Guided Workbook

Journey to Math Literacy

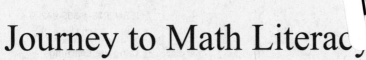

Richard Aufmann

Prepared by

Christi Verity
Sue Glascoe

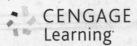

CENGAGE
Learning

Australia • Brazil • Mexico • Singapore • United Kingdom • United States

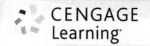
ISBN: 978-1-305-95229-4

Cengage Learning
20 Channel Center Street
Boston, MA 02210
USA

Cengage Learning is a leading provider of customiz learning solutions with office locations around the gl including Singapore, the United Kingdom, Australia, Mexico, Brazil, and Japan. Locate your local office a **www.cengage.com/global**.

Cengage Learning products are represented in Canada by Nelson Education, Ltd.

To learn more about Cengage Learning Solutions, visit **www.cengage.com**.

Purchase any of our products at your local college s at our preferred online store **www.cengagebrain.c**

Printed in the United States of America
Print Number: 01   Print Year: 2015

Table of Contents

Name: _____   Class: _____

Module A: AIM for Success: How to Succeed in This Course

Key Words and Concepts

Get Ready

You will have more success in math and in other courses you take if you actively _____.

Why are good math skills important? (Hint: think of reasons outside the classroom.) _____

_____

Motivate Yourself

Motivation alone won't lead to success. You'll need motivation *and* _____ in order to succeed.

_____ is the backbone of any successful endeavor—including math!

Do you think you will be successful in math if you just practice it once a week? Why or why not? _____

_____

Develop a "Can Do" Attitude Toward Math

Positive thinking towards math is very important to succeed in math.

Write the words "I can do math!" here: _____

Strategies for Success

Acknowledge your concerns, but make sure they don't prevent you from succeeding.

Start at a level that is appropriate for you.

Time Management

Create a time management plan of all the important activities and responsibilities in your life.

Be _____ about how much time you have.

　　　Note　　　To be successful, **attend class** and **participate in class**.

If you miss a class, what should you do? _____

_____

Why is it important to study and do homework in a timely fashion? _____

_____

1

Habits of Successful Students

List some habits of successful students that will work for you. _____

_____

Use the Interactive Method

How do the **Check your understanding** exercises related to the **Focus on** examples? _____

_____

Where can you find the solutions to the **Check your understanding** exercises? _____

Use a Strategy to Solve Word Problems

To help you solve word problems, which combine critical thinking skills to solve practical problems, you should:
    Read the problem.
    Write down what is known and what is unknown.
    Think of a method to find the unknown.
    Solve the equation.
    Check your solution.

In this text, the solution to every word problem is broken down into two steps, _____, which consists of the first three steps, and _____, which is the last two steps.

True or False: Math is a spectator sport.

Ace the Test

To prepare for a test, rework _____ exercises from each homework assignment.
         (one/two/three)

Ready, Set, Succeed!

Good luck! We wish you success.

Reflect on it
- You will have more success in math and in other courses you take, if you actively participate.
- You'll need motivation and learning in order to succeed.
- Practice is the backbone of any successful endeavor.
- Be realistic about how much time you have.
- In this text, the solution to every word problem is broken down into two steps, Strategy and Solution.
- Math is not a spectator sport.
- To prepare for a test, rework two exercises from each homework assignment

Application and Activities

Discussion and reflection questions:

1. How do you think this course (not just the math) will help you in your career?

2. It is sometimes popular for people to say things like "I can't do math", or "I am terrible at math". Discuss why you think people are encouraged to say negative things about math, but they might be embarrassed to say "I can't read", or "I am terrible at reading".

3. Have you ever been in a study group? How could you create a study group that would be most effective for you?

4. Which of the 5 steps to solve word problems do you have the most trouble with? How can you be more successful in that area?

5. Many students claim they are "bad test takers". Do you think that there is a way to get beyond feeling that way? If so, what steps do you think would help you the most?

Class Activity

Objectives for the lesson:

In this lesson you will use and enhance your understanding of the following:

- Strategies for success
- Habits of successful students

Use 3 whiteboards to create lists for the following as a class:

1. Discuss advantages and disadvantages of taking a math class only once per week, or taking it in a short period of time like summer school. Create a list on the board for advantages and a list for disadvantages.

2. Discuss how these advantages and disadvantages might be different for a student who math was difficult for, versus a student who was reviewing a course they took in high school.

3. "Practice" looks like many different things to people taking a math course. List as many way as you can think of to study for a math test. Get creative!

Group Activity (2-3 people)

Objectives for the lesson:

In this lesson you will use and enhance your understanding of the following:

- Listing possible concerns and a plan for addressing them
- Time management
- Setting reasonable goals and timelines for your education

Module A: AIM for Success: How to Succeed in This Course

1. List your concerns in each area and how you plan to address those concerns to be successful. Share with your group.
   a. Tuition

   b. Job

   c. Anxiety

   d. Child Care

   e. Time

   f. Degree Goals

   g. Other

2. Answer the following questions as honestly as possible, working alone. You will share the results from #2 - #5 with your group in activity #6.
   a. List the classes you are taking and what times/days they meet. How much time is this per day/week? Count 50 minute classes as 1 hour for ease.

   b. Do you have any online classes? How much time per day/week will you spend on that?

   c. Do you have a job? If so, how many hours per day on average do you work during the week? How many per weekend?

   d. How much time does it take you to drive to school and back every day? How much time does it take to drive to work?

3. Fill out the following table for 1 week, according to your answers in #2. If you prefer to put numbers like average work hours in on a weekly basis, then use the last column. Otherwise, total all the hours you wrote in per day in the last column. Sum the columns vertically as well.

TABLE 1

Day	Mon	Tues	Wed	Thurs	Fri	Sat	Sun	TOTAL/ WEEK
Class hours								
Online hours								
Job hours								
Driving hours								
TOTAL/DAY								

4. You filled out the non-negotiables in Table 1. Now it is time to think about the things you have in your life that make you successful, happy and healthy. Answer the following as honestly as possible:

   a. What grade would you like to earn in this course? How many hours per day/week do you plan on working on what you need to learn and do for this class? (Be specific for this activity, phrases like "as many as I need to" do not apply here.)

   b. How many hours total every day/week do you plan on working on all of your school work (outside of class time) to be successful in all of your courses?

   c. Do you have a spouse? Do you have children? How much time would you like to realistically spend with family each day? How much time would you like to dedicate to them on the weekends?

   d. How often do you exercise? List the amount of time you would like to exercise and which days per week.

   e. How much time would you like to spend on entertainment or eating out per week?

   f. Are you in any sports or any extra-curricular activities? How much time per day/week does that take up?

   g. What do you like to do to relax (i.e. read a book, watch TV)? How much time per day do you think you need in this area to help you re-energize?

   h. How many hours of sleep do you require each night to feel rested?

5. Fill out the following table for 1 week, according to your answers in #4. If you prefer to put numbers like average family hours in on a weekly basis, then use the last column. Otherwise, total all the hours you wrote in per day in the last column. Sum the columns vertically as well.

TABLE 2

Day	Mon	Tues	Wed	Thurs	Fri	Sat	Sun	TOTAL/WEEK
This class								
All classes								
Family								
Exercise								
Entertainment								
Sports								
Relax								
Sleep								
TOTAL/DAY								

6. Put the totals from both tables together in the following table to see if you need more than 24 hours in a day, or 168 hours in a week, to accomplish all of your goals!

TABLE 3

Day	Mon	Tues	Wed	Thurs	Fri	Sat	Sun	TOTAL/ WEEK
Totals from table #1								
Totals from Table # 2								
TOTAL/DAY								

Discuss the results with your group.

If you end up with more than 24 hours in a day or 168 hours in a week in Table 3, decide which areas you will need make sacrifices in, to make sure you fit in the things you need to be successful, happy, and healthy. Write down your ideas below and calculate your new results. Share your plan with your group.

7. How long do you think it will take you to finish your degree? Look back at your final table (table 3) and think about whether you can keep that schedule up for that amount of time. If it does not seem reasonable, then go adjust your schedule to create a manageable schedule that you can assure yourself at being successful with.

Objective 1.1A – Use inequality symbols with Integers

Key Words and Concepts

Mathematicians place objects with similar properties in groups called sets.

Define set. _____

The objects in a set are called the _____ **of the set**.

The **roster method** of writing a set encloses a list of the elements in braces.

   The set of **natural numbers** is the set {1, 2, 3, 4, …}.
   The set of **integers** is the set {…, –3, –2, –1, 0, 1, 2, 3, …}.

Each integer can be shown on a number line.
   The integers to the left of zero on the number line are called _____ **integers**.
                                                        (positive/negative)
   The integers to the right of zero are called _____**integers**, or natural numbers.
                                        (positive/negative)

   Note   Zero is neither a positive nor a negative integer.

The **graph** of an integer is shown by placing a heavy dot on the number line directly above the number.
The graphs of –4 and 3 are shown on the number line below.

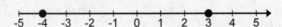

In mathematics, a letter of the alphabet can be used to stand for a number. Such a letter is called a **variable**.

   If a and b are two numbers and a is to the left of b on the number line, then a **is less than** b.
   This is written $a < b$.
For example,
   On the number line above, –4 is to the left of 3. So –4 is less than 3. This is written $-4 < 3$.

   If a and b are two numbers and a is to the right of b on the number line, then a **is greater than** b.
   This is written $a > b$.
For example,
   On the number line above, 3 is to the right of –4. So 3 is greater than –4. This is written $3 > -4$.

There are also inequality symbols for **is less than or equal to ($\leq$)** and **is greater than or equal to ($\geq$)**.

Example

Use the roster method to write the set of negative integers greater than –5

Solution
$A = \{-4,-3,-2,-1\}$

Try it

1. Use the roster method to write the set of positive integers less than or equal to 4.

Example

Which is the lesser number, –1 or –3?

Solution
Look at the numbers on the number line.

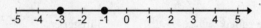

Since –3 is to the left of –1 on the number line, –3 is the lesser number.
$-3 < -1$

Try it

2. Which is the lesser number, –5 or –9?

Example

Given $B = \{-5,-3,-1, 3, 5\}$, which elements of set B are less than or equal to –1?

Solution
The elements –5, –3, and –1 are less than or equal to –1.

Try it

3. Given $B = \{-6,-4,-2, 4, 6\}$, which elements of set B are less than or equal to –2?

Reflect on it
- A set is a collection of objects.
- Integers to the left of zero on the number line are negative integers.
- Integers to the right of zero on the number line are positive integers.
- Inequality symbols are < (less than), > (greater than), ≤ (less than or equal to), and ≥ (greater than or equal to).
- Why do you think zero is neither negative nor positive?

Quiz Yourself 1.1A
1. Place the correct inequality symbol, < or >, between the numbers below.
   –13   –15

2. Place the correct inequality symbol, < or >, between the numbers below.
   –13   12

3. Place the correct inequality symbol, < or >, between the numbers below.
    0   −1

4. Use the roster method to write the set of positive integers less than 6.

5. Use the roster method to write the set of negative integers greater than or equal to −7.

6. Which is the lesser temperature, −8°F or −5°F?

7. Given $C = \{-8, -4, 0, 4, 8\}$, which elements of set C are greater than −4?

8. Given $C = \{-8, -4, 0, 4, 8\}$, which elements of set C are less than or equal to 0?

Practice Sheet 1.1A

Place the correct symbol, < or >, between the values of the two numbers.

1. −1   −3   2. −8   −5   3. 7   0   1. _____

                                                           2. _____

                                                           3. _____

4. −24   42   5. 53   −64   6. −55   53   4. _____

                                                           5. _____

                                                           6. _____

Answer True or False.

7. −5 > −2   8. 3 < −5   9. −12 < 14   7. _____

                                                           8. _____

                                                           9. _____

10. −34 > 47   11. 19 < −31   12. −10 > −7   10. _____

                                                           11. _____

                                                           12. _____

13. 53 < −71   14. −53 < −81   15. 84 > 73   13. _____

                                                           14. _____

                                                           15. _____

Use the roster method to write the set.

16. The natural numbers   17. The negative integers   18. The positive integers   16. _____
     less than 4   greater than or equal   less than 6
                                  to −3   17. _____

                                                                                       18. _____

19. Given A = {−6, −3, 1, 4}, 20. Given B = {−9, −5,   21. Given C = {−20, 0,   19. _____
     which elements of   −1, 3}, which elements   1, 18}, which elements
     set A are less than −1?   of set B are greater   of set C are less   20. _____
                                        than −6?   than 1?
                                                                                            21. _____

22. Given D = {−16, −5, 2,   23. Given E = {−26, −19,   24. Given F = {−35, −15,   22. _____
     3, 7, 18}, which   −7, −5, 2, 6, 9}, which   −1, 0, 3, 6}, which
     elements of set D are   elements of set E are   elements of set F 23. _____
     greater than or   less than or equal to 6?   are greater than −1?
     equal to 2?   24. _____

Answers

Try it 1.1A

1. $A = \{1, 2, 3, 4\}$
2. -9 is the lesser number, $-9 < -5$
3. $-6, -4$, and -2 are less than or equal to -2.

Quiz 1.1A

1. $-13 > -15$
2. $-13 < 12$
3. $0 > -1$
4. $A = \{1, 2, 3, 4, 5\}$
5. $A = \{-7, -6, -5, -4, -3, -2, -1\}$
6. $-8°F$ is a lesser temperature than $-5°F$.
7. The elements of C that are greater than -4 are $0, 4$, and 8.
8. The elements of C that are less than or equal to 0 are $-8, -4$, and 0.

Solutions to Practice Sheet 1.1A

1. $-1 > -3$

2. $-8 < -5$

3. $7 > 0$

4. $-24 < 42$

5. $53 > -64$

6. $-55 < 53$

7. $-5 > -2$ False

8. $3 < -5$ False

9. $-12 < 14$ True

10. $-34 > 47$ False

11. $19 < -31$ False

12. $-10 > -7$ False

13. $53 < -71$ False

14. $-53 < -81$ False

15. $84 > 73$ True

16. $\{1, 2, 3\}$

17. $\{-3, -2, -1\}$

18. $\{1, 2, 3, 4, 5\}$

19. $-6, -3$

20. $-5, -1, 3$

21. $-20, 0$

22. $2, 3, 7, 18$

23. $-26, -19, -7, -5, 2, 6$

24. $0, 3, 6$

Objective 1.1B – Simplify expressions with absolute value

Key Words and Concepts

Two numbers that are the same distance from zero on the number line but are on opposite sides of zero are **opposite numbers**, or **opposites**. The opposite of a number is also called its **additive inverse**.

Note   The negative sign can be read "the opposite of."

For example,

the opposite of 3 is –3.   This can be written as –(3) = –3

The opposite of –4 is 4.   This can be written as –(–4) = 4

The additive invers of 8 is –8.   This can be written as –(8) = –8

The **absolute value of a number** is its distance from _____ on the number line. Therefore, the absolute value of a number is a positive number or zero. The symbol for absolute value is two vertical bars, $|\ |$.

The absolute value of a positive number is _____.

The absolute value of a negative number is the opposite of the number.

The absolute value of zero is _____.

Using variables, the definition of absolute value is

$$|x| = \begin{cases} x, & x > 0 \\ 0, & x = 0 \\ -x, & x < 0 \end{cases}$$

Example

Evaluate $|-4|$.

   Solution
      Find the distance from –4 to 0.
      $|-4| = 4$

Try it

1. Evaluate $|-7|$.

Example

Evaluate $-|-5|$.

   Solution
      Find the *opposite* of the distance from –5 to 0.
      $-|-5| = -5$

Try it

2. Evaluate $-|-16|$.

Example
Evaluate $|21|$.

Solution
Find the distance from 21 to 0.
$$|21| = 21$$

Try it
3. Evaluate $|19|$.

Reflect on it
- Recall the meanings of opposite and additive inverse.
- The absolute value of
    a number is its distance from zero on the number line.
    a number is a positive number or zero
    a positive number is the number itself.
    a negative number is the opposite of the number.
    of zero is zero.

Quiz Yourself 1.1B
1. Evaluate $-(-2)$.

2. Find the additive inverse of -9.

3. Find the additive inverse of 8.

4. Find the opposite of 6.

5. Find the opposite of -17.

6. Evaluate $-(-23)$.

7. Evaluate $-|-23|$.

Practice Sheet 1.1B

Find the additive inverse.

1. 7	**2.** –4	**3.** –5
4. 34	**5.** –28	**6.** 66

Evaluate.

7. (–3)	**8.** – (7)	**9.** – (–5)
10. – (–13)	**11.** – | –4|	**12.** | 15 |
13. | – 16 |	**14.** – | 19 |	**15.** – | – 19|

Place the correct symbol, < or >, between the values of the two numbers.

16. – | 21 | | – 26 |	**17.** – | 22 | – | 31 |	**18.** – | 14 | – | – 35 |
19. – | – 33 | | 54 |	**20.** | 40 | | – 54 |	**21.** – | 28 | | 9 |
22. | 6 | – | 43 |	**23.** – | – 25 | | 18 |	**24.** | –15 | | 4 |

1. _____
2. _____
3. _____
4. _____
5. _____
6. _____
7. _____
8. _____
9. _____
10. _____
11. _____
12. _____
13. _____
14. _____
15. _____
16. _____
17. _____
18. _____
19. _____
20. _____
21. _____
22. _____
23. _____
24. _____

Answers

Try it 1.1B

1. $|-7| = 7$
2. $-|-16| = -16$
3. $|19| = 19$

Quiz 1.1B

1. $-(-2) = 2$
2. The additive inverse of –9 is 9.
3. The additive inverse of 8 is –8.
4. The opposite of 6 is –6.
5. The opposite of –17 is 17.
6. $-(-23) = 23$
7. $-|-23| = -23$

Solutions to Practice Sheet 1.1B

1. –7
2. 4
3. 5
4. –34
5. 28
6. –66
7. –3
8. –7
9. 5
10. 13
11. –4
12. 15
13. 16
14. –19
15. –19
16. $-|21| < |-26|$
17. $-|22| > -|31|$
18. $-|14| > -|-35|$
19. $-|-33| < |54|$
20. $|40| < |-54|$
21. $-|28| < |9|$
22. $|6| > -|43|$
23. $-|-25| < |18|$
24. $|-15| > |4|$

Objective 1.1C – Create stem-and-leaf diagrams

Key Words and Concepts

One way to analyze data is to draw a graph or display the data, such as using a **stem-and-leaf plot**.
Each data value is split into two parts, a **stem** and a **leaf**.

The leaf is usually the _____ digit of a number; the stem is the remaining digits.
                                     (leftmost/rightmost)

For example in the number 26, the stem is _____ and the leaf is _____.

For a three-digit number such as 349, the stem is 34 and the leaf is 9.

Data are placed into stem-and-leaf plots as a means of quickly showing the frequency of a particular event.
For example, both of the following lists show the round trip commuting distances for 29 employees of a digital art production company.

stem-and-leaf plot	raw data
0 | 7 9	
1 | 5 3 8	28 25 37 37 33
2 | 2 7 2 8 5	34 31 7 9 15
3 | 7 7 3 4 1 0 8	51 54 55 53 59
4 | 8 7 3 0 6 5	66 13 18 22 30
5 | 1 4 5 3 9	38 48 47 43 27
6 | 6	22 40 46 45

where 5|4 is 54.

Example

The data below show the final exam scores from last semester.
Draw a stem-and-leaf diagram for the data.

   78, 85, 69, 92, 77, 97, 83, 88, 86, 93, 79, 79, 56, 75, 63, 96, 86

Solution

The smallest value is 56 and the largest is 97. Therefore, stems from 5 (the first digit of 56) through 9 (the first digit of 97) are used. Draw a vertical line and list the stems to the left of the line as shown below.

```
5 |
6 |
7 |
8 |
9 |
```

Next, to the right of the vertical line, list the leaves.
To represent 78, the first data value, place an 8 to the right of 7.
Continue until you have used all the data value, placing the leaf values from smallest to largest.

```
5 | 6
6 | 3  9
7 | 5  7  8  9  9
8 | 3  5  6  6  8
9 | 2  3  6  7
```

Try it

1. The data below show the mid-term exam scores from last semester. Draw a stem-and-leaf diagram for the data.
83, 89, 77, 93, 58, 62, 68, 91, 93, 85, 74, 88, 95, 72, 88, 80, 96

Reflect on it
- In a stem-and-leaf plot, the leaf is usually the rightmost digit of a number; the stem is the remaining digits.
- What advantages are there to using a stem-and-leaf diagram to plot data?

Quiz Yourself 1.1C

1. The data below show the average number of days with wet weather for selected cities in the United States. Draw a stem-and-leaf plot for these data.

Orlando, FL	117	Columbus, OH	139
Louisville, KY	123	Rochester, NY	167
Birmingham, AL	117	St. Louis, MO	113
Nashville, TN	119	Chicago, IL	124
Detroit, MI	135	Indianapolis, IN	129
Virginia Beach, VA	117	Atlanta, GA	113
Miami, FL	135	Philadelphia, PA	118
Cincinatti, OH	137	Minneapolis, MN	117
Raleigh, NC	100	Portland, OR	164

Practice Sheet 1.1C

Solve.

1. The data below show the forecast of daily high temperatures for the next 15 days for   **1.** _____
    Sacramento, CA. Draw a stem-and-leaf diagram for the data.
    94, 97, 92, 90, 86, 81, 83, 86, 88, 85, 84, 82, 80, 79, 82

2. The data below show the forecast of daily humidity percentage levels for the next 15   **2.** _____
    days for Washington, DC. Draw a stem-and-leaf diagram for the data.
    50, 74, 58, 56, 62, 53, 51, 53, 57, 67, 73, 74, 69, 72, 70

3. The data below show the forecast of daily high temperatures for the next 15 days for   **3.** _____
    Hershey, PA. Draw a stem-and-leaf diagram for the data.
    50, 48, 42, 44, 39, 33, 33, 37, 42, 50, 45, 41, 42, 42, 43

4. The data below show the average number of days with wet weather for selected cities   **4.** _____
    in the United States. Draw a stem-and-leaf plot for these data.

Washington, DC	114	Houston, TX	104	Memphis, TN	108
Charlotte, NC	110	Tampa, FL	105	Buffalo, NY	167
New York, NY	122	Cleveland, OH	155	Richmond, VA	114
Baltimore, MD	116	New Orleans, LA	115	Hartford, CT	130
Seattle, WA	149	Jacksonville, FL	114	Providence, RI	100
Boston, MA	126	Pittsburgh, PA	151	Milwaukee, WI	127

Answers

Try it 1.1C

1. Mid-term exam scores

```
5 | 8
6 | 2  8
7 | 2  4  7
8 | 0  3  5  8  8  9
9 | 1  3  3  5  6
```

Quiz 1.1C

1. Average number of days with wet weather

```
10 | 0
11 | 3  3  7  7  7  7  8  9
12 | 3  4  9
13 | 5  5  7  9
16 | 4  7
```

Solutions to Practice Sheet 1.1C

1. Daily high temperatures

```
7 | 9
8 | 0  1  2  2  3  4  5  6  6  8
9 | 0  2  4  7
```

2. Humidity percentage levels

```
5 | 0  1  3  3  6  7  8
6 | 2  7  9
7 | 0  2  3  4  4
```

3. Daily low temperatures

```
3 | 3  3  7  9
4 | 1  2  2  2  2  3  4  5  8
5 | 0  0
```

4. Average number of days with wet weather

```
10 | 0  4  5  8
11 | 0  4  4  4  5  6
12 | 2  6  7
13 | 0
14 | 9
15 | 1  5
16 | 7
```

Objective 1.2A – Add integers

Key Words and Concepts

Addition is the process of finding the total of two numbers.
> The numbers being added are called **addends**.
> The total is called the **sum**.

To add two numbers with the same sign
> Add the absolute values of the numbers.
> Then attach the sign of the addends.

Example

Add: $-34 + (-56)$

> Solution
>> The numbers have the same sign.
>>> Add the absolute values of the numbers, $(34 + 56)$.
>>> Then attach the negative sign to the sum.
>>
>> $-34 + (-56) = -90$

Try it

1. Add: $-28 + (-52)$

To add two numbers with different signs
> Find the absolute value of each number.
> Subtract the smaller of the absolute values from the larger.
> Then attach the sign of the number with the larger absolute value.

Example

Add: $-34 + 56$

> Solution
>> The numbers have different signs.
>>
>> $|-34| = 34$ Find the absolute value of each number
>> $|56| = 56$
>>
>> $56 - 34 = 22$ Subtract the smaller from the larger
>>
>> $-34 + 56 = 22$ The sign of the larger addend (56) is positive, so the answer is positive.

Try it

2. Add: $-28 + 52$

Name some phrases which indicate addition. _____

_____

Example

Add: $-54+19+(-12)$

Solution

$$\begin{aligned} -54+19+(-12) \quad &\text{Copy the expression} \\ =-35+(-12) \quad &\text{Add } -54 \text{ and } 19 \\ =-47 \quad &\text{Add } -35 \text{ and } -12 \end{aligned}$$

Try it

3. Add: $-25+18+(-32)$

Example

Find -87 increased by 21.

Solution

$$-87+21=-66$$

Try it

4. Find -61 more than 37.

Reflect on it

- When adding two numbers with the same sign, the sum will have the same sign as the addends.
- When adding two numbers with different signs, the sum will have the same sign as the larger addend.
- Some examples of phrases which indicate addition are: added to, more than, the sum of, increased by, the total of, and plus.

Quiz Yourself 1.2A

1. Add: $-7+(-4)$

2. Add: $45+(-45)$

3. Find the sum of -13, 15, and 17.

4. What is 8 more than -5?

5. Add: $-12+15+(-18)$

6. Find -42 increased by 28.

Practice Sheet 1.2A

Add.

1. $2 + (-4)$	**2.** $-5 + 3$	**3.** $-3 + (-7)$	**1.** _____
			2. _____
			3. _____
4. $-16 + 23$	**5.** $9 + (-7)$	**6.** $-10 + (-8)$	**4.** _____
			5. _____
			6. _____
7. $-1 + (-6)$	**8.** $-7 + (-7)$	**9.** $5 + (-10)$	**7.** _____
			8. _____
			9. _____
10. $-9 + 3$	**11.** $-7 + 11$	**12.** $-13 + 9$	**10.** _____
			11. _____
			12. _____
13. $6 + (-4) + (-7)$	**14.** $-4 + (-10) + (-11)$	**15.** $11 + (-3) + (-13)$	**13.** _____
			14. _____
			15. _____
16. $-15 + (-5) + 30$	**17.** $10 + 54 + (-28)$	**18.** $-4 + (-9) + 9$	**16.** _____
			17. _____
			18. _____
19. $-36 + (-40) + (-16)$	**20.** $11 + (-20) + (-30)$	**21.** $-11 + (-6) + 8$	**19.** _____
			20. _____
			21. _____
22. $16 + (-18) + 5 + (-4)$	**23.** $-12 + (-4) + 6 + (-7)$	**24.** $-23 + 9 + 3 + (-16)$	**22.** _____
			23. _____
			24. _____
25. $6 + 8 + (-13) + (-4)$	**26.** $-12 + 15 + (-16) + (-10)$	**27.** $-21 + (-33) + 27 + 18$	**25.** _____
			26. _____
			27. _____

Answers

Try it 1.2A

1. -80
2. 24
3. -39
4. -24

Quiz 1.2A

1. -11
2. 0
3. 19
4. 3
5. -15
6. -14

Solutions to Practice Sheet 1.2A

1. $2 + (-4) = -2$

2. $-5 + 3 = -2$

3. $-3 + (-7) = -10$

4. $-16 + 23 = 7$

5. $9 + (-7) = 2$

6. $-10 + (-8) = -18$

7. $-1 + (-6) = -7$

8. $-7 + (-7) = -14$

9. $5 + (-10) = -5$

10. $-9 + 3 = -6$

11. $-7 + 11 = 4$

12. $-13 + 9 = -4$

13. $6 + (-4) + (-7) = 2 + (-7)$
$$= -5$$

14. $-4 + (-10) + (-11) = -14 + (-11)$
$$= -25$$

15. $11 + (-3) + (-13) = 8 + (-13)$
$$= -5$$

16. $-15 + (-5) + 30 = -20 + 30$
$$= 10$$

17. $10 + 54 + (-28) = 64 + (-28)$
$$= 36$$

18. $-4 + (-9) + 9 = -13 + 9$
$$= -4$$

19. $-36 + (-40) + (-16) = -76 + (-16)$
$$= -92$$

20. $11 + (-20) + (-30) = -9 + (-30)$
$$= -39$$

21. $-11 + (-6) + 8 = -17 + 8$
$$= -9$$

22. $16 + (-18) + 5 + (-4) = -2 + 5 + (-4)$
$$= 3 + (-4)$$
$$= -1$$

23. $-12 + (-4) + 6 + (-7) = -16 + 6 + (-7)$
$$= -10 + (-7)$$
$$= -17$$

24. $-23 + 9 + 3 + (-16) = -14 + 3 + (-16)$
$$= -11 + (-16)$$
$$= -27$$

25. $6 + 8 + (-13) + (-4) = 14 + (-13) + (-4)$
$$= 1 + (-4)$$
$$= -3$$

26. $-12 + 15 + (-16) + (-10) = 3 + (-16) + (-10)$
$$= -13 + (-10)$$
$$= -23$$

27. $-21 + (-33) + 27 + 18 = -54 + 27 + 18$
$$= -27 + 18$$
$$= -9$$

Objective 1.2B – Subtract integers

Key Words and Concepts

To **subtract** one number from another, add the opposite of the second number to the first number.

Give a definition of difference. _____

_____

Example

Subtract: $-18-15$

Solution

$-18-15$ Copy the expression

$=-18+(-15)$ Rewrite subtraction as
                addition of the opposite

$=-33$ Add

Try it

1. Subtract: $-61-13$

Example

Subtract: $-18-(-15)$

Solution

$-18-(-15)$ Copy the expression

$=-18+15$ Rewrite subtraction as
                addition of the opposite

$=-3$ Add

Try it

2. Subtract: $-26-(-21)$

Example

Subtract: $18-52$

Solution

$18-52$ Copy the expression

$=18+(-52)$ Rewrite subtraction as
                addition of the opposite

$=-34$ Add

Try it

3. Subtract: $34-61$

Name some phrases which indicate subtraction. _____

_____

Example
Find the difference between –5 and 6.

Solution

$-5-6$ Translate the expression

$=-5+(-6)$ Rewrite subtraction as
            addition of the opposite

$=-11$ Add

Try it
4. Find the difference between –12 and 9.

Example
Subtract –32 from –51.

Solution

$-51-(-32)$ Translate the expression

$=-51+32$ Rewrite subtraction as
            addition of the opposite

$=-19$ Add

Try it
5. Subtract –19 from –28.

Reflect on it
• A difference is the answer to a subtraction problem.
• Some examples of phrases which indicate subtraction are: minus, less, less than, the difference between, and decreased by.

Quiz Yourself 1.2B
1. Subtract: $20 - 14$

2. Subtract: $45-(-45)$

3. Subtract: $-12-15-(-18)$

4. What is 28 less than -42?

5. What is 8 decreased by -5?

Practice Sheet 1.2B

Subtract.

1. $13 - 7$ 2. $15 - 5$ 3. $5 - 15$

4. $-6 - 3$ 5. $-7 - 5$ 6. $6 - (-4)$

7. $2 - (-2)$ 8. $-9 - (-1)$ 9. $-3 - (-7)$

10. $-17 - 11$ 11. $-5 - 2 - 1$ 12. $-4 - 5 - 11$

13. $4 - 5 - 6$ 14. $10 - (-9) - 7$ 15. $-3 - 10 - (-6)$

16. $12 - 9 - 11$ 17. $-4 - (-5) - (-7)$ 18. $6 - 7 - (-2)$

19. $-28 - (-52) - 24 - 9$ 20. $44 - (-80) - 60 - 10$ 21. $-17 - 36 - 62 - 14$

22. $38 - (-21) - 62 - (-17)$ 23. $-36 - (-58) - 10 - 17$ 24. $-16 - 37 - (-64) - 35$

25. $-19 - (-31) - (-76) - 24$ 26. $21 - 33 - 57 - 14$ 27. $38 - (-22) - 63 - (-9)$

1. _____
2. _____
3. _____
4. _____
5. _____
6. _____
7. _____
8. _____
9. _____
10. _____
11. _____
12. _____
13. _____
14. _____
15. _____
16. _____
17. _____
18. _____
19. _____
20. _____
21. _____
22. _____
23. _____
24. _____
25. _____
26. _____
27. _____

Answers

Try it 1.2B
1. -74
2. -5
3. -27
4. -21
5. -9

Quiz 1.2B
1. 6
2. 90
3. -9
4. -70
5. 13

Solutions to Practice Sheet 1.2B

1. $13 - 7 = 6$

2. $15 - 5 = 10$

3. $5 - 15 = -10$

4. $-6 - 3 = -9$

5. $-7 - 5 = -12$

6. $6 - (-4) = 6 + 4 = 10$

7. $2 - (-2) = 2 + 2 = 4$

8. $-9 - (-1) = -9 + 1 = -8$

9. $-3 - (-7) = -3 + 7 = 4$

10. $-17 - 11 = -28$

11. $-5 - 2 - 1 = -7 - 1 = -8$

12. $-4 - 5 - 11 = -9 - 11 = -20$

13. $4 - 5 - 6 = -1 - 6 = -7$

14. $10 - (-9) - 7 = 10 + 9 - 7$
    $= 19 - 7$
    $= 12$

15. $-3 - 10 - (-6) = -13 - (-6)$
    $= -13 + 6$
    $= -7$

16. $12 - 9 - 11 = 3 - 11 = -8$

17. $-4 - (-5) - (-7) = -4 + 5 + 7$
    $= 1 + 7$
    $= 8$

18. $6 - 7 - (-2) = -1 - (-2)$
    $= -1 + 2$
    $= 1$

19. $-28 - (-52) - 24 - 9 = -28 + 52 - 24 - 9$
    $= 24 - 24 - 9$
    $= -9$

20. $44 - (-80) - 60 - 10 = 44 + 80 - 60 - 10$
    $= 124 - 60 - 10$
    $= 64 - 10$
    $= 54$

21. $-17 - 36 - 62 - 14 = -53 - 62 - 14$
    $= -115 - 14$
    $= -129$

22. $38 - (-21) - 62 - (-17) = 38 + 21 - 62 + 17$
    $= 59 - 62 + 17$
    $= -3 + 17$
    $= 14$

23. $-36 - (-58) - 10 - 17 = -36 + 58 - 10 - 17$
    $= 22 - 10 - 17$
    $= 12 - 17$
    $= -5$

24. $-16 - 37 - (-64) - 35 = -53 - (-64) - 35$
    $= -53 + 64 - 35$
    $= 11 - 35$
    $= -24$

25. $-19 - (-31) - (-76) - 24 = -19 + 31 + 76 - 24$
    $= 12 + 76 - 24$
    $= 88 - 24$
    $= 64$

26. $21 - 33 - 57 - 14 = -12 - 57 - 14$
    $= -69 - 14$
    $= -83$

27. $38 - (-22) - 63 - (-9) = 38 + 22 - 63 + 9$
    $= 60 - 63 + 9$
    $= -3 + 9$
    $= 6$

Objective 1.2C – Solve application problems

Some application problems can be solved with addition or subtraction.

Chemical Element	Boiling Point	Melting Point
Mercury	357	−39
Radon	−62	−71
Xenon	−108	−112

The table above shows the boiling point and the melting point, in degrees Celsius, of three chemical elements.

Example

What is the difference between the boiling point of mercury and the melting point of mercury?

  Solution

  Subtract the melting point of mercury (−39°C) from its boiling point (357°C).

  $357 - (-39)$ Translate the expression

  $= 357 + 39$ Rewrite subtraction as
                     addition of the opposite

  $= 396$ Add

The difference between the boiling point and melting point is 396°C.

Try it

1. What is the difference between the boiling point of xenon and the melting point of xenon?

Quiz Yourself 1.2C

The table below shows the average temperatures, in degrees Fahrenheit, at different cruising altitudes for airplanes.

Cruising Altitude	Average Temperature
12,000 ft	16°
20,000 ft	−12°
30,000 ft	−48°
40,000 ft	−70°
50,000 ft	−70°

1. What is the difference between the average temperatures at 12,000 ft and at 40,000 ft?

2. What is the difference between the average temperatures at 40,000 ft and 50,000 ft?

3. How much colder is the average temperature at 30,000 ft than at 12,000 ft?

Practice Sheet 1.2C

Solve.

1. The highest elevation in Asia is Mt. Everest with an elevation of 8850 m. The lowest point is the Dead Sea with an elevation of –400 m. What is the difference between the highest and lowest points in Asia?

2. The average high temperature in Fargo, ND during the month of July is 89.6°F. The average low temperature in February is –17.6°F. Find the difference between the two average temperatures.

1. _____

2. _____

3. When a certain amount of salt is added to water, the freezing point of the salt water mixture is –4°C. The boiling point of the mixture is 103°C. What is the difference between the boiling point and the freezing point of the saltwater mixture?

4. The temperature at which a rare element boils is 350°C. This element freezes at –29°C. Determine the difference between the temperature at which it boils and the temperature at which it freezes.

3. _____

4. _____

5. When the temperature is –10°F and the wind is blowing at 15 mph, the wind chill factor is –32°F. For the same temperature of –10°F and the wind blowing at 30 mph, the wind chill factor is –39°F. Find the difference between the wind chill factors.

6. The temperature at which Xenon boils is –107°C. Xenon melts at –112°C. Determine the difference between the temperatures at the boiling point and the melting point.

5. _____

6. _____

Answers

Try it 1.2C
1. 4°C`

Quiz 1.2C
1. 86°F
2. 0°F
3. 64°F

Solutions to Practice Sheet 1.2C

1. STRATEGY
   To find the difference, subtract the elevation of the Dead Sea (–400 m) from the elevation of Mt Everest (8850 m).

   SOLUTION
   $8850 – (–400) = 8850 + 400 = 9250$
   The difference in elevation is 9250 m.

2. STRATEGY
   To find the difference, subtract the temperature of February (–17.6°F) from the temperature of July (89.6°F).

   SOLUTION
   $89.6 – (–17.6) = 89.6 + 17.6 = 107.2$
   The difference in temperature is 107.2°F.

3. STRATEGY
   To find the difference, subtract the temperature of freezing point (–4°C) from the temperature of boiling point (103°C).

   SOLUTION
   $103 – (–4) = 103 + 4 = 107$
   The difference in temperature is 107°C.

4. STRATEGY
   To find the difference, subtract the temperature of freezing point (–29°C) from the temperature of boiling point (350°C).

   SOLUTION
   $350 – (–29) = 350 + 29 = 379$
   The difference in temperature is 379°C.

5. STRATEGY
   To find the difference, subtract the temperature of 30 mph (–39°F) from the temperature of 15 mph (–32°F).

   SOLUTION
   $–32 – (–39) = –32 + 39 = 7$
   The difference in temperature is 7°F.

6. STRATEGY
   To find the difference, subtract the temperature of melting point (–112°C) from the temperature of boiling point (–107°C).

   SOLUTION
   $–107 – (–112) = –107 + 112 = 5$
   The difference in temperature is 5°C.

Objective 1.3A – Multiply integers

Key Words and Concepts

Several different symbols are used to indicate multiplication.

The numbers being multiplied are called _____. The result is called the _____.

> *Note* When parentheses are used and there is no operation symbol, the operation is multiplication.

To multiply two numbers with the *same* sign,
multiply the absolute values of the numbers.   The product is _____.
(positive/negative)

To multiply two numbers with *different* signs,
multiply the absolute values of the numbers.   The product is _____.
(positive/negative)

Example
Multiply: $-15(8)$

Solution
The signs are different. The product is negative.
$$-15(8) = -120$$

Try it
1. Multiply: $-13(4)$

Example
Multiply: $(-15)(-8)$

Solution
The signs are the same. The product is positive.
$$(-15)(-8) = 120$$

Try it
2. Multiply: $(-3)(-24)$

Name some phrases which indicate multiplication. _____

_____

Example

Find the product of $-3, -4,$ and $-5.$

Try it

3. Find the product of $-6, -2,$ and $-10.$

Solution

To multiply more than two numbers, multiply the first
    two numbers, -3 and $-4.$
Then multiply that product by the third number.
Continue until all the numbers are multiplied.

$$(-3)(-4)(-5) \qquad \text{Write the expression}$$
$$= 12(-5) \qquad \text{Mulitply } (-3)(-4)$$
$$= -60 \qquad \text{Mulitply } 12(-5)$$

Reflect on it

- The numbers being multiplied are called factors. The result is called the product.
- Some examples of phrases which indicate subtraction are: times, the product of, multiplied by, and twice.
- The product of an even number of negative factors is positive.
- The product of an odd number of negative factors is negative.

Quiz Yourself 1.3A

1. Multiply: $(-14)(7)$

2. Multiply: $14(-7)$

3. Multiply: $8(-9)(-10)$

4. Multiply: $(15)3(-5)$

5. Find the product of $-2, -4, -6,$ and $-8.$

33

Name: Class:
Module 1: Problem Solving with Integers   Section 1.3: Multiplication and Division of Integers

Practice Sheet 1.3A

Multiply.

1. 11×4 **2.** -2×4 **3.** $3(-4)$

4. $-7(-2)$ **5.** $(8)(-2)$ **6.** $4(-6)$

7. $(-6)(7)$ **8.** $(-4)(5)$ **9.** $(-4)(-7)$

10. $(-2)(-9)$ **11.** $(0)(-6)$ **12.** -21×3

13. $18(-5)$ **14.** $3(-19)$ **15.** $-4(-13)$

16. $-4(27)$ **17.** $9(-18)$ **18.** $-6(-32)$

19. $-5(36)$ **20.** $3(-7)(2)$ **21.** $(-3)(5)(8)$

22. $6 \times 6 \times (-3)$ **23.** $(-8)(3)(-2)$ **24.** $-7(-6)(-5)$

25. $-4(7)(-2)$ **26.** $(-5)(4)(9)$ **27.** $4(-3)(4)$

1. _____
2. _____
3. _____
4. _____
5. _____
6. _____
7. _____
8. _____
9. _____
10. _____
11. _____
12. _____
13. _____
14. _____
15. _____
16. _____
17. _____
18. _____
19. _____
20. _____
21. _____
22. _____
23. _____
24. _____
25. _____
26. _____
27. _____

Answers

Try it 1.3A
1. -52
2. 72
3. -120

Quiz 1.3A
1. -98
2. -98
3. 720
4. -225
5. 384

Solutions to Practice Sheet 1.3A

1. $11 \times 4 = 44$

2. $-2 \times 4 = -8$

3. $3(-4) = -12$

4. $-7(-2) = 14$

5. $(8)(-2) = -16$

6. $4(-6) = -24$

7. $(-6)(7) = -42$

8. $(-4)(5) = -20$

9. $(-4)(-7) = 28$

10. $(-2)(-9) = 18$

11. $(0)(-6) = 0$

12. $-21 \times 3 = -63$

13. $18(-5) = -90$

14. $3(-19) = -57$

15. $-4(-13) = 52$

16. $-4(27) = -108$

17. $9(-18) = -162$

18. $-6(-32) = 192$

19. $-5(36) = -180$

20. $3(-7)(2) = -21(2) = -42$

21. $(-3)(5)(8) = -15(8) = -120$

22. $6 \times 6 \times (-3) = 36 \times (-3) = -108$

23. $(-8)(3)(-2) = -24(-2) = 48$

24. $-7(-6)(-5) = 42(-5) = -210$

25. $-4(7)(-2) = -28(-2) = 56$

26. $(-5)(4)(9) = -20(9) = -180$

27. $4(-3)(4) = -12(4) = -48$

Objective 1.3B – Divide integers

Key Words and Concepts

To divide two numbers with the *same* sign,

      divide the absolute values of the numbers.   The quotient is _____.

                                                    (positive/negative)

To divide two numbers with *different* signs,

      divide the absolute values of the numbers.   The quotient is _____.

                                                      (positive/negative)

Example

Divide: $(-21) \div 7$

    Solution

        The signs are different. The quotient is negative.

$$(-21) \div 7 = -3$$

Try it

1. Divide: $(-56) \div 8$

Example

Divide: $(-21) \div (-7)$

    Solution

        The signs are the same. The quotient is positive.

$$(-21) \div (-7) = 3$$

Try it

2. Divide: $(-63) \div (-7)$

Example

Divide: $\dfrac{64}{-4}; \dfrac{-64}{4}; -\dfrac{64}{4}$

    Solution

$$\frac{64}{-4} = -16$$

$$\frac{-64}{4} = -16$$

$$-\frac{64}{4} = -16$$

Try it

3. Divide: $\dfrac{81}{-3}; \dfrac{-81}{3}; -\dfrac{81}{3}$

In general,

    If a and b are integers, and $b \neq 0$, then $\dfrac{-a}{b} = \dfrac{a}{-b} = -\dfrac{a}{b}$.

      Note   The symbol $\neq$ is read "is not equal to."

Properties of Zero and One in Division

1. If $a \neq 0$, $\dfrac{0}{a} = 0$.   Zero divided by any number other than zero is zero.

2. If $a \neq 0$, $\dfrac{a}{a} = 1$.   Any number other than zero divided by itself is 1.

3. $\dfrac{a}{1} = a$ A number divided by 1 is the number.

4. $\dfrac{a}{0}$ is undefined.   Division by 0 is not defined.

Name some phrases which indicate division. _____

_____

Reflect on it
- The division of two numbers with the same sign has a positive quotient.
- The division of two numbers with different signs has a negative quotient.
- Some examples of phrases which indicate subtraction are: divided by, the quotient of, and the ratio of.

Quiz Yourself 1.3B

1. Divide: $(-54) \div 6$

2. Divide: $(-54) \div (-6)$

3. Divide: $\dfrac{72}{-3}$

4. Divide: $0 \div (-28)$

5. What is the quotient of -94 and 2?

Practice Sheet 1.3B

Divide.

1. $18 \div (-6)$ 2. $(-36) \div (6)$ 3. $(-32) \div (-4)$

4. $7 \div (-7)$ 5. $-42 \div 6$ 6. $40 \div (-8)$

7. $-30 \div 5$ 8. $-48 \div 6$ 9. $-72 \div (-8)$

10. $-42 \div (-7)$ 11. $-42 \div (-6)$ 12. $-50 \div 5$

13. $-55 \div (-11)$ 14. $-84 \div (-7)$ 15. $-60 \div (-5)$

16. $-75 \div (-5)$ 17. $105 \div 7$ 18. $68 \div (-4)$

19. $72 \div (-9)$ 20. $(-128) \div 8$ 21. $192 \div (-12)$

22. $154 \div (-22)$ 23. $-182 \div (-13)$ 24. $205 \div (-5)$

25. $-120 \div (-8)$ 26. $135 \div 15$ 27. $168 \div (-12)$

1. _____
2. _____
3. _____
4. _____
5. _____
6. _____
7. _____
8. _____
9. _____
10. _____
11. _____
12. _____
13. _____
14. _____
15. _____
16. _____
17. _____
18. _____
19. _____
20. _____
21. _____
22. _____
23. _____
24. _____
25. _____
26. _____
27. _____

Answers

Try it 1.3B
1. -7
2. 9
3. $-27; -27; -27$

Quiz 1.3B
1. -9
2. 9
3. -24
4. 0
5. -47

Solutions to Practice Sheet 1.3B

1. $18 \div (-6) = -3$

2. $(-36) \div (6) = -6$

3. $(-32) \div (-4) = 8$

4. $7 \div (-7) = -1$

5. $-42 \div 6 = -7$

6. $40 \div (-8) = -5$

7. $-30 \div 5 = -6$

8. $-48 \div 6 = -8$

9. $-72 \div (-8) = 9$

10. $-42 \div (-7) = 6$

11. $-42 \div (-6) = 7$

12. $-50 \div 5 = -10$

13. $-55 \div (-11) = 5$

14. $-84 \div (-7) = 12$

15. $-60 \div (-5) = 12$

16. $-75 \div (-5) = 15$

17. $105 \div 7 = 15$

18. $68 \div (-4) = -17$

19. $72 \div (-9) = -8$

20. $(-128) \div 8 = -16$

21. $192 \div (-12) = -16$

22. $154 \div (-22) = -7$

23. $-182 \div (-13) = 14$

24. $205 \div (-5) = -41$

25. $-120 \div (-8) = 15$

26. $135 \div 15 = 9$

27. $168 \div (-12) = -14$

Objective 1.3C – Solve application problems

Counting Principle

To find the number of possible outcomes when making a series of choices, multiply the number of options available for each choice.

Example

A password is formed by first choosing three numbers from 1, 2, 3, 4, 5, and then choosing three letters from A, B, C, D. How many passwords can be formed if a number or a letter can be used more than once?

> Solution
>
> There are five number options and four letter options. Because the password must contain three numbers and three letters, use the counting principle to find the number of possible passwords.

$$\underbrace{5 \cdot 5 \cdot 5}_{\text{number options}} \cdot \underbrace{4 \cdot 4 \cdot 4}_{\text{letter options}} = 8000$$

> There are 8000 possible passwords, using a letter or number more than once.

The phrase _____ replacement means used more than once.
              (with/without)

Try it

1. An identification tag for a package is created by choosing three numbers from 3, 4, 5, 6, 7, 8, and two letters from E, F, G, H. How many different identification tags can be formed if a letter or number can be used more than once?

Example

A password is formed by first choosing three numbers from 1, 2, 3, 4, 5, and then choosing three letters from A, B, C, D. How many passwords can be formed if a number or a letter can be used only once?

> Solution
>
> There are five number options and four letter options. Because the password must contain three numbers and three letters, use the counting principle to find the number of possible passwords.

$$\underbrace{5 \cdot 4 \cdot 3}_{\text{number options}} \cdot \underbrace{4 \cdot 3 \cdot 2}_{\text{letter options}} = 1440$$

> There are 1440 possible passwords, using a letter or number only once.

The phrase _____ replacement means used more only once.
              (with/without)

Try it

2. An identification tag for a package is created by choosing five numbers from 4, 5, 6, 7, 8, 9, and one letter from K, L, M, N. How many different identification tags can be formed if a letter or number can be used only once?

Example

The daily low temperatures (in degrees Celsius) for five days in Windsor, Ontario, Canada were $-7°, -3°, -3°, 1°,$ and $2°$. Find the average low temperature.

Solution

To find the **average** low temperature
Add the five temperature readings.
Divide the sum by 5.

$$\frac{-7 + (-3) + (-3) + 1 + 2}{5} = \frac{-10}{5} = -2$$

The average low temperature was $-2°C$.

Try it

3. The daily low temperatures (in degrees Fahrenheit) for six days in Fargo, Minnesota were $-10°, -2°, -1°, 3°, -5°,$ and $-3°$. Find the average low temperature.

Example

A credit to an account can be represented with a positive number, and a debit from an account can be represented with a negative number. If 5 credits of $30 and 8 debits of $40 are processed for and against an account with an initial balance of $340, what is the ending balance in the account?

Solution

The ending balance can be determined by computing
$$340 + 5(30) + 8(-40) = 340 + 150 + (-320)$$
$$= 490 + (-320)$$
$$= 170$$

The ending balance in the account is $170.

Try it

4. A credit to an account can be represented with a positive number, and a debit from an account can be represented with a negative number. If 3 credits of $50 and 6 debits of $35 are processed for and against an account with an initial balance of $430, what is the ending balance in the account?

Reflect on it

- The phrase with replacement means used more than once.
- The phrase without replacement means used more only once.

Quiz Yourself 1.3C

In Exercises 1 and 2, use the counting principle to determine the number of ways in which the task can be completed.

1. Three digits are selected without replacement from the digits 2, 3, 4, 5, 6.

2. Three digits are selected with replacement from the digits 2, 3, 4, 5, 6.

3. The daily high temperatures, in degrees Celsius, for a city in northern Canada were $-10°, -12°, -15°$, and $-11°$. Find the average daily high temperature for this period.

4. The daily low temperatures, in degrees Fahrenheit, for a city in Alaska were $-8°, -3°, 4°, 2°, 0°$, and $-1°$. Find the average daily low temperature for this period.

5. To discourage random guessing on a multiple-choice exam, an instructor assigns 5 points for a correct answer, -2 points for an incorrect answer, and -1 points for leaving a question blank. What is the score for a student who had 15 correct answers, 6 incorrect answers, and 4 answers left blank?

6. To discourage random guessing on a multiple-choice exam, an instructor assigns 6 points for a correct answer, -4 points for an incorrect answer, and -2 points for leaving a question blank. What is the score for a student who had 16 correct answers, 7 incorrect answers, and 2 answers left blank?

Practice Sheet 1.3C

Solve.

1. Find the temperature after a rise of 4°C from –2°C.

2. Find the temperature after a rise of 6°C from –10°C.

1. _____

2. _____

3. During a card game of Hearts, you have a score of 14 points before your opponent "shoots the moon," subtracting a score of 26 from your total. What is your score after your opponent "shoots the moon"?

4. In a card game of Hearts, you have a score of –15 before you "shoot the moon," entitling you to add 26 points to your score. What is your score after you "shoot the moon"?

3. _____

4. _____

5. The daily low temperatures during one week were recorded as follows: 4°, –6°, 8°, –2°, –9°, –11°, and –5°. Find the average daily low temperature for the week.

6. The daily high temperatures during one week were recorded as follows: –5°, –8°, 6°, 8°, 0°, –6°, and –2°. Find the average daily high temperature for the week.

5. _____

6. _____

7. One golfer had a score of six under par (–6) while a second golfer had a score of 8 over par (+8). Find the difference between their scores.

8. A four-member golf team had a combined score of 18 under par (–18). Another team had a combined score of 2 over par (+2). Find the difference between their scores.

7. _____

8. _____

9. During the last week in June a stock rose 15 points. The stock then fell 9 points during the next week. Find the net change in the value of the stock for the two week period.

10. During the first week in May a stock fell 11 points. Then the stock rose 6 points during the next week. Find the net change in the value of the stock for the two week period.

9. _____

10. _____

Use the counting principle to determine the number of ways in which the task can be completed.

11. Three letters are selected without replacement from A, B, C, D.

12. Three letters are selected with replacement from A, B, C, D.

11. _____

12. _____

13. Five numbers are selected without replacement from 2, 3, 4, 5, 6, 7.

14. Five numbers are selected with replacement from 2, 3, 4, 5, 6, 7.

13. _____

14. _____

Answers

Try it 1.3C
1. 3456 ways
2. 2880 ways
3. –3°F
4. $370

Quiz 1.3C
1. 60 ways
2. 125 ways
3. The average daily high temperature for this period is –12°F.
4. The average daily low temperature for this period is –1°F.
5. The student's score is 59 points.
6. The student's score is 64 points.

Solutions to Practice Sheet 1.3C

1. STRATEGY
   To find the temperature, add the rise (4°C) to the temperature (–2°C).

   SOLUTION
   $-2 + 4 = 2$
   The temperature is 2°C.

2. STRATEGY
   To find the temperature, add the rise (6°C) to the temperature (–10°C).

   SOLUTION
   $-10 + 6 = -4$
   The temperature is –4°C.

3. STRATEGY
   To find the score, subtract 26 from your score (14).

   SOLUTION
   $14 - 26 = -12$
   Your score is –12 points.

4. STRATEGY
   To find the score, add 26 to your score (–15).

   SOLUTION
   $-15 + 26 = 11$
   Your score is 11 points.

5. STRATEGY
   To find the average daily low temperature:
   Add the seven temperature readings
   Divide the sum by 7.

SOLUTION
$4 + (-6) + 8 + (-2) + (-9) + (-11) + (-5) = -21$
$-21 \div 7 = -3$
The average daily low temperature was –3°.

6. STRATEGY
   To find the average daily high temperature:
   Add the seven temperature readings
   Divide the sum by 7.

   SOLUTION
   $-5 + (-8) + 6 + 8 + 0 + (-6) + (-2) = -7$
   $-7 \div 7 = -1$
   The average daily high temperature was –1°.

7. STRATEGY
   To find the difference, subtract the first score (–6) from the second score (+8).

   SOLUTION
   $8 - (-6) = 8 + 6 = 14$
   The difference in scores is 14.

8. STRATEGY
   To find the difference, subtract the first score (–18) from the second score (+2).

   SOLUTION
   $2 - (-18) = 2 + 18 = 20$
   The difference in scores is 20.

9. STRATEGY
   To find the net change, add the rise (15) to the fall (–9).

   SOLUTION
   $15 + (-9) = 6$
   The net change is 6 points.

10. STRATEGY
    To find the net change, add the fall (–11) to the rise (6).

    SOLUTION
    $-11 + 6 = -5$
    The net change is –5 points.

11. $4 \cdot 3 \cdot 2 = 24$ ways
12. $4 \cdot 4 \cdot 4 = 64$ ways
13. $6 \cdot 5 \cdot 4 \cdot 3 \cdot 2 = 720$ ways
14. $6 \cdot 6 \cdot 6 \cdot 6 \cdot 6 = 7776$ ways

Objective 1.4A – Simplify expressions containing exponents

Key Words and Concepts

Repeated multiplication of the same factor can be written using an exponent.

The **exponent** indicates how many times the factor, which is called the **base**, occurs in the multiplication.

For example, in the expression 3^6, 6 is the exponent and 3 is the base. $3^6 = 3 \cdot 3 \cdot 3 \cdot 3 \cdot 3 \cdot 3$.

The expression 3^6 is in **exponential form**. The expression $3 \cdot 3 \cdot 3 \cdot 3 \cdot 3 \cdot 3$ is in **factored form**.

3^1 is read "3 to the first power" or just "3." Usually the exponent 1 is not written.

3^2 is read "3 to the second power" or "3 squared."

3^3 is read "3 to the third power" or "3 cubed."

To evaluate an exponential expression
   Write each factor as many times as indicated by the exponent.
   Then multiply.

Example

Evaluate $(-3)^4$.

   Solution
      Write −3 as a factor 4 times. Then multiply.
      $$(-3)^4 = (-3)(-3)(-3)(-3)$$
      $$= 81$$

Try it

1. Evaluate $(-4)^4$.

Example

Evaluate -3^4.

   Solution
      Write 3 as a factor 4 times. Then multiply.
      $$-3^4 = -(3 \cdot 3 \cdot 3 \cdot 3)$$
      $$= -81$$

Try it

2. Evaluate -4^4.

Example

Evaluate $(-1)^3 (-2)^5 (-3)^2$.

Try it

3. Evaluate $(-1)^2 (-2)^3 (-3)^4$.

Solution

Write out all the factors. Then multiply.

$(-1)^3 (-2)^5 (-3)^2$

$= (-1)(-1)(-1) \cdot (-2)(-2)(-2)(-2)(-2) \cdot (-3)(-3)$

$= -1 \cdot (-32) \cdot 9 \quad$ Since $(-1)^3 = -1;\ (-2)^5 = -32$

$\qquad\qquad\qquad$ and $(-3)^2 = 9$

$= 32 \cdot 9$

$= 288$

Reflect on it

• To evaluate an exponential expression, write each factor as many times as indicated by the exponent. Then multiply.

Quiz Yourself 1.4A

1. Evaluate -4^2.

2. Evaluate $(-4)^2$.

3. Evaluate $(-3)^5$.

4. Evaluate $(-3)^6$.

5. Evaluate $(-1)^8 (-4)^3 (-2)^2$.

6. Evaluate $(-1)^{11} \cdot 3^3 \cdot (-2)^4$

Practice Sheet 1.4A

Evaluate.

1. 4^2

2. 5^3

3. -5^2

4. -3^4

5. $(-2)^4$

6. $(-3)^3$

7. $(-5)^2$

8. $(-4)^2$

9. $-(-2)^3$

10. $-(-3)^3$

11. $(12)^3$

12. $(-2)^2 \cdot 4^3$

13. $(-1)^3 \cdot 2^2$

14. $(-1)^2 \cdot 2^3$

15. $(-2)^4 \cdot 5^2$

16. $(-5) \cdot 4^3$

17. $2^2 \cdot 3^2 \cdot (-4)^2$

18. $(-3) \cdot 2^4$

19. $(-3) \cdot (-2)^2$

20. $(-3) \cdot (-2)^3$

21. $2^2 \cdot 3^2 \cdot (-5)$

22. $(-2)^2 \cdot 3^2 \cdot 6$

23. $(-6) \cdot 2^3 \cdot 3^2$

24. $(-2) \cdot 3^2 \cdot (-2)^2$

1. _____

2. _____

3. _____

4. _____

5. _____

6. _____

7. _____

8. _____

9. _____

10. _____

11. _____

12. _____

13. _____

14. _____

15. _____

16. _____

17. _____

18. _____

19. _____

20. _____

21. _____

22. _____

23. _____

24. _____

Answers

Try it 1.4A
1. 256
2. −256
3. −648

Quiz 1.4A
1. −16
2. 16
3. −243
4. 729
5. −256
6. −432

Solutions to Practice Sheet 1.4A

1. $4^2 = 4 \cdot 4 = 16$

2. $5^3 = 5 \cdot 5 \cdot 5 = 125$

3. $-5^2 = -(5 \cdot 5) = -25$

4. $-3^4 = -(3 \cdot 3 \cdot 3 \cdot 3) = -81$

5. $(-2)^4 = (-2)(-2)(-2)(-2) = 16$

6. $(-3)^3 = (-3)(-3)(-3) = -27$

7. $(-5)^2 = (-5)(-5) = 25$

8. $(-4)^2 = (-4)(-4) = 16$

9. $-(-2)^3 = -(-2)(-2)(-2) = 8$

10. $-(-3)^3 = -(-3)(-3)(-3) = 27$

11. $12^3 = 12 \cdot 12 \cdot 12 = 1728$

12. $(-2)^2 \cdot 4^3 = (-2)(-2)(4 \cdot 4 \cdot 4)$
$= 4(64)$
$= 256$

13. $(-1)^3 \cdot 2^2 = (-1)(-1)(-1)(2 \cdot 2)$
$= -1(4)$
$= -4$

14. $(-1)^2 \cdot 2^3 = (-1)(-1)(2 \cdot 2 \cdot 2)$
$= 1(8)$
$= 8$

15. $(-2)^4 \cdot 5^2 = (-2)(-2)(-2)(-2)(5 \cdot 5)$
$= 16(25)$
$= 400$

16. $-5 \cdot 4^3 = -5(4 \cdot 4 \cdot 4)$
$= -5(64)$
$= -320$

17. $2^2 \cdot 3^2 \cdot (-4)^2 = (2 \cdot 2)(3 \cdot 3)(-4)(-4)$
$= 4(9)(16)$
$= 36(16)$
$= 576$

18. $(-3) \cdot 2^4 = (-3)(2 \cdot 2 \cdot 2 \cdot 2)$
$= -3(16)$
$= -48$

19. $(-3) \cdot (-2)^2 = (-3)(-2)(-2)$
$= -3(4)$
$= -12$

20. $(-3) \cdot (-2)^3 = (-3)(-2)(-2)(-2)$
$= -3(-8)$
$= 24$

21. $2^2 \cdot 3^2 \cdot (-5) = (2 \cdot 2)(3 \cdot 3)(-5)$
$= 4(9)(-5)$
$= 36(-5)$
$= -180$

22. $(-2)^2 \cdot 3^2 \cdot 6 = (-2)(-2)(3 \cdot 3)(6)$
$= 4(9)(6)$
$= 36(6)$
$= 216$

23. $(-6) \cdot 2^3 \cdot 3^2 = (-6)(2 \cdot 2 \cdot 2)(3 \cdot 3)$
$= -6(8)(9)$
$= -48(9)$
$= -432$

24. $(-2) \cdot 3^2 \cdot (-2)^2 = (-2)(3 \cdot 3)(-2)(-2)$
$= -2(9)(4)$
$= -18(4)$
$= -72$

Objective 1.4B – Use the Order of Operations Agreement to simplify expressions

Key Words and Concepts

To prevent there being more than one answer when simplifying a numerical expression, an Order of Operations Agreement has been established.

The Order of Operations Agreement

Step 1: Perform operations inside grouping symbols. Grouping symbols include parentheses ( ), brackets [ ], braces { }, the absolute value symbol | |, and the fraction bar.

Step 2: Simplify exponential expressions.

Step 3: Do multiplication and division as they occur from left to right.

Step 4: Do addition and subtraction as they occur from left to right.

Example

Evaluate $24 \div \left[-10-(1-3)\right]+2^2$.

Solution

$24 \div \left[-10-(1-3)\right]+2^2$ Copy the expression

$= 24 \div \left[-10-(-2)\right]+2^2$ Perform operations inside parentheses
$(1-3)=-2$

$= 24 \div (-8)+2^2$ Perform operations inside brackets
$-10-(-2)=-8$

$= 24 \div (-8)+4$ Simplify exponential expression
$2^2=4$

$= -3+4$ Do division from left to right
$24 \div (-8)=-3$

$= 1$ Do addition from left to right
$-3+4=1$

Try it

1. Evaluate $15 \div \left[-8-(6-9)\right]+3^2$.

Example

Evaluate $5 - \dfrac{4-20}{2-(-2)^2} - 3$.

Try it

2. Evaluate $7 + \dfrac{8-20}{3-(-3)^2} - 3$.

Solution

$5 - \dfrac{4-20}{2-(-2)^2} - 3$ Copy the expression

$= 5 - \dfrac{4-20}{2-4} - 3$ Simplify exponential expression
$(-2)^2 = 4$

$= 5 - \dfrac{-16}{-2} - 3$ Do addition and subtraction
$(4-20) = -16$ and $(2-4) = -2$

$= 5 - 8 - 3$ Do division
$-16 \div (-2) = 8$

$= -3 - 3$ Do addition and subtraction left to right
$5 - 8 = -3$

$= -6$

Reflect on it

- One or more of the steps listed may not be needed to evaluate an expression. In that case, proceed to the next step in the Order of Operations Agreement.

Quiz Yourself 1.4B

1. Evaluate $-5^2 + 5 \cdot 2$.

2. Evaluate $12 \div (-6) \cdot 2 + 4$.

3. Evaluate $-28 \div 4 - 2^2 - (-3)^3$.

4. Evaluate $12 \div (3-5)^2 \cdot (-2)$.

5. Evaluate $9 + \dfrac{(-3)^2 + 1}{-1 - 2^2} \div (-2)$.

6. Evaluate $-8^2 + 3\left[4 - (1-3)^2\right]$.

Practice Sheet 1.4B

Evaluate.

1. $9 - 3 \times 2$

2. $3^2 \cdot 2 - 4$

3. $3(5-1) - (-2)^2$

4. $15 - 16 \div 2^2$

5. $21 - 15 \div 3 + 4$

6. $10 - (-2)^3 - (-1)$

7. $12 + 18 \div (-3) - 1$

8. $10 - 2^3 - (2 - 5)$

9. $5 - 3[9 - (5 - 3)]$

10. $-3^2 + 2[8 \div (1 + 3)]$

11. $20 \div 5 - 3^2 + (-2)^3$

12. $25 \div (14 - 3^2) + (-5)$

13. $8(-9) + [3(7 - 3)^2]$

14. $11 + \dfrac{19 - 3}{3^2 - 1} - 7$

15. $32 \div \dfrac{4^2}{7 - 5} - (-6)$

16. $48 \div 4[11 - (6 - 1)] - (-2)^2$

17. $3 \cdot [15 - (8 - 5)] \div 6$

18. $\dfrac{(-9) + (-7)}{5^2 - 17} \div (3 - 5)$

19. $-3^2 + 10[20 \div (2 - 7)]$

20. $5 \cdot [20 - (9 - 4)] \div 25$

21. $6 \cdot [14 - (11 - 9)] \div 3^2$

22. $15 + 21 \div (-7) - 3$

23. $12 - 3^2 - (2 - 7)$

24. $(2.54 - 2.09)^2 \div 0.5 + 0.9$

1. _____

2. _____

3. _____

4. _____

5. _____

6. _____

7. _____

8. _____

9. _____

10. _____

11. _____

12. _____

13. _____

14. _____

15. _____

16. _____

17. _____

18. _____

19. _____

20. _____

21. _____

22. _____

23. _____

24. _____

Answers

Try it 1.4B
1. 6
2. 6

Quiz 1.4B
1. -15
2. 0
3. 16
4. -6
5. 10
6. -64

Solutions to Practice Sheet 1.4B

1. $9 - 3 \times 2 = 9 - 6 = 3$

2. $3^2 \cdot 2 - 4 = 9 \cdot 2 - 4$
   $= 18 - 4$
   $= 14$

3. $3(5 - 1) - (-2)^2 = 3(4) - 4$
   $= 12 - 4$
   $= 8$

4. $15 - 16 \div 2^2 = 15 - 16 \div 4$
   $= 15 - 4$
   $= 11$

5. $21 - 15 \div 3 + 4 = 21 - 5 + 4$
   $= 16 + 4$
   $= 20$

6. $10 - (-2)^3 - (-1) = 10 - (-8) - (-1)$
   $= 10 + 8 + 1$
   $= 19$

7. $12 + 18 \div (-3) - 1 = 12 - 6 - 1$
   $= 6 - 1$
   $= 5$

8. $10 - 2^3 - (2 - 5) = 10 - 8 - (-3)$
   $= 2 + 3$
   $= 5$

9. $5 - 3[9 - (5 - 3)] = 5 - 3[9 - 2]$
   $= 5 - 3[7]$
   $= 5 - 21$
   $= -16$

10. $-3^2 + 2[8 \div (1 + 3)] = -9 + 2[8 \div 4]$
    $= -9 + 2[2]$
    $= -9 + 4$
    $= -5$

11. $20 \div 5 - 3^2 + (-2)^3 = 20 \div 5 - 9 - 8$
    $= 4 - 9 - 8$
    $= -5 - 8$
    $= -13$

12. $25 \div (14 - 3^2) + (-5) = 25 \div (14 - 9) + (-5)$
    $= 25 \div (5) + (-5)$
    $= 5 + (-5)$
    $= 0$

13. $8(-9) + [3(7 - 3)^2] = -72 + [3(4)^2]$
    $= -72 + 3(16)$
    $= -72 + 48$
    $= -24$

14. $11 + \dfrac{19 - 3}{3^2 - 1} - 7 = 11 + \dfrac{16}{9 - 1} - 7$
    $= 11 + \dfrac{16}{8} - 7$
    $= 11 + 2 - 7$
    $= 6$

15. $32 \div \dfrac{4^2}{7 - 5} - (-6) = 32 \div \dfrac{16}{2} + 6$
    $= 32 \div 8 + 6$
    $= 4 + 6$
    $= 10$

16. $48 \div 4[11 - (6 - 1)] - (-2)^2$
    $= 48 \div 4[11 - 5] - 4$
    $= 48 \div 4[6] - 4$
    $= 12[6] - 4$
    $= 72 - 4$
    $= 68$

17. $3 \cdot [15 - (8 - 5)] \div 6 = 3 \cdot [15 - 3] \div 6$
    $= 3 \cdot [12] \div 6$
    $= 36 \div 6$
    $= 6$

18. $\dfrac{(-9) + (-7)}{5^2 - 17} \div (3 - 5) = \dfrac{-16}{25 - 17} \div (-2)$
    $= \dfrac{-16}{8} \div (-2)$
    $= -2 \div (-2)$
    $= 1$

19. $-3^2 + 10[20 \div (2 - 7)] = -9 + 10[20 \div (-5)]$
    $= -9 + 10[-4]$
    $= -9 + (-40)$
    $= -49$

20. $5 \cdot [20 - (9 - 4)] \div 25 = 5 \cdot [20 - 5] \div 25$
    $= 5 \cdot [15] \div 25$
    $= 75 \div 25$
    $= 3$

21. $6 \cdot [14 - (11 - 9)] \div 3^2 = 6 \cdot [14 - 2] \div 9$
$$= 6 \cdot [12] \div 9$$
$$= 72 \div 9$$
$$= 8$$

22. $15 + 21 \div (-7) - 3 = 15 + (-3) - 3$
$$= 9$$

23. $12 - 3^2 - (2 - 7) = 12 - 9 - (-5)$
$$= 3 + 5$$
$$= 8$$

24. $(2.54 - 2.09)^2 \div 0.5 + 0.9 = (0.45)^2 \div 0.5 + 0.9$
$$= 0.2025 \div 0.5 + 0.9$$
$$= 0.405 + 0.9$$
$$= 1.305$$

Objective 1.5A – Use inductive reasoning

Key Words and Concepts

The type of reasoning in which a conclusion is formed by examining specific examples is called *inductive reasoning*.

Define conjecture. _____

_____

Inductive Reasoning

Inductive reasoning is the process of reaching a general conclusion by examining specific examples.

Example

Use inductive reasoning to predict the next number in the list.

1, 5, 9, 13, 17, ?

Solution
Each successive number is 4 larger than the preceding number.
Thus, we predict that the next number in the list is 4 larger than 17, which is 21.

Try it

1. Use inductive reasoning to predict the next number in the list.
   2, 7, 12, 17, 22, ?

Example

Consider the following procedure: Pick a number. Multiply the number by 4, add 8 to the product, divide the sum by 2, and subtract 4. Complete the procedure for several different numbers. Use inductive reasoning to make a conjecture about the relationship between the size of the resulting number and the size of the original number.

Solution
Suppose set pick 5 as our original number.

Original number: 5
Multiply by 4: $5 \times 4 = 20$
Add 8: $20 + 8 = 28$
Divide by 2: $28 \div 2 = 14$
Subtract 5: $14 - 4 = 10$

Starting with 6 as our original number produces a final result of 12. Starting with 10 produces a final result of 20. Starting with 100 produces a final result of 200.

We *conjecture* that the given procedure produces a number that is two times the original number.

Try it

2. Consider the following procedure: Pick a number. Multiply the number by 5, add 15 to the product, divide the sum by 5, and subtract 3.

   Complete the procedure for several different numbers. Use inductive reasoning to make a conjecture about the relationship between the size of the resulting number and the size of the original number.

In mathematics, a statement is considered a true statement if and only if it is true in all cases. If you can find even one case for which a statement is not true, called a **counterexample**, then the statement is false.

Example
Find a counterexample to the statement "All numbers ending in 7 can be divided evenly by 7."

Solution

The statement is true for some numbers that end in 7.

For instance, 77 ends in 7, and $77 \div 7 = 11$.
However, 47 ends in 7, and 7 does not divide into 47 evenly.

47 is a counterexample to the statement.

Since are other counterexamples, such as 37 and 57, answers will vary.

Try it
3. Find a counterexample to the statement "Any number squared is greater than the original number."

Reflect on it
- The conclusion reached through inductive reasoning is often called conjecture, since it may or may not be correct.
- To prove a statement false, you need only one counterexample.

Quiz Yourself 1.5A
1. Use inductive reasoning to predict the next number in the list: 2, 7, 17, 32, 52, 77, ?

2. Use inductive reasoning to predict the next number in the list: 2, –1, 5, 2, 8, 5, 11, 8, ?

3. Consider the following procedure: pick a number. Multiply the number by 25, add 15 to the product, divide the sum by 5, and subtract 3 from the quotient. Complete the above procedure for several different numbers. Use inductive reasoning to make a conjecture about the relationship between the resulting number and the original number.

4. Consider the following procedure: pick a number. Multiply the number by 100, subtract 40 from the product, divide the difference by 20, and add 2 to the quotient. Complete the above procedure for several different numbers. Use inductive reasoning to make a conjecture about the relationship between the resulting number and the original number.

5. Find a counterexample to the statement "All numbers ending in 3 can be divided evenly by three."

Practice Sheet 1.5A

Solve.

1. Use inductive reasoning to predict the next number in the list.
2, 12, 32, 62, 102, 152, ?

2. Use inductive reasoning to predict the next number in the list.
3, 7, 15, 27, 43, 63, ?

1. _____

2. _____

3. Use inductive reasoning to predict the next number in the list.
7, 9, 3, 5, –1, 1, ?

4. Consider the following procedure: Pick a counting number, multiply the number by 6, add 24 to the product, divide the sum by 3, and subtract 8 from the quotient.
Complete the above procedure for several different numbers. Use inductive reasoning to make a conjecture about the relationship between the resulting number and the original number.

3. _____

4. _____

5. Consider the following procedure: Pick a counting number, multiply the number by 12, subtract 8 from the product, divide the difference by 4, and add 2 to the quotient.
Complete the above procedure for several different numbers. Use inductive reasoning to make a conjecture about the relationship between the resulting number and the original number.

6. Find a counterexample to the statement:
"Twice any number is greater than the original number."

5. _____

6. _____

7. Find a counterexample to the statement:
"Any number squared is greater than the original number."

8. Find a counterexample to the statement:
"A number divided by itself equals one."

7. _____

8. _____

Answers

Try it 1.5A
1. 27
2. The resulting number is the same as the original number.
3. Answers will vary. One counterexample is 0.

Quiz 1.5A
1. 107
2. 14
3. The resulting number is 5 times the original number.
4. The resulting number is 5 times the original number.
5. Answers will vary. One counterexample is 23.

Solutions to Practice Sheet 1.5A

1. 212

2. 87

3. −5   (add 2, subtract 6)

4. The resulting number is twice the original number.

5. The resulting number is three times the original number.

6. Answers will vary. One counterexample is −10.

7. Answers will vary. One counterexample is 0.5.

8. Answers will vary. Zero is the counterexample.

Objective 1.5B – Use deductive reasoning

Key Words and Concepts

Another type of reasoning is called *deductive reasoning*.

Deductive Reasoning

Deductive reasoning is the process of reaching a conclusion by applying general assumptions, procedures, or principles.

Example

Use deductive reasoning to create a true statement from the following information:
Breakfast is served in the cafeteria between 8AM and 9:30AM. Bill eats breakfast in the cafeteria.

Solution
From the given information, we can deduce that
Bill eats breakfast between 8AM and 9:30AM.

Try it

1. Use deductive reasoning to create a true statement from the following information:
All football players wear shoulder pads. Tony is a football player.

Example

Use deductive reasoning to complete the statement: If JJJJJ = HH and GGGG = HH, then how many J's = GGGG

Solution
Because five "J" symbols are equal to HH and four "G" symbols are also equal to HH, we can use deductive reasoning to state that
five J symbols are equal to four G symbols.

Try it

2. Use deductive reasoning to complete the statement:
If NN = LLLL and MMMMMM = LLLL, then how many N's = MMM?

Example

Each of three friends, Frances, Jack, and Taylor, eats a different food for lunch: sandwich, pizza, or taco. From the following clues, determine which food each friend eats.

1. Taylor does not eat a sandwich or pizza.
2. Frances eats after the friend who eats the taco.
3. The friend who eats the pizza eats before Taylor.

Solution

Create a table to help solve the problem. Make a row for each friend and a column for each food.

	Sandwich	Pizza	Taco
Frances			
Jack			
Taylor			

From the first clue, we know that Taylor does not eat a sandwich or pizza. Put "X1" in the sandwich and pizza columns of Taylor's row. From the table, we can see that Taylor must eat the taco, so put a check in that spot.

	Sandwich	Pizza	Taco
Frances			
Jack			
Taylor	X1	X1	√

Now since Taylor eats the taco, that means that neither Frances nor Jack can eat the taco. Put "X1" in the taco column of both Frances and Jack's rows.

	Sandwich	Pizza	Taco
Frances			X1
Jack			X1
Taylor	X1	X1	√

From the second clue, we know that Frances eats *after* Taylor, who eats the taco. From the third clue we know that the friend who eats the pizza eats before Taylor. That means that Jack must eat the pizza, because Frances eats after Taylor.

	Sandwich	Pizza	Taco
Frances			X1
Jack	X3	√	X1
Taylor	X1	X1	√

Therefore, Frances must eat the sandwich.

	Sandwich	Pizza	Taco
Frances	√	X2	X1
Jack	X3	√	X1
Taylor	X1	X1	√

From the completed table, Frances eats the sandwich, Jack eats the pizza, and Taylor eats the taco.

Try it

3. Three football players, Tom, Bob, and Sam, each have a different numbered jersey: 10, 45, and 51. From the following clues, determine which numbered jersey each football player wears.

1. Sam does not wear an odd-numbered jersey.
2. Bob runs faster than the player who wears 45.

Reflect on it
- Take your time to think through the logic puzzles.

Quiz Yourself 1.5A

1. Use deductive reasoning to create a true statement from the following information: The entire 8AM chemistry lab wore protective eyewear in lab today. Alexis is a student in the 8AM chemistry lab.

2. Use deductive reasoning to create a true statement from the following information: Dan earned $3 per delivery for a pizza shop. Last evening, Dan made 12 pizza deliveries.

3. Use deductive reasoning to complete the statement: If AAA = BB and BB = CCCCCC, then A = _?_ C's.

4. Use deductive reasoning to answer the following question: If five X's are equal to ten P's and two P's equal three Q's, how many Q's are equal to one X?

5. Is the following argument a correct example of deductive reasoning? Troy likes to solve equations. All the students in Joliet's algebra class like to solve equations. Therefore Troy is a student in Joliet's algebra class.

6. Is the following argument a correct example of deductive reasoning? Rene is a member of the choir at his college. All of the members of the choir like to practice yoga. Therefore, Rene likes to practice yoga.

7. Use deductive reasoning to solve the following logic problem. Each of three students, Edward, Tyler, and Samantha, take a different form of transportation to the park: walk, bike, or gets a ride. Use the following clues to determine which student takes each form of transportation.
   1. Tyler gets to the park before Edward, who does not get a ride.
   2. Edwards gets to the park after the person who rides a bike.
   3. The person who rides a bike gets to the park after Samantha.

Practice Sheet 1.5B

Solve.

1. Use deductive reasoning to create a true statement from the following information:
All Girl Scouts like cookies. Tori is a Girl Scout.

2. Use deductive reasoning to create a true statement from the following information:
All pokers players do not like to smile. Miles is a poker player.

1. _____

2. _____

3. Use deductive reasoning to create a true statement from the following information:
When mail is posted in the mailbox, the red flag is raised. Mail is posted in the mailbox.

4. Use deductive reasoning to answer the following question:
If AAA = BBBBBB, and BB = CCC, then A = _?_ C.

3. _____

4. _____

5. Use deductive reasoning to answer the following question:
If three X's are equal to four O's, and two O's are equal to three Q's, how many Q's are equal to one X?

6. Use deductive reasoning to answer the following question:
If five X's are equal to nine O's, and three O's are equal to five Q's, how many Q's are equal to one X?

5. _____

6. _____

7. Is the following argument a correct example of deductive reasoning?
Stephanie likes to eat raw cookie dough. All cookie bakers like to eat raw cookie dough. Therefore, Stephanie is a cookie baker.

8. Use deductive reasoning to solve the following logic problem.
Each of three college roommates Sarah, Abby, and Grace, is from a different state: California, Minnesota, and Pennsylvania. Use the following clues to determine the home state of each roommate.
1. Sarah arrives before Abby who is not from Pennsylvania.
2. Abby arrives immediately after the person who is from California.
3. The roommate who is from California arrives to the room after Grace.

7. _____

8. _____

Answers

Try it 1.5B
1. Tony wears shoulder pads.
2. One
3. Tom wears 45, Bob wears 51, and Sam wears 10

Quiz 1.5B
1. Alexis wore protective eyewear in the chemistry lab today.
2. Last evening, Dan earned $36 making pizza deliveries.
3. Two
4. Three Q's
5. No
6. Yes
7. Edward walks, Tyler rides the bike, Samantha gets a ride.

Solutions to Practice Sheet 1.5B
1. Tori likes cookies.

2. Miles does not like to smile.

3. The red flag is up.

4. CCC

5. Two Q's

6. Three Q's

7. No

8. Sarah is from California, Abby is from Minnesota, Grace is from Pennsylvania.

Objective 1.5C – Choose between induction and deductive reasoning

Example

Determine whether the argument is an example of inductive reasoning or deductive reasoning.
Every Monday there is a flight to Chicago. Tomorrow is Monday. Thus, tomorrow there is a flight to Chicago.

Solution
Because the conclusion is a specific case of a general assumption, this argument is an example of deductive reasoning.

Try it

1. Determine whether the argument is an example of inductive reasoning or deductive reasoning.
The crossing guard helped 200 children safely cross the street today. Therefore, the crossing guard will help children safely cross the street tomorrow.

Quiz Yourself 1.5C

In Exercises 1–5, determine whether the argument is based on inductive or deductive reasoning.

1. All miles are exactly 5280 ft. The school track is a mile. Therefore, the school track is 5280 ft.

2. Every French chef likes to make desserts. Jules is a French chef. Jules likes to make desserts.

3. Jason wore contact lenses six days in a row. Therefore, Jason will wear contact lenses tomorrow.

4. The CTA subway line to O'Hare airport has never been more than 10 minutes late all week. Therefore, the CTA subway line will arrive within 10 minutes of its scheduled time next week.

5. All book club members have read this month's book. Andrea is in the book club. Therefore, Andrea has read this month's book.

Practice Sheet 1.5C

Determine whether the argument is based on inductive or deductive reasoning.

1. All trapezoids have exactly four sides. Figure D is a trapezoid. Therefore, Figure D has four sides.

2. All delivery trucks have a warning back-up signal. Ted has a delivery truck. Therefore, Ted's truck has a warning back-up signal.

1. _____

2. _____

3. The Cubs have not won a World Series over 100 years. Therefore, the Cubs will not win a World Series next year.

4. All Presidents are sworn into office. Ronald Reagan was a President. Ronald Reagan was sworn into office.

3. _____

4. _____

5. The Girl Scout sold twenty boxes of cookies today. Therefore, the Girl Scout will sell a box of cookies tomorrow.

6. Books checked out from the public library are due in three weeks. Joy check out a library book today. Therefore, Joy's library book will be due in three weeks.

5. _____

6. _____

7. The garage door opener opened and closed the garage door twenty times last week. Therefore, the garage door opener will open and close the garage tomorrow.

8. Yesterday, the delivery truck driver signaled every time he changed lanes while he drove on the highway. Therefore, the delivery truck driver will signal his lane change while he drives today.

7. _____

8. _____

Answers

Try it 1.5C
  1. Inductive

Quiz 1.5C
  1. Deductive
  2. Deductive
  3. Inductive
  4. Inductive
  5. Deductive

Solutions to Practice Sheet 1.5C

  1. Deductive

  2. Deductive

  3. Inductive

  4. Deductive

  5. Inductive

  6. Deductive

  7. Inductive

  8. Inductive

Objective 1.6A – Write sets and use Venn diagrams

Key Words and Concepts

A collection of objects is called a set, and the objects in the set are called elements, or members, of the set.

What is the universal set? _____

_____

Note   The letter U is used to denote the universal set.

The Complement of a Set

The complement of a set A, denoted by A' is the set of all elements of the universal set U that are not elements of A.

Example

Let $U = \{5, 10, 15, 20, 25, 30\}$, $S = \{10, 20, 30\}$. Find S'.

Solution
   From the universal set, exclude the elements of S.

   $$S' = \{5,\ 15,\ 25\}$$

Try it

1. Let $U = \{3, 6, 9, 12, 15, 18\}$, $P = \{6, 12, 18\}$. Find P'.

What is the empty set? _____

_____

Note   The symbol $\varnothing$ is used to denote the empty set.

The Complement of the Universal Set and the Complement of the Empty Set

   $U' = \varnothing$ and $\varnothing' = U$

A Subset of a Set

Set A is a subset of set B, denoted by $A \subseteq B$, if and only if every element of A is also an element of B.

Example

Determine whether the statement is true or false.

   $\{4,\ 7,\ 11,\ 14\} \subseteq \{2,\ 4,\ 7,\ 9,\ 11,\ 13,\ 14\}$

Solution
   True. Every element in the first set is an element in the second set.

Try it

2. Determine whether the statement is true or false.

   $\{2,\ 9,\ 13\} \subseteq \{2,\ 5,\ 13\}$

Reflect on it
- The set of all elements that are being considered for a particular problem is called the universal set.
- The empty set is the set containing no elements.
- In a Venn diagram, the universal set is represented by a rectangular region, and subsets of the universal set are generally represented by oval or circular regions drawn inside the rectangle.

Quiz Yourself 1.6A

In Exercises 1 and 2, find the complement of the set given that $U = \{0, 2, 4, 5, 6, 7, 9\}$.

   1. $\{0, 5, 9\}$

   2. $\{0, 2, 4, 5, 6, 7, 9\}$

In Exercises 3 to 5, let $U = \{a, b, c, d, e\}$, $P = \{a, b, c\}$, $Q = \{d, e\}$, and $R = \{a, b, c, d, e\}$. Determine whether each statement is true or false.

   3. $Q \subseteq R$

   4. $P' \subseteq Q$

   5. $R' \subseteq P$

Practice Sheet 1.6A

For Exercises 1–4, find the complement of each set given $U = \{1, 3, 5, 7, 9, 11, 13\}$.

1. $\{3, 7, 11\}$

2. $\{1, 3, 5, 7, 9\}$

1. _____

2. _____

3. $\{1, 3, 5, 7, 9, 11, 13\}$

4. $\{9, 11, 13\}$

3. _____

4. _____

For Exercises 5–10, let $U = \{a, b, c, d, e, f, g\}$, $K = \{a, b, d, e, f, g\}$, $M = \{b, d, f\}$, $N = \{a, g\}$. **Determine whether each statement is true or false.**

5. $N \subseteq K$

6. $M \subseteq N'$

5. _____

6. _____

7. $K' \subseteq M$

8. $N' \subseteq K$

7. _____

8. _____

9. $M' \subseteq N$

10. $K \subseteq U$

9. _____

10. _____

Answers

Try it 1.6A

1. $P' = \{3,\ 9,\ 15\}$
2. False

Quiz 1.6A

1. $\{2, 4, 6, 7\}$
2. $\varnothing$
3. True
4. True
5. True

Solutions to Practice Sheet 1.6A

1. $\{1, 5, 9, 13\}$

2. $\{11, 13\}$

3. $\varnothing$

4. $\{1, 3, 5, 7\}$

5. True

6. True

7. False

8. False

9. False

10. True

Objective 1.6B – Find the intersections and union of sets

Key Words and Concepts

Intersection of Sets

The intersection of sets A and B, denoted by $A \cap B$, is the set of elements common to both A and B.

$$A \cap B = \{x \mid x \in A \text{ and } x \in B\}$$

Example

Let $P = \{1, 2, 5, 8\}$, $Q = \{2, 4, 6, 8\}$. Find $P \cap Q$.

Solution

The elements common to P and Q are 2 and 8.

$$P \cap Q = \{1,\ 2,\ 5,\ 8\} \cap \{2,\ 4,\ 6,\ 8\}$$
$$= \{2,\ 8\}$$

Try it

1. Let $M = \{3, 5, 6, 9\}$ and $N = \{1, 2, 5, 9\}$. Find $M \cap N$.

What does it mean if two sets are **disjoint**? _____

_____

Union of Sets

The union of sets A and B, denoted by $A \cup B$, is the set that contains all the elements that belong to A or to B, or to both.

$$A \cup B = \{x \mid x \in A \text{ or } x \in B\}$$

Example

Let $W = \{6, 7, 8, 9\}$ and $Z = \{1, 3, 7, 9\}$. Find $W \cup Z$.

Solution

The elements of W, 6, 7, 8, 9, with the elements of Z that have not already been listed, 1 and 3.

$$W \cup Z = \{6,\ 7,\ 8,\ 9\} \cup \{1,\ 3,\ 7,\ 9\}$$
$$= \{1,\ 3,\ 6,\ 7,\ 8,\ 9\}$$

Try it

2. Let $G = \{2, 4, 6, 8\}$ and $H = \{1, 3, 5, 7\}$. Find $G \cup H$.

Example

Let $U = \{2, 4, 6, 8, 10, 12, 14, 16, 18\}$, $D = \{4, 8, 12, 16\}$, $E = \{4, 6, 8, 10, 12, 14, 18\}$, and $F = \{4, 14, 16, 18\}$.
Find $D \cup (E \cap F)$.

   Solution

   First find $E \cap F$, which is $\{4, 14, 18\}$.

   Then find $D \cup (E \cap F)$.

   $$D \cup (E \cap F) = D \cup \{4, 14, 18\}$$
   $$= \{4, 8, 12, 14, 16, 18\}$$

Try it

3. Let
   $U = \{1, 4, 7, 10, 13, 16, 19, 22, 25\}$,
   $A = \{7, 16, 22, 25\}$,
   $B = \{1, 7, 10, 16, 25\}$, and
   $C = \{1, 4, 13, 19, 22\}$.
   Find $(A \cup C) \cap B$.

Reflect on it

- Intersection uses the keyword "and".
- Union uses the keyword "or".

Quiz Yourself 1.6B

In Exercises 1 to 5, let $U = \{1, 3, 5, 7, 9, 11, 13, 15, 17, 19, 21, 23, 25\}$, $A = \{3, 9, 15, 21\}$, $B = \{7, 9, 11, 15, 19\}$, $C = \{1, 17, 25\}$. Find each of the following.

   1. $A \cup B$

   2. $A \cap B'$

   3. $A \cap (B \cap C)$

   4. $A' \cup (B \cap C)$

   5. $(A \cap B') \cap C'$

Practice Sheet 1.6B

Let $U = \{2, 4, 6, 8, 10, 12, 14, 16\}$, $A = \{2, 6, 10\}$, $B = \{2, 8, 10, 12\}$, $C = \{6, 10, 14, 16\}$. **Find each of the following.**

1. $A \cup B$ 2. $A' \cap B$ 1. _____

2. _____

3. $A \cap C$ 4. $A \cap (B \cap C)$ 3. _____

4. _____

5. $A' \cap (B \cap C)$ 6. $A' \cup (B \cap C)$ 5. _____

6. _____

7. $(A \cup B) \cap C'$ 8. $(A' \cup B') \cap C$ 7. _____

8. _____

9. $A' \cup B'$ 10. $B' \cap C'$ 9. _____

10. _____

Answers

Try it 1.6B

1. $M \cap N = \{5, 9\}$
2. $G \cup H = \{1, 2, 3, 4, 5, 6, 7, 8\}$
3. $(A \cup C) \cap B = \{1, 7, 16, 25\}$

Quiz 1.6B

1. $A \cup B = \{3, 7, 9, 11, 15, 19, 21\}$
2. $A \cap B' = \{3, 21\}$
3. $A \cap (B \cap C) = \varnothing$
4. $A' \cup (B \cap C) = \{1, 5, 7, 11, 13, 17, 19, 23, 25\}$
5. $(A \cap B') \cap C' = \{3, 21\}$

Solutions to Practice Sheet 1.6B

1. $A \cup B = \{2, 6, 8, 10, 12\}$

2. $A' \cap B = \{8, 12\}$

3. $A \cap C = \{6, 10\}$

4. $A \cap (B \cap C) = \{10\}$

5. $A' \cap (B \cap C) = \varnothing$

6. $A' \cup (B \cap C) = \{4, 8, 10, 12, 14, 16\}$

7. $(A \cup B) \cap C' = \{2, 8, 12\}$

8. $(A' \cup B') \cap C = \{6, 14, 16\}$

9. $A' \cup B' = \{4, 6, 8, 12, 14, 16\}$

10. $B' \cap C' = \{4\}$

Application and Activities

Discussion and reflection questions:

1. What would you like your net worth to be (realistically) by the time you are 50?

2. What percent of your income will you try to save each year towards retirement?

3. If you made $100,000 net each year, how expensive of a house do you think you could realistically afford?

Class Activity:

Objectives for the lesson:

In this lesson you will use and enhance your understanding of the following:

- Adding integers
- Solving applied problems

1. Carl and Lisa want to start a budget. They will start by figuring out their net worth. There are two categories, assets and debts. Assets are positive values for things you own or money you have saved, while debts are negative values and include money you owe on loans and credit cards. To find their net worth, you will need to add the assets and the debts.

   Carl and Lisa own a house worth $420,000, but still owe $250,000 on the mortgage. They took out a home equity loan of $40,000 a few years ago to fix up their backyard and kitchen. Carl and Lisa have invested $50,000 in mutual funds to help pay for their children's education. They have both been working for many years and have $200,000 saved in retirement accounts (a 401(k) and an IRA). They estimated that they own about $50,000 worth of jewelry and furnishings for their house. They have two cars that are paid off; one is worth about $10,000 and the other is currently worth about $6,000. They recently bought their daughter a used car and took out a loan for $15,000. Carl and Lisa have 2 credit cards. Their Visa has $5,000 on it, while their American Express card has $2,000 on it. Carl and Lisa keep $10,000 in a savings account for emergencies. Calculate their net worth after filling in the chart containing assets and debts.

Assets descriptions	$$ (positive values)	Debts descriptions	$$ (negative values)

   Carl and Lisa's net worth is $_____.

Name: Class:
Module 1: Problem Solving with Integers

Group Activity (2-3 people):

Objectives for the lesson:

In this lesson you will use and enhance your understanding of the following:

- Adding integers
- Solving applied problems
- Multiplying integers
- Dividing integers
- Simplifying expressions containing exponents
- Using order of operations to simplify expressions
- Writing sets
- Inductive reasoning
- Venn diagrams
- Finding the intersection of sets

1. Lisa is creating a spreadsheet to track some of their spending habits to begin the process of making a budget. Money spent, called expenses, will be written as negative numbers, while income will be written as positive numbers in her spreadsheet.
   Looking at several past bank statements, she calculates that they spend about $700 each month on groceries, $400 on gas, $275 on electricity, $60 for water/sewer, $150 for cable TV, $80 for internet, $300 for cell phones, $500 on clothing, $1,750 for their mortgage payment, $450 for their daughter's car payments, $65 for dog food/vet bills, $300 on eating out, $190 for entertainment (movies, bowling, etc.…), $120 on gifts, $800 for health insurance, $500 in doctor visits/medicine, and about $300 on other miscellaneous items each month. Create a table for listing their expenses, leaving a few blank rows at the bottom. Write all of their expenses in the table, but do not calculate the total yet.

2. Carl chose the car insurance payment plan of 4 months on, two months off. The months he pays, the bill is $450 for the month. Find his average monthly payment to use for their budget, and add that to the table you created for expenses. Add all the expenses and find their total monthly expenses. Remember, this is a negative value.

3. Both Carl and Lisa have their retirement savings taken directly out of their paychecks before taxes, along with the money they save for their children's college. After retirement, college savings, taxes and social security are taken out, Carl takes home $42,500 yearly and Lisa takes home $50,700 yearly. They have no other income. Find their average monthly net income, and round to the nearest dollar.

4. Find Carl and Lisa's net monthly take home by adding their income and their expenses (remember, expenses are negative numbers).

5. Carl and Lisa want to start saving 10% of their total monthly take home for long term savings. They hope to one day buy a sailboat. How much money do they want to save every month?

   Do they have enough money left over every month after their expenses to save 10% of their net monthly income?   If not, how much more do they need each month?

   Discuss where they might save money in their monthly expenses so they can save enough for their future dream of owning a sailboat.

6. Use the following formula and a calculator to find out how much money they will have saved over 15 years, putting 10% of their net income every month into an account that earns 5% annual interest:

$$P = M\left(\frac{(1+i)^n - 1}{i}\right),$$

   where P is the amount of money they will have in the future if they make n total monthly payments of M dollars each month, and earn an interest rate of i % each month (interest rate divided by 12).

   Carl and Lisa want to have $200,000 saved in an account in 15 years to be able to buy a sailboat. By putting 10% of their net income each month into this savings account, will they have the amount they want in 15 years?

   $M =$ _____   (your amount from #6)
   $i = 0.05$ divided by 12 months $=$ _____ (round to 5 decimal places)
   $n = 15$ years x 12 payments per year $=$ _____

   Find P and round to the nearest penny. $P =$ _____

   How much money will they have in the account after 15 years?

7.   Carl looked at his net salary over the past 4 years and saw he took home $37,820, $39,380, $40,940, and this year was $42,500. Using inductive reasoning, how much will Carl most likely net next year?

8.   Carl and Lisa want the following options on their sailboat: autopilot, 3 cabins, furling sails, air conditioning, 2 helms, a watermaker, a spinnaker and a bow thruster.
     Looking at sailboats, they found 3 used sailboats within their budget, with the following options:

     Beneteau: autopilot, furling sails, 3 cabins, air conditioning, 2 helms
     Hunter: autopilot, 3 cabins, furling sails, spinnaker
     Catalina: 3 cabins, spinnaker, watermaker

     a.   Write the information in set notation, using B for Beneteau, H for Hunter, and C for Catalina.
     b.   Draw a Venn diagram containing all of the options they wanted as the Universe, and all 3 sets.
     c.   Find $B \cap C$
     d.   Find $H \cap B$
     e.   Find $B \cup H \cup C$ and explain what it represents.

Objective 2.1A – Factor numbers and find the prime factorization of numbers

Key Words and Concepts

Natural number _____ of a number divide that number evenly (there is no remainder).

The following rules are helpful in finding the factors of a number.
   2 is a factor of a number if the digit in the ones place of the number is 0, 2, 4, 6, or 8.
   3 is a factor of a number if the sum of the digits of the number is divisible by 3.
   4 is a factor of a number if the last two digits of the number are divisible by 4.
   5 is a factor of a number if the ones digit of the number is 0 or 5.

Example
Find all the factors of 48.

   Solution
      Look for all factors of 48 until the factors start to repeat.
      $48 \div 1 = 48$ 1 and 48 are factors of 48.
      $48 \div 2 = 24$ 2 and 24 are factors of 48.
      $48 \div 3 = 16$ 3 and 16 are factors of 48.
      $48 \div 4 = 12$ 4 and 12 are factors of 48.
      $48 \div 5$ 5 will not divide 48 evenly.
      $48 \div 6 = 8$ 6 and 8 are factors of 48.
      $48 \div 7$ 7 will not divide 48 evenly.
      $48 \div 8 = 6$ 8 and 6 are factors of 48.
            The factors are repeating.
            All the factors of 48 have been found.
      The factors of 48 are 1, 2, 3, 4, 6, 8, 12, 16, 24, and 48.

Try it
1. Find all the factors of 56.

A **prime number** is a natural number greater than 1 that has exactly two natural number factors. What are those two natural number factors? _____

If a number is not prime, what kind of number is it? _____

The **prime factorization** of a number is the expression of the number as a product of its prime factors.

Example
Find the prime factorization of 90.

   Solution
      Look for all prime factors starting with 2, including repeating factors, until 90 is completely factored.

Try it
2. Find the prime factorization of 60.

$$2)\overline{90} \quad \begin{array}{c}15\\3)\overline{45}\\2)\overline{90}\end{array} \quad \begin{array}{c}5\\3)\overline{15}\\3)\overline{45}\\2)\overline{90}\end{array}$$

      The prime factorization of 90 is $2 \cdot 3^2 \cdot 5$.

Finding the prime factorization of larger numbers can be more difficult.

Try each prime number as a trial divisor.

Stop when the square of the trial divisor is greater than the number being factored.

Example

Find the prime factorization of 355.

Solution

355 is not divisible by 2 or by 3 but is divisible by 5.
Divide 355 by 5:

$$\begin{array}{r} 71 \\ 5\overline{)355} \end{array}$$

71 cannot be divided evenly by 7 or 11.
Prime numbers greater than 11 need not be tried

because $11^2 = 121$ and $121 > 71$.
The prime factorization of 355 is $5 \cdot 71$.

Try it

3. Find the prime factorization of 455.

Reflect on it
• Look for all factors of a number until the factors start to repeat.
• A prime number is a natural number greater than 1 that has exactly two natural number factors, 1 and the number itself.
• If a number is not prime, then it is a composite number.

Quiz Yourself 2.1A

1. Find all the factors of 18.

2. Find all the factors of 96.

3. Find all the factors of 97.

4. Find the prime factorization of 108.

5. Find the prime factorization of 109.

6. Find the prime factorization of 2520.

Practice Sheet 2.1A

Find all the factors of the number.

1. 15	**2.** 23	**3.** 40	**1.** _____
			2. _____
			3. _____
4. 62	**5.** 81	**6.** 51	**4.** _____
			5. _____
			6. _____
7. 88	**8.** 63	**9.** 35	**7.** _____
			8. _____
			9. _____
10. 45	**11.** 78	**12.** 55	**10.** _____
			11. _____
			12. _____

Find the prime factorization.

13. 8	**14.** 30	**15.** 53	**13.** _____
			14. _____
			15. _____
16. 11	**17.** 20	**18.** 82	**16.** _____
			17. _____
			18. _____
19. 35	**20.** 44	**21.** 69	**19.** _____
			20. _____
			21. _____
22. 72	**23.** 88	**24.** 94	**22.** _____
			23. _____
			24. _____

Answers

Try it 2.1A

1. 1, 2, 4, 7, 8, 14, 28, 56
2. $2^2 \cdot 3 \cdot 5$
3. $5 \cdot 7 \cdot 13$

Quiz 2.1A

1. 1, 2, 3, 6, 9, 18
2. 1, 2, 3, 4, 6, 8, 12, 16, 24, 32, 48, 96
3. 1, 97
4. $2^2 \cdot 3^3$
5. Prime
6. $2^3 \cdot 3^2 \cdot 5 \cdot 7$

Solutions to Practice Sheet 2.1A

1. $15 \div 1 = 15$
   $15 \div 3 = 5$
   $15 \div 5 = 3$
   $15 \div 15 = 1$
   The factors of 15 are 1, 3, 5, and 15.

2. $23 \div 1 = 23$
   $23 \div 23 = 1$
   The factors of 23 are 1 and 23.

3. $40 \div 1 = 40$
   $40 \div 2 = 20$
   $40 \div 4 = 10$
   $40 \div 5 = 8$
   $40 \div 8 = 5$
   $40 \div 10 = 4$
   $40 \div 20 = 2$
   $40 \div 40 = 1$
   The factors of 40 are 1, 2, 4, 5, 8, 10, 20, and 40.

4. $62 \div 1 = 62$
   $62 \div 2 = 31$
   $62 \div 31 = 2$
   $62 \div 62 = 1$
   The factors of 62 are 1, 2, 31, and 62.

5. $81 \div 1 = 81$
   $81 \div 3 = 27$
   $81 \div 9 = 9$
   $81 \div 27 = 3$
   $81 \div 81 = 1$
   The factors of 81 are 1, 3, 9, 27, and 81.

6. $51 \div 1 = 51$
   $51 \div 3 = 17$
   $51 \div 17 = 3$
   $51 \div 51 = 1$
   The factors of 51 are 1, 3, 17, and 51.

7. $88 \div 1 = 88$
   $88 \div 2 = 44$
   $88 \div 4 = 22$
   $88 \div 8 = 11$
   $88 \div 11 = 8$
   $88 \div 22 = 4$
   $88 \div 44 = 2$
   $88 \div 88 = 1$
   The factors of 40 are 1, 2, 4, 8, 11, 22, 44, and 88.

8. $63 \div 1 = 63$
   $63 \div 3 = 21$
   $63 \div 7 = 9$
   $63 \div 9 = 7$
   $63 \div 21 = 3$
   $63 \div 63 = 1$
   The factors of 63 are 1, 3, 7, 9, 21, and 63.

9. $35 \div 1 = 35$
   $35 \div 5 = 7$
   $35 \div 7 = 5$
   $35 \div 35 = 1$
   The factors of 35 are 1, 5, 7, and 35.

10. $45 \div 1 = 45$
    $45 \div 3 = 15$
    $45 \div 5 = 9$
    $45 \div 9 = 5$
    $45 \div 15 = 3$
    $45 \div 45 = 1$
    The factors of 45 are 1, 3, 5, 9, 15, and 45.

11. $78 \div 1 = 78$
    $78 \div 2 = 39$
    $78 \div 3 = 26$
    $78 \div 6 = 13$
    $78 \div 13 = 6$
    $78 \div 26 = 3$
    $78 \div 39 = 2$
    $78 \div 78 = 1$
    The factors of 40 are 1, 2, 3, 6, 13, 26, 39, and 78.

12. $55 \div 1 = 55$
    $55 \div 5 = 11$
    $55 \div 11 = 5$
    $55 \div 55 = 1$
    The factors of 62 are 1, 5, 11, and 55.

13.
$$\begin{array}{r} 2 \\ 2\overline{)4} \\ 2\overline{)8} \end{array}$$

$8 = 2 \cdot 2 \cdot 2 = 2^3$

14.
$$\begin{array}{r} 5 \\ 3\overline{)15} \\ 2\overline{)30} \end{array}$$

$30 = 2 \cdot 3 \cdot 5$

15. 53 is a prime number.

16. 11 is a prime number.

17.
$$\begin{array}{r} 5 \\ 2\overline{)10} \\ 2\overline{)20} \end{array}$$

$20 = 2 \cdot 2 \cdot 5 = 2^2 \cdot 5$

18.
$$\begin{array}{r} 41 \\ 2\overline{)82} \end{array}$$

$82 = 2 \cdot 41$

19.
$$\begin{array}{r} 7 \\ 5\overline{)35} \end{array}$$

$35 = 5 \cdot 7$

20.
$$\begin{array}{r} 11 \\ 2\overline{)22} \\ 2\overline{)44} \end{array}$$

$44 = 2 \cdot 2 \cdot 11 = 2^2 \cdot 11$

21.
$$\begin{array}{r} 23 \\ 3\overline{)69} \end{array}$$

$69 = 3 \cdot 23$

22.
$$\begin{array}{r} 3 \\ 3\overline{)9} \\ 2\overline{)18} \\ 2\overline{)36} \\ 2\overline{)72} \end{array}$$

$72 = 2 \cdot 2 \cdot 2 \cdot 3 \cdot 3 = 2^3 \cdot 3^2$

23.
$$\begin{array}{r} 11 \\ 2\overline{)22} \\ 2\overline{)44} \\ 2\overline{)88} \end{array}$$

$88 = 2 \cdot 2 \cdot 2 \cdot 11 = 2^3 \cdot 11$

24.
$$\begin{array}{r} 47 \\ 2\overline{)94} \end{array}$$

$94 = 2 \cdot 47$

Objective 2.1B – Find the least common multiple (LCM)

Key Words and Concepts

The **multiples of a number** are the products of that number and the numbers 1, 2, 3, 4, 5, …
A number that is a multiple of two or more numbers is a _____ **multiple** of those numbers.

Define the least common multiple. _____

_____

Listing the multiples of each number is one way to find the LCM.

Another way to find the LCM uses the prime factorization of each number.
    Find the prime factorization of each number and write the factorization of each number in a table.
    Mark the greatest product in each column.
    The LCM is the product of the marked numbers.

Example

Find the LCM of 126 and 220.

    Solution
        First, write the prime factorization of 126 and 220.

	2	3	5	7	11
126 =	2	3·3		7	
220 =	2·2		5		11

        Next mark the greatest product in each column.

	2	3	5	7	11
126 =	2	3·3		7	
220 =	2·2		5		11

        The LCM is the product of the marked numbers.

        The LCM = $2 \cdot 2 \cdot 3 \cdot 3 \cdot 5 \cdot 7 \cdot 11 = 13,860$.

Try it

1. Find the LCM of 75 and 60.

Example

Find the LCM of 9, 21, and 28.

    Solution
        First, write the prime factorization of 9, 21, and 28.

	2	3	7
9 =		3·3	
21 =		3	7
28 =	2·2		7

        Next, mark the greatest product in each column.

	2	3	7
9 =		3·3	
21 =		3	7
28 =	2·2		7

        The LCM is the product of the marked numbers.

        The LCM = $2 \cdot 2 \cdot 3 \cdot 3 \cdot 7 = 252$.

Try it

2. Find the LCM of 165, 660, and 264.

Reflect on it
- A number that is a multiple of two or more numbers is a common multiple of those numbers.
- The least common multiple (LCM) is the smallest common multiple of two or more numbers.
- Why is listing the multiples of each number not a very efficient way to find the LCM?

Quiz Yourself 2.1B

1. Find the LCM of 6 and 9.

2. Find the LCM of 12 and 15.

3. Find the LCM of 16 and 32.

4. Find the LCM of 106 and 108.

5. Find the LCM of 8, 10, and 12.

6. Find the LCM of 14, 16, and 52.

Name:

Class:

Module 2: Problem Solving with Fractions and Decimals

Section 2.1: The LCM and GCF

Practice Sheet 2.1B

Find the LCM.

1. 3, 4
2. 3, 7
3. 6, 9

4. 8, 10
5. 4, 8
6. 9, 12

7. 4, 9
8. 6, 15
9. 16, 24

10. 15, 25
11. 28, 32
12. 4, 18

13. 72, 108
14. 84, 126
15. 32, 128

16. 3, 7, 9
17. 6, 12, 27
18. 3, 7, 11

19. 9, 12, 24
20. 10, 25, 40
21. 4, 7, 21

22. 2, 7, 11
23. 28, 32, 56
24. 16, 20, 40

25. 8, 27, 36
26. 6, 12, 18
27. 2, 16, 32

1. _____
2. _____
3. _____
4. _____
5. _____
6. _____
7. _____
8. _____
9. _____
10. _____
11. _____
12. _____
13. _____
14. _____
15. _____
16. _____
17. _____
18. _____
19. _____
20. _____
21. _____
22. _____
23. _____
24. _____
25. _____
26. _____
27. _____

Answers

Try it 2.1B
1. 300
2. 1320

Quiz 2.1B
1. 18
2. 60
3. 32
4. 5724
5. 120
6. 1456

Solutions to Practice Sheet 2.1B

1.
$$3 = $$
$$4 = $$
$$LCM = 2 \cdot 2 \cdot 3 = 12$$

2.
$$3 = $$
$$7 = $$
$$LCM = 3 \cdot 7 = 21$$

3.
$$6 = $$
$$9 = $$
$$LCM = 2 \cdot 3 \cdot 3 = 18$$

4.
$$8 = $$
$$10 = $$
$$LCM = 2 \cdot 2 \cdot 2 \cdot 5 = 40$$

5.
$$4 = $$
$$8 = $$
$$LCM = 2 \cdot 2 \cdot 2 = 8$$

6.
$$9 = $$
$$12 = $$
$$LCM = 2 \cdot 2 \cdot 3 \cdot 3 = 36$$

7.
$$4 = $$
$$9 = $$
$$LCM = 2 \cdot 2 \cdot 3 \cdot 3 = 36$$

8.
$$6 = $$
$$15 = $$
$$LCM = 2 \cdot 3 \cdot 5 = 30$$

9.
$$16 = $$
$$24 = $$
$$LCM = 2 \cdot 2 \cdot 2 \cdot 2 \cdot 3 = 48$$

10.
$$15 = $$
$$25 = $$
$$LCM = 3 \cdot 5 \cdot 5 = 75$$

11.
$$28 = $$
$$32 = $$
$$LCM = 2 \cdot 2 \cdot 2 \cdot 2 \cdot 2 \cdot 7 = 224$$

12.
$$4 = $$
$$18 = $$
$$LCM = 2 \cdot 2 \cdot 3 \cdot 3 = 36$$

13.
$$72 = $$
$$108 = $$
$$LCM = 2 \cdot 2 \cdot 2 \cdot 3 \cdot 3 \cdot 3 = 216$$

14.
$$84 = $$
$$126 = $$
$$LCM = 2 \cdot 2 \cdot 3 \cdot 3 \cdot 7 = 252$$

15.
$$32 = $$
$$128 = $$
$$LCM = 2 \cdot 2 \cdot 2 \cdot 2 \cdot 2 \cdot 2 \cdot 2 = 128$$

16.
$$3 = $$
$$7 = $$
$$9 = $$
$$LCM = 3 \cdot 3 \cdot 7 = 63$$

17.

	2	3
6 =	2	3
12 =	2·2	3
27 =		3·3·3

LCM = 2·2·3·3·3 = 108

18.

	3	7	11
3 =	3		
7 =		7	
11 =			11

LCM = 3·7·11 = 231

19.

	2	3
9 =		3·3
12 =	2·2	3
24 =	2·2·2	3

LCM = 2·2·2·3·3 = 72

20.

	2	5
10 =	2	5
25 =		5·5
40 =	2·2·2	5

LCM = 2·2·2·5·5 = 200

21.

	2	3	7
4 =	2·2		
7 =			7
21 =		3	

LCM = 2·2·3·7 = 84

22.

	2	7	11
2 =	2		
7 =		7	
11 =			11

LCM = 2·7·11 = 154

23.

	2	7
28 =	2·2	7
32 =	2·2·2·2·2	
56 =	2·2·2	7

LCM = 2·2·2·2·2·7 = 224

24.

	2	5
16 =	2·2·2·2	
20 =	2·2	5
40 =	2·2·2	5

LCM = 2·2·2·2·5 = 80

25.

	2	3
8 =	2·2·2	
27 =		3·3·3
36 =	2·2	3·3

LCM = 2·2·2·3·3·3 = 216

26.

	2	3
6 =	2	3
12 =	2·2	3
18 =	2	3·3

LCM = 2·2·3·3 = 36

27.

	2
2 =	2
16 =	2·2·2·2
32 =	2·2·2·2·2

LCM = 2·2·2·2·2 = 32

Objective 2.1C – Find the greatest common factor (GCF)

Key Words and Concepts

A number that is a factor of two or more numbers is a _____ **factor** of those numbers.

Define the greatest common factor. _____

_____

Listing the factors of each number is one way to find the GCF.

Another way to find the GCF is to use the prime factorization of each number.
    Find the prime factorization of each number and write the factorization of each number in a table.
    Mark the least product in each column that does not have a blank.
    The GCF is the product of the marked numbers.

Example
Find the GCF of 126 and 220.

    Solution
        First, write the prime factorization of 126 and 220.

	2	3	5	7	11
126 =	2	3·3		7	
220 =	2·2		5		11

        Next mark the least product in each column.

	2	3	5	7	11
126 =	[2]	3·3		7	
220 =	2·2		5		11

        The GCF is the product of the marked numbers.

        The GCF = 2.

Try it
1. Find the GCF of 75 and 60.

Example
Find the GCF of 98, 1008, and 1176.

    Solution
        First, write the prime factorization of 98, 1008, and 1176.

	2	3	7
98 =	2		7·7
1008 =	2·2·2·2	3·3	7
1176 =	2·2·2	3	7·7

        Next, mark the least product in each column.

	2	3	7
98 =	[2]		7·7
1008 =	2·2·2·2	3·3	[7]
1176 =	2·2·2	3	7·7

        The GCF is the product of the marked numbers.

        The GCF = $2 \cdot 7 = 14$.

Try it
2. Find the LCM of 165, 264, and 660.

Reflect on it
- A number that is a factor of two or more numbers is a common factor of those numbers.
- The greatest common factor (GCF) is the largest common factor of two or more numbers.
- Why is listing the factors of each number not a very efficient way to find the GCF?

Quiz Yourself 2.1C
1. Find the GCF of 12 and 20.

2. Find the GCF of 75 and 375.

3. Find the GCF of 144 and 360.

4. Find the GCF of 60, 72, and 96.

5. Find the GCF of 52, 78, and 156.

6. Find the GCF of 90, 91, and 93.

Practice Sheet 2.1C

Find the GCF.

1. 3, 7	**2.** 3, 6	**3.** 9, 16	**1.** _____
			2. _____
			3. _____
4. 8, 18	**5.** 10, 15	**6.** 30, 65	**4.** _____
			5. _____
			6. _____
7. 15, 30	**8.** 36, 56	**9.** 18, 27	**7.** _____
			8. _____
			9. _____
10. 21, 35	**11.** 15, 20	**12.** 30, 50	**10.** _____
			11. _____
			12. _____
13. 48, 64	**14.** 39, 52	**15.** 37, 67	**13.** _____
			14. _____
			15. _____
16. 4, 8, 10	**17.** 3, 5, 7	**18.** 3, 9, 12	**16.** _____
			17. _____
			18. _____
19. 5, 11, 13	**20.** 10, 25, 30	**21.** 16, 40, 80	**19.** _____
			20. _____
			21. _____
22. 16, 20, 32	**23.** 24, 32, 40	**24.** 18, 27, 81	**22.** _____
			23. _____
			24. _____
25. 28, 44, 56	**26.** 17, 68, 85	**27.** 30, 75, 150	**25.** _____
			26. _____
			27. _____

Answers

Try it 2.1C
1. 15
2. 33

Quiz 2.1C
1. 4
2. 75
3. 72
4. 12
5. 26
6. 1

Solutions to Practice Sheet 2.1C

1.
	3	7
3 =	3	
7 =		7

GCF = 1

2.
	2	3
3 =		3
6 =	2	3

GCF = 3

3.
	2	3
9 =		3·3
16 =	2·2·2·2	

GCF = 1

4.
	2	3
8 =	2·2·2	
18 =	2	3·3

GCF = 2

5.
	2	3	5
10 =	2		5
15 =		3	5

GCF = 5

6.
	2	3	5	13
30 =	2	3	5	
65 =			5	13

GCF = 5

7.
	2	3	5
15 =		3	5
30 =	2	3	5

GCF = 3·5 = 15

8.
	2	3	7
36 =	2·2	3·3	
56 =	2·2·2		7

GCF = 2·2 = 4

9.
	2	3
18 =	2	3·3
27 =		3·3·3

GCF = 3·3 = 9

10.
	3	5	7
21 =	3		7
35 =		5	7

GCF = 7

11.
	2	3	5
15 =		3	5
20 =	2·2		5

GCF = 5

12.
	2	3	5
30 =	2	3	5
50 =	2		5·5

GCF = 2·5 = 10

13.
	2	3
48 =	2·2·2·2	3
64 =	2·2·2·2·2·2	

GCF = 2·2·2·2 = 16

14.
	2	3	13
39 =		3	13
52 =	2·2		13

GCF = 13

15.
	37	67
37 =	37	
67 =		67

GCF = 1

16.
	2	5
4 =	2·2	
8 =	2·2·2	
10 =	2	5

GCF = 2

17.
	3	5	7
3 =	3		
5 =		5	
7 =			7

GCF = 1

18.

	2	3
3 =		**3**
9 =		3·3
12 =	2·2	3

GCF = 3

26.

	2	5	17
17 =			**17**
68 =	2·2		17
85 =		5	17

GCF = 17

19.

	5	11	13
5 =	5		
11 =		11	
13 =			13

GCF = 1

27.

	2	3	5
30 =	2	**3**	**5**
75 =		3	5·5
150 =	2	3	5·5

GCF = 3·5 = 15

20.

	2	3	5
10 =	2		5
25 =			5·5
30 =	2	3	**5**

GCF = 5

21.

	2	5
16 =	2·2·2·2	
40 =	**2·2·2**	5
80 =	2·2·2·2	5

GCF = 2·2·2 = 8

22.

	2	5
16 =	2·2·2·2	
20 =	**2·2**	5
32 =	2·2·2·2·2	

GCF = 2·2 = 4

23.

	2	3	5
24 =	**2·2·2**	3	
32 =	2·2·2·2·2		
40 =	2·2·2		5

GCF = 2·2·2 = 8

24.

	2	3
18 =	2	**3·3**
27 =		3·3·3
81 =		3·3·3·3

GCF = 3·3 = 9

25.

	2	7	11
28 =	**2·2**	7	
44 =	2·2		11
56 =	2·2·2	7	

GCF = 2·2 = 4

Objective 2.2A – Write an improper fraction as a mixed number or a whole number, and a mixed number as an improper fraction

Key Words and Concepts

Note from the diagram that the mixed number $1\frac{2}{5}$ and the improper fraction $\frac{7}{5}$ both represent the shaded portion of the circles, so $1\frac{2}{5}=\frac{7}{5}$.

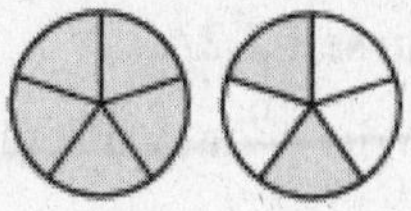

Note An improper fraction can be written as a mixed number or a whole number.

Example

Write $\frac{7}{5}$ as a mixed number.

Solution

Divide the numerator by the denominator.

$$\begin{array}{r} 1 \\ 5\overline{)7} \\ \underline{5} \\ 2 \end{array}$$

To write the fractional part of the mixed number, write the remainder over the divisor.

$$\frac{7}{5}=1\frac{2}{5}$$

Try it

1. Write $\frac{9}{4}$ as a mixed number.

Example

Write $\frac{27}{9}$ as a whole number.

Solution

$$\frac{27}{9}=27\div 9=3$$

Try it

2. Write $\frac{32}{4}$ as a whole number.

To write a mixed number as an improper fraction,
multiply the denominator of the fractional part by the whole-number part.
The sum of this product and the numerator of the fractional part is the numerator of the improper fraction.
The denominator remains the same.

Example

Write $8\frac{3}{5}$ as an improper fraction.

Solution

$$8\frac{3}{5}=\frac{(5\times 8)+3}{5}$$ Multiply the denominator by the whole-number part and add to the numerator

$$=\frac{40+3}{5}$$

$$=\frac{43}{5}$$

Try it

3. Write $6\frac{4}{7}$ as an improper fraction.

Reflect on it
- Can you think of a situation in which an improper fraction is more useful than a mixed number?
- Can you think of a situation in which a mixed number is more useful than an improper fraction?

Quiz Yourself 2.2A

1. Write $\dfrac{19}{7}$ as a mixed or whole number.

2. Write $\dfrac{84}{6}$ as a mixed or whole number.

3. Write $5\dfrac{7}{9}$ as an improper fraction.

4. Write $12\dfrac{4}{13}$ as an improper fraction.

Practice Sheet 2.2A

Write the improper fraction as a mixed number or whole number.

1. $\dfrac{11}{3}$ 2. $\dfrac{30}{5}$ 3. $\dfrac{17}{9}$

4. $\dfrac{9}{2}$ 5. $\dfrac{24}{9}$ 6. $\dfrac{19}{4}$

7. $\dfrac{36}{12}$ 8. $\dfrac{11}{6}$ 9. $\dfrac{8}{8}$

10. $\dfrac{64}{8}$ 11. $\dfrac{14}{1}$ 12. $\dfrac{70}{14}$

1. _____
2. _____
3. _____
4. _____
5. _____
6. _____
7. _____
8. _____
9. _____
10. _____
11. _____
12. _____

Write the mixed number as an improper fraction.

13. $2\dfrac{1}{5}$ 14. $7\dfrac{2}{3}$ 15. $4\dfrac{7}{9}$

16. $2\dfrac{7}{8}$ 17. $5\dfrac{2}{7}$ 18. $6\dfrac{3}{5}$

19. $10\dfrac{3}{4}$ 20. $9\dfrac{7}{10}$ 21. $12\dfrac{5}{6}$

22. $7\dfrac{13}{15}$ 23. $11\dfrac{1}{3}$ 24. $16\dfrac{3}{8}$

13. _____
14. _____
15. _____
16. _____
17. _____
18. _____
19. _____
20. _____
21. _____
22. _____
23. _____
24. _____

99

Answers

Try it 2.2A

1. $2\frac{1}{4}$

2. 8

3. $\frac{46}{7}$

Quiz 2.2A

1. $2\frac{5}{7}$

2. 14

3. $\frac{52}{9}$

4. $\frac{160}{13}$

Solutions to Practice Sheet 2.2A

1. $3\overline{)11}$ $\begin{array}{r}3\\-9\\\hline 2\end{array}$ $\frac{11}{3}=3\frac{2}{3}$

2. $5\overline{)30}$ $\begin{array}{r}6\\-30\\\hline 0\end{array}$ $\frac{30}{5}=6$

3. $9\overline{)17}$ $\begin{array}{r}1\\-9\\\hline 8\end{array}$ $\frac{17}{9}=1\frac{8}{9}$

4. $2\overline{)9}$ $\begin{array}{r}4\\-8\\\hline 1\end{array}$ $\frac{9}{2}=4\frac{1}{2}$

5. $9\overline{)24}$ $\begin{array}{r}2\\-18\\\hline 6\end{array}$ $\frac{24}{9}=2\frac{6}{9}$

6. $4\overline{)19}$ $\begin{array}{r}4\\-16\\\hline 3\end{array}$ $\frac{19}{4}=4\frac{3}{4}$

7. $12\overline{)36}$ $\begin{array}{r}3\\-36\\\hline 0\end{array}$ $\frac{36}{12}=3$

8. $6\overline{)11}$ $\begin{array}{r}1\\-6\\\hline 5\end{array}$ $\frac{11}{6}=1\frac{5}{6}$

9. $8\overline{)8}$ $\begin{array}{r}1\\-8\\\hline 0\end{array}$ $\frac{8}{8}=1$

10. $8\overline{)64}$ $\begin{array}{r}8\\-64\\\hline 0\end{array}$ $\frac{64}{8}=8$

11. $1\overline{)14}$ $\begin{array}{r}14\\-14\\\hline 0\end{array}$ $\frac{14}{1}=14$

12. $14\overline{)70}$ $\begin{array}{r}5\\-70\\\hline 0\end{array}$ $\frac{70}{14}=5$

13. $2\frac{1}{5}=\frac{10+1}{5}=\frac{11}{5}$

14. $7\frac{2}{3}=\frac{21+2}{3}=\frac{23}{3}$

15. $4\frac{7}{9}=\frac{36+7}{9}=\frac{43}{9}$

16. $2\frac{7}{8}=\frac{16+7}{8}=\frac{23}{8}$

17. $5\frac{2}{7}=\frac{35+2}{7}=\frac{37}{7}$

18. $6\frac{3}{5}=\frac{30+3}{5}=\frac{33}{5}$

19. $10\frac{3}{4}=\frac{40+3}{4}=\frac{43}{4}$

20. $9\frac{7}{10}=\frac{90+7}{10}=\frac{97}{10}$

21. $12\frac{5}{6}=\frac{72+5}{6}=\frac{77}{6}$

22. $7\frac{13}{15}=\frac{105+13}{15}=\frac{118}{15}$

23. $11\frac{1}{3}=\frac{33+1}{3}=\frac{34}{3}$

24. $16\frac{3}{8}=\frac{128+3}{8}=\frac{131}{8}$

Objective 2.2B – Write a fraction in simplest form

Key Words and Concepts

Writing the **simplest form** of a fraction means writing it so that the numerator and denominator have no common factors other than 1.

The _____ Property of One can be used to write fractions in simplest form.

Write the numerator and denominator of the given fraction as a product of factors.

Write factors common to both the numerator and denominator as an improper fraction equivalent to 1.

$$\frac{9}{12} = \frac{3 \cdot 3}{2 \cdot 2 \cdot 3} = \frac{3}{2 \cdot 2} \cdot \frac{3}{3} = \frac{3}{2 \cdot 2} \cdot 1 = \frac{3}{4}$$

The process of eliminating common factors is displayed with slashes through the common factors as shown below:

$$\frac{9}{12} = \frac{3 \cdot \overset{1}{\cancel{3}}}{2 \cdot 2 \cdot \underset{1}{\cancel{3}}} = \frac{3}{2 \cdot 2} = \frac{3}{4}$$

To write a fraction in simplest form, eliminate the common factors.

An improper fraction can be changed to a mixed number.

Example

Write $\frac{26}{39}$ in simplest form.

Solution

Write the numerator and denominator as a product of factors, then eliminate the common factors.

$$\frac{26}{39} = \frac{2 \cdot \overset{1}{\cancel{13}}}{3 \cdot \underset{1}{\cancel{13}}} = \frac{2}{3}$$

Try it

1. Write $\frac{34}{51}$ in simplest form.

Example

Write $\frac{15}{45}$ in simplest form.

Solution

Write the numerator and denominator as a product of factors, then eliminate the common factors.

$$\frac{15}{45} = \frac{\overset{1}{\cancel{3}} \cdot \overset{1}{\cancel{5}}}{\underset{1}{\cancel{3}} \cdot 3 \cdot \underset{1}{\cancel{5}}} = \frac{1}{3}$$

Try it

2. Write $\frac{14}{56}$ in simplest form.

Example

Write $\dfrac{10}{21}$ in simplest form.

Solution

$$\dfrac{10}{21} = \dfrac{2 \cdot 5}{3 \cdot 7} = \dfrac{10}{21}$$

$\dfrac{10}{21}$ is already in simplest form because there are no common

factors in the numerator and denominator.

Try it

3. Write $\dfrac{25}{28}$ in simplest form.

Example

Write $\dfrac{26}{8}$ as a fraction in simplest form.

Solution

$$\dfrac{26}{8} = \dfrac{\overset{1}{\cancel{2}} \cdot 13}{\underset{1}{\cancel{2}} \cdot 2 \cdot 2} = \dfrac{13}{4}$$

Note that $\dfrac{13}{4}$ is a fraction in simplest form.

This fraction is equivalent to the mixed number $3\dfrac{1}{4}$.

Try it

4. Write $\dfrac{28}{6}$ as a fraction in simplest form.

Reflect on it

*   The Multiplication Property of One can be used to write fractions in simplest form.
*   To write a fraction in simplest form, eliminate the common factors.
*   How many different ways can you write an improper fraction equivalent to one?

Quiz Yourself 2.2B

1. Write $\dfrac{32}{80}$ in simplest form.

2. Write $\dfrac{9}{27}$ in simplest form.

3. Write $\dfrac{20}{27}$ in simplest form.

4. Write $\dfrac{42}{32}$ as a fraction in simplest form.

Practice Sheet 2.2B

Reduce the fraction to simplest form.

1. $\dfrac{8}{12}$

2. $\dfrac{15}{25}$

3. $\dfrac{2}{16}$

4. $\dfrac{28}{49}$

5. $\dfrac{54}{81}$

6. $\dfrac{40}{64}$

7. $\dfrac{75}{90}$

8. $\dfrac{36}{27}$

9. $\dfrac{0}{15}$

10. $\dfrac{60}{96}$

11. $\dfrac{18}{36}$

12. $\dfrac{7}{18}$

13. $\dfrac{25}{100}$

14. $\dfrac{84}{144}$

15. $\dfrac{39}{13}$

16. $\dfrac{169}{234}$

17. $\dfrac{112}{126}$

18. $\dfrac{59}{177}$

19. $\dfrac{85}{65}$

20. $\dfrac{34}{238}$

21. $\dfrac{16}{34}$

22. $\dfrac{104}{240}$

23. $\dfrac{69}{150}$

24. $\dfrac{26}{75}$

25. $\dfrac{112}{160}$

26. $\dfrac{143}{182}$

27. $\dfrac{92}{23}$

1. _____
2. _____
3. _____
4. _____
5. _____
6. _____
7. _____
8. _____
9. _____
10. _____
11. _____
12. _____
13. _____
14. _____
15. _____
16. _____
17. _____
18. _____
19. _____
20. _____
21. _____
22. _____
23. _____
24. _____
25. _____
26. _____
27. _____

103

Answers

Try it 2.2B

1. $\dfrac{2}{3}$

2. $\dfrac{1}{4}$

3. $\dfrac{25}{28}$

4. $\dfrac{14}{3}$, or $4\dfrac{2}{3}$

Quiz 2.2B

1. $\dfrac{2}{5}$

2. $\dfrac{1}{3}$

3. $\dfrac{20}{27}$

4. $\dfrac{21}{16}$

Solutions to Practice Sheet 2.2B

1. $\dfrac{8}{12} = \dfrac{\cancel{2}\cdot\cancel{2}\cdot 2}{\cancel{2}\cdot\cancel{2}\cdot 3} = \dfrac{2}{3}$

2. $\dfrac{15}{25} = \dfrac{3\cdot\cancel{5}}{\cancel{5}\cdot 5} = \dfrac{3}{5}$

3. $\dfrac{2}{16} = \dfrac{\cancel{2}}{\cancel{2}\cdot 2\cdot 2\cdot 2} = \dfrac{1}{8}$

4. $\dfrac{28}{49} = \dfrac{2\cdot 2\cdot\cancel{7}}{\cancel{7}\cdot 7} = \dfrac{4}{7}$

5. $\dfrac{54}{81} = \dfrac{\cancel{3}\cdot\cancel{3}\cdot\cancel{3}\cdot 2}{\cancel{3}\cdot\cancel{3}\cdot\cancel{3}\cdot 3} = \dfrac{2}{3}$

6. $\dfrac{40}{64} = \dfrac{\cancel{2}\cdot\cancel{2}\cdot\cancel{2}\cdot 5}{\cancel{2}\cdot\cancel{2}\cdot\cancel{2}\cdot 2\cdot 2\cdot 2} = \dfrac{5}{8}$

7. $\dfrac{75}{90} = \dfrac{\cancel{3}\cdot\cancel{5}\cdot 5}{2\cdot\cancel{3}\cdot 3\cdot\cancel{5}} = \dfrac{5}{6}$

8. $\dfrac{36}{27} = \dfrac{2\cdot 2\cdot\cancel{3}\cdot\cancel{3}}{\cancel{3}\cdot\cancel{3}\cdot 3} = \dfrac{4}{3} = 1\dfrac{1}{3}$

9. $\dfrac{0}{15} = 0$

10. $\dfrac{60}{96} = \dfrac{\cancel{2}\cdot\cancel{2}\cdot\cancel{3}\cdot 5}{\cancel{2}\cdot\cancel{2}\cdot 2\cdot 2\cdot 2\cdot\cancel{3}} = \dfrac{5}{8}$

11. $\dfrac{18}{36} = \dfrac{\cancel{2}\cdot\cancel{3}\cdot\cancel{3}}{2\cdot\cancel{2}\cdot\cancel{3}\cdot\cancel{3}} = \dfrac{1}{2}$

12. $\dfrac{7}{18}$ is in simplest form.

13. $\dfrac{25}{100} = \dfrac{\cancel{5}\cdot\cancel{5}}{2\cdot 2\cdot\cancel{5}\cdot\cancel{5}} = \dfrac{1}{4}$

14. $\dfrac{84}{144} = \dfrac{\cancel{2}\cdot\cancel{2}\cdot\cancel{3}\cdot 7}{2\cdot 2\cdot\cancel{2}\cdot\cancel{2}\cdot\cancel{3}\cdot 3} = \dfrac{7}{12}$

15. $\dfrac{39}{13} = \dfrac{3\cdot\cancel{13}}{\cancel{13}} = \dfrac{3}{1} = 3$

16. $\dfrac{169}{234} = \dfrac{\cancel{13}\cdot 13}{2\cdot 3\cdot 3\cdot\cancel{13}} = \dfrac{13}{18}$

17. $\dfrac{112}{126} = \dfrac{2\cdot 2\cdot 2\cdot\cancel{2}\cdot\cancel{7}}{\cancel{2}\cdot 3\cdot 3\cdot\cancel{7}} = \dfrac{8}{9}$

18. $\dfrac{59}{177} = \dfrac{\cancel{59}}{3\cdot\cancel{59}} = \dfrac{1}{3}$

19. $\dfrac{85}{65} = \dfrac{\cancel{5}\cdot 17}{\cancel{5}\cdot 13} = \dfrac{17}{13} = 1\dfrac{4}{13}$

20. $\dfrac{34}{238} = \dfrac{\cancel{2}\cdot\cancel{17}}{\cancel{2}\cdot 7\cdot\cancel{17}} = \dfrac{1}{7}$

21. $\dfrac{16}{34} = \dfrac{\cancel{2} \cdot 2 \cdot 2 \cdot 2}{\cancel{2} \cdot 17} = \dfrac{8}{17}$

22. $\dfrac{104}{240} = \dfrac{\cancel{2} \cdot \cancel{2} \cdot \cancel{2} \cdot 13}{\cancel{2} \cdot \cancel{2} \cdot \cancel{2} \cdot 2 \cdot 3 \cdot 5} = \dfrac{13}{30}$

23. $\dfrac{69}{150} = \dfrac{\cancel{3} \cdot 23}{2 \cdot \cancel{3} \cdot 5 \cdot 5} = \dfrac{23}{50}$

24. $\dfrac{26}{75}$ is in simplest form.

25. $\dfrac{112}{160} = \dfrac{\cancel{2} \cdot \cancel{2} \cdot \cancel{2} \cdot \cancel{2} \cdot 7}{2 \cdot \cancel{2} \cdot \cancel{2} \cdot \cancel{2} \cdot \cancel{2} \cdot 5} = \dfrac{7}{10}$

26. $\dfrac{143}{182} = \dfrac{11 \cdot \cancel{13}}{2 \cdot 7 \cdot \cancel{13}} = \dfrac{11}{14}$

27. $\dfrac{92}{23} = \dfrac{4 \cdot \cancel{23}}{\cancel{23}} = 4$

Objective 2.2C – Find equivalent fractions by raising to higher terms

Key Words and Concepts

Equal fractions with different denominators are called _____ **fractions**.

The Multiplication Property of One states that the product of a number and 1 is the number.

This property can be used to write equivalent fractions.

Example

Write $\frac{2}{5}$ as an equivalent fraction with denominator 30.

Solution
Divide the larger denominator by the smaller.
$30 \div 5 = 6$
Multiply the numerator and denominator of the given fraction by the quotient (6).

$$\frac{2}{5} = \frac{2 \cdot 6}{5 \cdot 6} = \frac{12}{30}$$

$\frac{2}{5}$ is equivalent to $\frac{12}{30}$.

Try it

1. Write $\frac{3}{7}$ as an equivalent fraction with denominator 42.

Example

Write $\frac{5}{8}$ as an equivalent fraction with denominator 96.

Solution
Divide the larger denominator by the smaller.
$96 \div 8 = 12$
Multiply the numerator and denominator of the given fraction by the quotient (12).

$$\frac{5}{8} = \frac{5 \cdot 12}{8 \cdot 12} = \frac{60}{96}$$

$\frac{5}{8}$ is equivalent to $\frac{60}{96}$.

Try it

2. Write $\frac{7}{12}$ as an equivalent fraction with denominator 84.

Example

Write 9 as a fraction with denominator 12.

Solution
Write 9 as $\frac{9}{1}$.
$12 \div 1 = 12$
$$\frac{9}{1} = \frac{9 \cdot 12}{1 \cdot 12} = \frac{108}{12}$$
9 is equivalent to $\frac{108}{12}$.

Try it

3. Write 6 as a fraction with denominator 11.

Reflect on it
- Equal fractions with different denominators are called equivalent fractions.

Quiz Yourself 2.2C

1. Write $\frac{5}{6}$ as an equivalent fraction with denominator 18.

2. Write $\frac{4}{7}$ as an equivalent fraction with denominator 28.

3. Write $\frac{11}{15}$ as an equivalent fraction with denominator 180.

4. Write 5 as a fraction with denominator 7.

Practice Sheet 2.2C

Build an equivalent fraction with the given denominator.

1. $\dfrac{2}{3} = \dfrac{}{39}$

2. $\dfrac{5}{6} = \dfrac{}{18}$

3. $\dfrac{4}{7} = \dfrac{}{56}$

4. $\dfrac{1}{5} = \dfrac{}{25}$

5. $\dfrac{3}{4} = \dfrac{}{28}$

6. $\dfrac{6}{10} = \dfrac{}{30}$

7. $\dfrac{1}{2} = \dfrac{}{30}$

8. $\dfrac{4}{5} = \dfrac{}{45}$

9. $\dfrac{2}{9} = \dfrac{}{54}$

10. $\dfrac{8}{12} = \dfrac{}{48}$

11. $\dfrac{11}{13} = \dfrac{}{65}$

12. $\dfrac{5}{10} = \dfrac{}{70}$

13. $\dfrac{6}{7} = \dfrac{}{84}$

14. $6 = \dfrac{}{9}$

15. $8 = \dfrac{}{15}$

16. $\dfrac{2}{4} = \dfrac{}{36}$

17. $\dfrac{1}{3} = \dfrac{}{102}$

18. $\dfrac{5}{9} = \dfrac{}{180}$

19. $\dfrac{10}{15} = \dfrac{}{90}$

20. $\dfrac{17}{24} = \dfrac{}{96}$

21. $\dfrac{4}{11} = \dfrac{}{121}$

22. $\dfrac{8}{20} = \dfrac{}{240}$

23. $\dfrac{15}{16} = \dfrac{}{336}$

24. $\dfrac{7}{33} = \dfrac{}{165}$

25. $\dfrac{5}{8} = \dfrac{}{312}$

26. $\dfrac{14}{19} = \dfrac{}{437}$

27. $\dfrac{21}{25} = \dfrac{}{1075}$

1. _____

2. _____

3. _____

4. _____

5. _____

6. _____

7. _____

8. _____

9. _____

10. _____

11. _____

12. _____

13. _____

14. _____

15. _____

16. _____

17. _____

18. _____

19. _____

20. _____

21. _____

22. _____

23. _____

24. _____

25. _____

26. _____

27. _____

Name:

Class:

Module 2: Problem Solving with Fractions and Decimals

Section 2.2: Writing Equivalent Fractions

Answers

Try it 2.2C

1. $\dfrac{18}{42}$

2. $\dfrac{49}{84}$

3. $\dfrac{66}{11}$

Quiz 2.2C

1. $\dfrac{15}{18}$

2. $\dfrac{16}{28}$

3. $\dfrac{132}{180}$

4. $\dfrac{35}{7}$

Solutions to Practice Sheet 2.2C

1. $39 \div 3 = 13$

$\dfrac{2}{3} = \dfrac{2 \cdot 13}{3 \cdot 13} = \dfrac{26}{39}$

2. $18 \div 6 = 3$

$\dfrac{5}{6} = \dfrac{5 \cdot 3}{6 \cdot 3} = \dfrac{15}{18}$

3. $56 \div 7 = 8$

$\dfrac{4}{7} = \dfrac{4 \cdot 8}{7 \cdot 8} = \dfrac{32}{56}$

4. $25 \div 5 = 5$

$\dfrac{1}{5} = \dfrac{1 \cdot 5}{5 \cdot 5} = \dfrac{5}{25}$

5. $28 \div 4 = 7$

$\dfrac{3}{4} = \dfrac{3 \cdot 7}{4 \cdot 7} = \dfrac{21}{28}$

6. $30 \div 10 = 3$

$\dfrac{6}{10} = \dfrac{6 \cdot 3}{10 \cdot 3} = \dfrac{18}{30}$

7. $30 \div 2 = 15$

$\dfrac{1}{2} = \dfrac{1 \cdot 15}{2 \cdot 15} = \dfrac{15}{30}$

8. $45 \div 5 = 9$

$\dfrac{4}{5} = \dfrac{4 \cdot 9}{5 \cdot 9} = \dfrac{36}{45}$

9. $54 \div 9 = 6$

$\dfrac{2}{9} = \dfrac{2 \cdot 6}{9 \cdot 6} = \dfrac{12}{54}$

10. $48 \div 12 = 4$

$\dfrac{8}{12} = \dfrac{8 \cdot 4}{12 \cdot 4} = \dfrac{32}{48}$

11. $65 \div 13 = 5$

$\dfrac{11}{13} = \dfrac{11 \cdot 5}{13 \cdot 5} = \dfrac{55}{65}$

12. $70 \div 10 = 7$

$\dfrac{5}{10} = \dfrac{5 \cdot 7}{10 \cdot 7} = \dfrac{35}{70}$

13. $84 \div 7 = 12$

$\dfrac{6}{7} = \dfrac{6 \cdot 12}{7 \cdot 12} = \dfrac{72}{84}$

14. $\dfrac{6}{1} = \dfrac{6 \cdot 9}{1 \cdot 9} = \dfrac{54}{9}$

15. $\dfrac{8}{1} = \dfrac{8 \cdot 15}{1 \cdot 15} = \dfrac{120}{15}$

16. $36 \div 4 = 9$

$\dfrac{2}{4} = \dfrac{2 \cdot 9}{4 \cdot 9} = \dfrac{18}{36}$

17. $102 \div 3 = 34$

$\dfrac{1}{3} = \dfrac{1 \cdot 34}{3 \cdot 34} = \dfrac{34}{102}$

18. $180 \div 9 = 20$

$\dfrac{5}{9} = \dfrac{5 \cdot 20}{9 \cdot 20} = \dfrac{100}{180}$

19. $90 \div 15 = 6$

$\dfrac{10}{15} = \dfrac{10 \cdot 6}{15 \cdot 6} = \dfrac{60}{90}$

20. $96 \div 24 = 4$

$\dfrac{17}{24} = \dfrac{17 \cdot 4}{24 \cdot 4} = \dfrac{68}{96}$

21. $121 \div 11 = 11$

$\dfrac{4}{11} = \dfrac{4 \cdot 11}{11 \cdot 11} = \dfrac{44}{121}$

22. $240 \div 20 = 12$

$\dfrac{8}{20} = \dfrac{8 \cdot 12}{20 \cdot 12} = \dfrac{96}{240}$

23. $336 \div 16 = 21$

$\dfrac{15}{16} = \dfrac{15 \cdot 21}{16 \cdot 21} = \dfrac{315}{336}$

24. $165 \div 33 = 5$

$\dfrac{7}{33} = \dfrac{7 \cdot 5}{33 \cdot 5} = \dfrac{35}{165}$

25. $312 \div 8 = 39$

$\dfrac{5}{8} = \dfrac{5 \cdot 39}{8 \cdot 39} = \dfrac{195}{312}$

© 2017 Cengage Learning. All Rights Reserved. May not be scanned, copied or duplicated, or posted to a publicly accessible website, in whole or in part.

26. $437 \div 19 = 23$

$$\frac{14}{19} = \frac{14 \cdot 23}{19 \cdot 23} = \frac{322}{437}$$

27. $1075 \div 25 = 43$

$$\frac{21}{25} = \frac{21 \cdot 43}{25 \cdot 43} = \frac{903}{1075}$$

Objective 2.2D – Identify the order relation between two fractions

Key Words and Concepts

Recall that whole numbers can be graphed as points on the number line. Fractions can also be graphed as points on the number line. The number line can be used to determine the order relation between two fractions.

A fraction that appears to the left of a given fraction is _____ than the given fraction.
(greater/less)

A fraction that appears to the _____ of a given fraction is greater than the given fraction.
(left/right)

For example,

$$\frac{4}{7} < \frac{6}{7}$$

To find the order relation between two fractions with the same denominator,
   compare the numerators.
   The fraction that has the smaller numerator is the smaller fraction.

For example, $\frac{4}{7} < \frac{6}{7}$ since $4 < 6$.

When the denominators are different,
   begin by writing equivalent fractions with a common denominator;
   then compare the numerators.

Example	Try it
Find the order relation between $\frac{5}{6}$ and $\frac{7}{9}$.	1. Find the order relation between $\frac{5}{12}$ and $\frac{9}{20}$.

Solution
   The LCD of the fractions is 18.
   $$\frac{5}{6} = \frac{15}{18} \leftarrow \text{Larger numerator}$$
   $$\frac{7}{9} = \frac{14}{18} \leftarrow \text{Smaller numerator}$$

   Therefore, $\frac{5}{6} > \frac{7}{9}$ or $\frac{7}{9} < \frac{5}{6}$.

Reflect on it
- A fraction that appears to the left of a given fraction is less than the given fraction.
- A fraction that appears to the right of a given fraction is greater than the given fraction.

Quiz Yourself 2.2D

1. Place the correct symbol, < or >, between the two numbers.

    $\dfrac{9}{19}$ $\dfrac{14}{19}$

2. Place the correct symbol, < or >, between the two numbers.

    $\dfrac{3}{4}$ $\dfrac{5}{7}$

3. Place the correct symbol, < or >, between the two numbers.

    $\dfrac{6}{11}$ $\dfrac{8}{13}$

4. Place the correct symbol, < or >, between the two numbers.

    $\dfrac{7}{15}$ $\dfrac{11}{23}$

Practice Sheet 2.2D

Place the correct symbol, < or >, between the two numbers.

1. $\dfrac{11}{17}$ $\dfrac{14}{17}$ 2. $\dfrac{4}{5}$ $\dfrac{5}{9}$ 3. $\dfrac{7}{10}$ $\dfrac{2}{3}$

4. $\dfrac{7}{12}$ $\dfrac{11}{15}$ 5. $\dfrac{5}{7}$ $\dfrac{3}{4}$ 6. $\dfrac{2}{3}$ $\dfrac{3}{5}$

7. $\dfrac{3}{10}$ $\dfrac{1}{6}$ 8. $\dfrac{7}{9}$ $\dfrac{9}{14}$ 9. $\dfrac{4}{11}$ $\dfrac{1}{2}$

10. $\dfrac{1}{4}$ $\dfrac{4}{15}$ 11. $\dfrac{13}{20}$ $\dfrac{5}{7}$ 12. $\dfrac{8}{13}$ $\dfrac{11}{18}$

13. $\dfrac{19}{24}$ $\dfrac{25}{36}$ 14. $\dfrac{8}{15}$ $\dfrac{23}{35}$ 15. $\dfrac{5}{9}$ $\dfrac{3}{5}$

16. $\dfrac{5}{22}$ $\dfrac{3}{8}$ 17. $\dfrac{1}{4}$ $\dfrac{5}{26}$ 18. $\dfrac{5}{12}$ $\dfrac{4}{9}$

19. $\dfrac{9}{16}$ $\dfrac{10}{17}$ 20. $\dfrac{19}{22}$ $\dfrac{39}{46}$ 21. $\dfrac{23}{30}$ $\dfrac{17}{20}$

22. $\dfrac{5}{6}$ $\dfrac{3}{4}$ 23. $\dfrac{11}{14}$ $\dfrac{15}{19}$ 24. $\dfrac{7}{24}$ $\dfrac{8}{15}$

25. $\dfrac{4}{5}$ $\dfrac{16}{21}$ 26. $\dfrac{7}{15}$ $\dfrac{5}{8}$ 27. $\dfrac{21}{25}$ $\dfrac{27}{35}$

1. _____
2. _____
3. _____
4. _____
5. _____
6. _____
7. _____
8. _____
9. _____
10. _____
11. _____
12. _____
13. _____
14. _____
15. _____
16. _____
17. _____
18. _____
19. _____
20. _____
21. _____
22. _____
23. _____
24. _____
25. _____
26. _____
27. _____

Answers

Try it 2.2D

1. $\dfrac{5}{12} < \dfrac{9}{20}$

Quiz 2.2D

1. $\dfrac{9}{19} < \dfrac{14}{19}$

2. $\dfrac{3}{4} > \dfrac{5}{7}$

3. $\dfrac{6}{11} < \dfrac{8}{13}$

4. $\dfrac{7}{15} < \dfrac{11}{23}$

Solutions to Practice Sheet 2.2D

1. $\dfrac{11}{17} < \dfrac{14}{17}$

2. $\dfrac{4}{5} = \dfrac{36}{45}; \ \dfrac{5}{9} = \dfrac{25}{45}$
   $\dfrac{4}{5} > \dfrac{5}{9}$

3. $\dfrac{7}{10} = \dfrac{21}{30}; \ \dfrac{2}{3} = \dfrac{20}{30}$
   $\dfrac{7}{10} > \dfrac{2}{3}$

4. $\dfrac{7}{12} = \dfrac{35}{60}; \ \dfrac{11}{15} = \dfrac{44}{60}$
   $\dfrac{7}{12} < \dfrac{11}{15}$

5. $\dfrac{5}{7} = \dfrac{20}{28}; \ \dfrac{3}{4} = \dfrac{21}{28}$
   $\dfrac{5}{7} < \dfrac{3}{4}$

6. $\dfrac{2}{3} = \dfrac{10}{15}; \ \dfrac{3}{5} = \dfrac{9}{15}$
   $\dfrac{2}{3} > \dfrac{3}{5}$

7. $\dfrac{3}{10} = \dfrac{9}{30}; \ \dfrac{1}{6} = \dfrac{5}{30}$
   $\dfrac{3}{10} > \dfrac{1}{6}$

8. $\dfrac{7}{9} = \dfrac{98}{126}; \ \dfrac{9}{14} = \dfrac{81}{126}$
   $\dfrac{7}{9} > \dfrac{9}{14}$

9. $\dfrac{4}{11} = \dfrac{8}{22}; \ \dfrac{1}{2} = \dfrac{11}{22}$
   $\dfrac{4}{11} < \dfrac{1}{2}$

10. $\dfrac{1}{4} = \dfrac{15}{60}; \ \dfrac{4}{15} = \dfrac{16}{60}$
    $\dfrac{1}{4} < \dfrac{4}{15}$

11. $\dfrac{13}{20} = \dfrac{91}{140}; \ \dfrac{5}{7} = \dfrac{100}{140}$
    $\dfrac{13}{20} < \dfrac{5}{7}$

12. $\dfrac{8}{13} = \dfrac{144}{234}; \ \dfrac{11}{18} = \dfrac{143}{234}$
    $\dfrac{8}{13} > \dfrac{11}{18}$

13. $\dfrac{19}{24} = \dfrac{57}{72}; \ \dfrac{25}{36} = \dfrac{50}{72}$
    $\dfrac{19}{24} > \dfrac{25}{36}$

14. $\dfrac{8}{15} = \dfrac{56}{105}; \ \dfrac{23}{35} = \dfrac{69}{105}$
    $\dfrac{8}{15} < \dfrac{23}{35}$

15. $\dfrac{5}{9} = \dfrac{25}{45}; \ \dfrac{3}{5} = \dfrac{27}{45}$
    $\dfrac{5}{9} < \dfrac{3}{5}$

16. $\dfrac{5}{22} = \dfrac{20}{88}; \ \dfrac{3}{8} = \dfrac{33}{88}$
    $\dfrac{5}{22} < \dfrac{3}{8}$

17. $\dfrac{1}{4} = \dfrac{13}{52}; \ \dfrac{5}{26} = \dfrac{10}{52}$
    $\dfrac{1}{4} > \dfrac{5}{26}$

18. $\dfrac{5}{12} = \dfrac{15}{36}; \ \dfrac{4}{9} = \dfrac{16}{36}$
    $\dfrac{5}{12} < \dfrac{4}{9}$

19. $\dfrac{9}{16} = \dfrac{153}{272}; \ \dfrac{10}{17} = \dfrac{160}{272}$
    $\dfrac{9}{16} < \dfrac{10}{17}$

20. $\dfrac{19}{22} = \dfrac{437}{506}; \ \dfrac{39}{46} = \dfrac{429}{506}$
    $\dfrac{19}{22} > \dfrac{39}{46}$

21. $\dfrac{23}{30} = \dfrac{46}{60}$; $\dfrac{17}{20} = \dfrac{51}{60}$

$\dfrac{23}{30} < \dfrac{17}{20}$

22. $\dfrac{5}{6} = \dfrac{10}{12}$; $\dfrac{3}{4} = \dfrac{9}{12}$

$\dfrac{5}{6} > \dfrac{3}{4}$

23. $\dfrac{11}{14} = \dfrac{209}{266}$; $\dfrac{15}{19} = \dfrac{210}{266}$

$\dfrac{11}{14} < \dfrac{15}{19}$

24. $\dfrac{7}{24} = \dfrac{35}{120}$; $\dfrac{8}{15} = \dfrac{64}{120}$

$\dfrac{7}{24} < \dfrac{8}{15}$

25. $\dfrac{4}{5} = \dfrac{84}{105}$; $\dfrac{16}{21} = \dfrac{80}{105}$

$\dfrac{4}{5} > \dfrac{16}{21}$

26. $\dfrac{7}{15} = \dfrac{56}{120}$; $\dfrac{5}{8} = \dfrac{75}{120}$

$\dfrac{7}{15} < \dfrac{5}{8}$

27. $\dfrac{21}{25} = \dfrac{147}{175}$; $\dfrac{27}{35} = \dfrac{135}{175}$

$\dfrac{21}{25} > \dfrac{27}{35}$

Objective 2.3A – Multiply and divide positive and negative fractions

Key Words and Concepts

The product of two fractions is the product of the numerators over the product of the denominators.

$$\frac{a}{b} \cdot \frac{c}{d} = \frac{ac}{bd} \quad \text{where } b \neq 0 \text{ and } d \neq 0$$

The sign rules for multiplying positive and negative fractions are the same rules used to multiply integers.

The product of two numbers with the same sign is positive.

The product of two numbers with different signs is negative.

Example

Multiply $-\dfrac{7}{9} \cdot \dfrac{3}{5}$.

Solution

The signs are different. The product is negative.

$\begin{aligned}
-\frac{7}{9} \cdot \frac{3}{5} &= -\frac{7 \cdot 3}{9 \cdot 5} & \text{Multiply the numerators.} \\
&& \text{Multiply the denominators.} \\
&= -\frac{7 \cdot 3}{3 \cdot 3 \cdot 5} & \text{Write in factored form.} \\
&= -\frac{7}{15} & \text{Simplify}
\end{aligned}$

Try it

1. Multiply $-\dfrac{5}{6} \cdot \dfrac{3}{10}$.

The _____ of a fraction is the fraction with numerator and denominator interchanged. Division is defined as the multiplication by the reciprocal.

To divide two fractions, multiply by the reciprocal of the divisor.

$$\frac{a}{b} \div \frac{c}{d} = \frac{a}{b} \cdot \frac{d}{c} \quad \text{where } b \neq 0, \ c \neq 0, \text{ and } d \neq 0$$

The sign rules for dividing positive and negative fractions are the same rules used to divide integers.

The quotient of two numbers with the same sign is positive.

The quotient of two numbers with different signs is negative.

Example

Divide $-\dfrac{2}{3} \div \left(-\dfrac{6}{9}\right)$.

Solution

The signs are the same. The quotient is positive.

$\begin{aligned}
-\frac{2}{3} \div \left(-\frac{6}{9}\right) &= \frac{2}{3} \cdot \frac{9}{6} & \begin{array}{l}\text{Rewrite division as multiplication} \\ \text{by the reciprocal.}\end{array} \\
&= \frac{2 \cdot 9}{3 \cdot 6} & \begin{array}{l}\text{Multiply the numerators.} \\ \text{Multiply the denominators.}\end{array} \\
&= \frac{2 \cdot 3 \cdot 3}{3 \cdot 2 \cdot 3} & \text{Write in factored form.} \\
&= 1 & \text{Simplify}
\end{aligned}$

Try it

2. Divide $-\dfrac{6}{8} \div \left(-\dfrac{3}{4}\right)$.

Example

Multiply $-\dfrac{2}{5}\left(\dfrac{7}{10}\right)\left(-\dfrac{25}{7}\right)$

    Solution

$$-\frac{2}{5}\left(\frac{7}{10}\right)\left(-\frac{25}{7}\right) = \frac{2\cdot 7 \cdot 25}{5 \cdot 10 \cdot 7}$$
$$= \frac{1}{1} = 1$$

Try it

3. Multiply $-\dfrac{6}{11}\left(\dfrac{4}{12}\right)\left(-\dfrac{88}{16}\right)$

Example

Evaluate xy for $x = -1\dfrac{1}{5}$ and $y = -3\dfrac{1}{3}$.

    Solution

xy

$$\left(-1\frac{1}{5}\right)\left(-3\frac{1}{3}\right) = \left(-\frac{6}{5}\right)\left(-\frac{10}{3}\right)$$
$$= \frac{2\cdot 3 \cdot 2 \cdot 5}{5 \cdot 3}$$
$$= \frac{4}{1} = 4$$

Try it

4. Evaluate xy for

$$x = -2\frac{1}{4} \text{ and } y = -6\frac{2}{3}.$$

Example

Divide $-5\dfrac{3}{7} \div \left(-\dfrac{7}{10}\right)$

    Solution

        The signs are the same. The quotient is positive.

$$-5\frac{3}{7} \div \left(-\frac{7}{10}\right) = -\frac{38}{7}\cdot\left(-\frac{7}{10}\right)$$
$$= \frac{2 \cdot 19 \cdot 7}{7 \cdot 2 \cdot 5}$$
$$= \frac{19}{5} = 3\frac{4}{5}$$

Try it

5. Divide $-2\dfrac{5}{6} \div \left(-\dfrac{8}{9}\right)$

Reflect on it

- To multiply two fractions, multiply the numerators and multiply the denominators.
- The reciprocal of a fraction is the fraction with numerator and denominator interchanged.
- To divide two fractions, multiply by the reciprocal of the divisor.

Quiz Yourself 2.3A

1. Find the product of $-\frac{2}{3}$ and $\frac{6}{7}$.

2. Multiply: $-1\frac{3}{5}\cdot\left(-\frac{10}{13}\right)$

3. What is $-2\frac{2}{9}$ times $-1\frac{4}{5}$?

4. Find the quotient of $-\frac{2}{3}$ and $\frac{6}{7}$.

5. Divide: $-4\div\left(-\frac{6}{11}\right)$

6. Divide: $-7\frac{1}{8}\div\left(-3\frac{1}{2}\right)$

7. Evaluate $x\div y$ for $x=9\frac{1}{2}$ and $y=-3$.

Name: Class:

Module 2: Problem Solving with Fractions and Decimals   Section 2.3: Operations on Positive/Negative Fractions

Practice Sheet 2.3A

Multiply or divide.

1. $\dfrac{15}{16} \cdot \dfrac{4}{9}$

2. $-\dfrac{a}{10} \cdot \left(-\dfrac{b}{6}\right)$

3. $\dfrac{2}{3} \cdot \dfrac{3}{8} \cdot \dfrac{4}{9}$

4. $\dfrac{5}{6} \cdot \left(-\dfrac{2}{3}\right) \cdot \dfrac{3}{25}$

5. $24 \cdot \left(-\dfrac{3}{8}\right)$

6. $3\dfrac{1}{2} \cdot 5\dfrac{3}{7}$

7. $-2\dfrac{1}{2} \cdot 4$

8. $\dfrac{0}{1} \div \dfrac{1}{9}$

9. $\dfrac{3}{4} \div (-6)$

10. $\left(-\dfrac{4}{9}\right) \div (-6)$

11. $-1\dfrac{3}{5} \div 3\dfrac{1}{10}$

12. $6\dfrac{8}{9} \div \left(-\dfrac{31}{36}\right)$

13. $-16 \div 1\dfrac{1}{3}$

14. $2\dfrac{4}{13} \div 1\dfrac{5}{26}$

15. $3\dfrac{3}{8} \div 2\dfrac{7}{16}$

16. Find $5\dfrac{1}{3}$ multiplied by $\dfrac{4}{27}$.

17. What is $5\dfrac{1}{3}$ times $\dfrac{3}{16}$?

18. Find the quotient of $\dfrac{9}{10}$ and $\dfrac{3}{4}$.

19. Find $-\dfrac{3}{8}$ divided by $2\dfrac{1}{4}$.

Evaluate the variable expression *xyz* for the given values *x*, *y*, and *z*.

20. $x = \dfrac{3}{8}, y = \dfrac{2}{8}, z = \dfrac{4}{5}$

21. $x = 4, y = \dfrac{0}{8}, z = 1\dfrac{5}{9}$

22. $x = \dfrac{5}{6}, y = -3, z = 1\dfrac{7}{15}$

Evaluate the variable expression *x* ÷ *y* for the given values *x* and *y*.

23. $x = -\dfrac{14}{3}, y = -\dfrac{7}{9}$

24. $x = -\dfrac{1}{2}, y = -3\dfrac{5}{8}$

25. $x = -5\dfrac{2}{5}, y = -9$

1. _____
2. _____
3. _____
4. _____
5. _____
6. _____
7. _____
8. _____
9. _____
10. _____
11. _____
12. _____
13. _____
14. _____
15. _____
16. _____
17. _____
18. _____
19. _____
20. _____
21. _____
22. _____
23. _____
24. _____
25. _____

Answers

Try it 2.3A

1. $-\dfrac{1}{4}$

2. 1

3. 1

4. 15

5. $3\dfrac{3}{16}$

Quiz 2.3A

1. $-\dfrac{4}{7}$

2. $\dfrac{16}{13} = 1\dfrac{3}{13}$

3. 4

4. $-\dfrac{7}{9}$

5. $\dfrac{22}{3} = 7\dfrac{1}{3}$

6. $\dfrac{57}{28} = 2\dfrac{1}{28}$

7. $\dfrac{-19}{6} = -3\dfrac{1}{6}$

Solutions to Practice Sheet 2.3A

1. $\dfrac{15}{16} \cdot \dfrac{4}{9} = \dfrac{3 \cdot 5 \cdot 2 \cdot 2}{2 \cdot 2 \cdot 2 \cdot 2 \cdot 3 \cdot 3} = \dfrac{5}{12}$

2. $-\dfrac{a}{10} \cdot \left(-\dfrac{b}{6}\right) = \dfrac{ab}{60}$

3. $\dfrac{2}{3} \cdot \dfrac{3}{8} \cdot \dfrac{4}{9} = \dfrac{2 \cdot 3 \cdot 2 \cdot 2}{3 \cdot 2 \cdot 2 \cdot 2 \cdot 3 \cdot 3} = \dfrac{1}{9}$

4. $\dfrac{5}{6} \cdot \left(-\dfrac{2}{3}\right) \cdot \dfrac{3}{25} = -\dfrac{5 \cdot 2 \cdot 3}{2 \cdot 3 \cdot 3 \cdot 5 \cdot 5} = -\dfrac{1}{15}$

5. $24 \cdot \left(-\dfrac{3}{8}\right) = -\dfrac{2 \cdot 2 \cdot 2 \cdot 3 \cdot 3}{2 \cdot 2 \cdot 2} = -9$

6. $3\dfrac{1}{2} \cdot 5\dfrac{3}{7} = \dfrac{7}{2} \cdot \dfrac{38}{7} = \dfrac{7 \cdot 2 \cdot 19}{2 \cdot 7} = 19$

7. $-2\dfrac{1}{2} \cdot 4 = -\dfrac{5}{2} \cdot \dfrac{4}{1} = -\dfrac{5 \cdot 2 \cdot 2}{2} = -10$

8. $\dfrac{0}{1} \div \dfrac{1}{9} = \dfrac{0}{1} \cdot \dfrac{9}{1} = 0$

9. $\dfrac{3}{4} \div (-6) = -\dfrac{3}{4} \cdot \dfrac{1}{6} = -\dfrac{3}{2 \cdot 2 \cdot 2 \cdot 3} = -\dfrac{1}{8}$

10. $\left(-\dfrac{4}{9}\right) \div (-6) = \dfrac{4}{9} \cdot \dfrac{1}{6} = \dfrac{2 \cdot 2}{3 \cdot 3 \cdot 2 \cdot 3} = \dfrac{2}{27}$

11. $-1\dfrac{3}{5} \div 3\dfrac{1}{10} = -\dfrac{8}{5} \div \dfrac{31}{10} = -\dfrac{8}{5} \cdot \dfrac{10}{31}$

    $= -\dfrac{2 \cdot 2 \cdot 2 \cdot 2 \cdot 5}{5 \cdot 31}$

    $= -\dfrac{16}{31}$

12. $6\dfrac{8}{9} \div \left(-\dfrac{31}{36}\right) = \dfrac{62}{9} \div \left(-\dfrac{31}{36}\right)$

    $= -\dfrac{62}{9} \cdot \dfrac{36}{31}$

    $= -\dfrac{2 \cdot 31 \cdot 2 \cdot 2 \cdot 3 \cdot 3}{3 \cdot 3 \cdot 31}$

    $= -8$

13. $-16 \div 1\dfrac{1}{3} = -16 \div \dfrac{4}{3} = -\dfrac{16}{1} \cdot \dfrac{3}{4}$

    $= -\dfrac{2 \cdot 2 \cdot 2 \cdot 2 \cdot 3}{2 \cdot 2}$

    $= -12$

14. $2\dfrac{4}{13} \div 1\dfrac{5}{26} = \dfrac{30}{13} \div \dfrac{31}{26}$

    $= \dfrac{30}{13} \cdot \dfrac{26}{31}$

    $= \dfrac{2 \cdot 3 \cdot 5 \cdot 2 \cdot 13}{13 \cdot 31}$

    $= \dfrac{60}{31} = 1\dfrac{29}{31}$

15. $3\dfrac{3}{8} \div 2\dfrac{7}{16} = \dfrac{27}{8} \div \dfrac{39}{16}$

    $= \dfrac{27}{8} \cdot \dfrac{16}{39}$

    $= \dfrac{3 \cdot 3 \cdot 3 \cdot 2 \cdot 2 \cdot 2 \cdot 2}{2 \cdot 2 \cdot 2 \cdot 3 \cdot 13}$

    $= \dfrac{18}{13}$

    $= 1\dfrac{5}{13}$

16. $5\dfrac{1}{3} \cdot \dfrac{4}{27} = \dfrac{16}{3} \cdot \dfrac{4}{27} = \dfrac{64}{81}$

17. $5\dfrac{1}{3} \cdot \dfrac{3}{16} = \dfrac{16}{3} \cdot \dfrac{3}{16} = 1$

18. $\dfrac{9}{10} \div \dfrac{3}{4} = \dfrac{9}{10} \cdot \dfrac{4}{3} = \dfrac{3 \cdot 3 \cdot 2 \cdot 2}{2 \cdot 5 \cdot 3} = \dfrac{6}{5} = 1\dfrac{1}{5}$

19. $-\dfrac{3}{8} \div 2\dfrac{1}{4} = -\dfrac{3}{8} \div \dfrac{9}{4}$

$= -\dfrac{3}{8} \cdot \dfrac{4}{9}$

$= -\dfrac{3 \cdot 2 \cdot 2}{2 \cdot 2 \cdot 2 \cdot 3 \cdot 3}$

$= -\dfrac{1}{6}$

20. xyz

$\dfrac{3}{8} \cdot \dfrac{2}{8} \cdot \dfrac{4}{5} = \dfrac{3 \cdot 2 \cdot 2 \cdot 2}{2 \cdot 2 \cdot 2 \cdot 2 \cdot 2 \cdot 2 \cdot 5} = \dfrac{3}{40}$

21. xyz

$4 \cdot \dfrac{0}{8} \cdot 1\dfrac{5}{9} = 0$

22. xyz

$\dfrac{5}{6} \cdot (-3) \cdot 1\dfrac{7}{15} = -\dfrac{5}{6} \cdot \dfrac{3}{1} \cdot \dfrac{22}{15}$

$= -\dfrac{5 \cdot 3 \cdot 2 \cdot 11}{2 \cdot 3 \cdot 3 \cdot 5}$

$= -\dfrac{11}{3} = -3\dfrac{2}{3}$

23. $x \div y$

$-\dfrac{14}{3} \div \left(-\dfrac{7}{9}\right) = \dfrac{14}{3} \cdot \dfrac{9}{7} = \dfrac{2 \cdot 7 \cdot 3 \cdot 3}{3 \cdot 7} = 6$

24. $x \div y$

$-\dfrac{1}{2} \div \left(-3\dfrac{5}{8}\right) = -\dfrac{1}{2} \div \left(-\dfrac{29}{8}\right)$

$= \dfrac{1}{2} \cdot \dfrac{8}{29}$

$= \dfrac{2 \cdot 2 \cdot 2}{2 \cdot 29}$

$= \dfrac{4}{29}$

25. $x \div y$

$-5\dfrac{2}{5} \div (-9) = -\dfrac{27}{5} \div (-9)$

$= \dfrac{27}{5} \cdot \dfrac{1}{9}$

$= \dfrac{3 \cdot 3 \cdot 3}{5 \cdot 3 \cdot 3}$

$= \dfrac{3}{5}$

Objective 2.3B – Add and subtract positive and negative fractions

Key Words and Concepts

To add a fraction with a negative sign,
    rewrite the fraction with the negative sign in the numerator,
    rewrite each fraction as equivalent fractions with common denominators, if necessary,
    add the numerators,
    place the sum over the common denominator.

Example

Add $\frac{1}{3}$ to $-\frac{5}{6}$.

Solution

$-\frac{5}{6} + \frac{1}{3} = \frac{-5}{6} + \frac{1}{3}$ Rewrite the first fraction with negative sign in the numerator

$= \frac{-5}{6} + \frac{2}{6}$ Rewrite the fractions with a common denominator

$= \frac{-3}{6}$ Add

$= \frac{-1}{2}$ Simplify

$= -\frac{1}{2}$

Although the answer could have been left as $\frac{-1}{2}$,

we will write all negative fractions with the negative sign in front of the fraction.

Try it

1. Add $\frac{1}{4}$ to $-\frac{7}{12}$.

To subtract fractions with negative signs, first rewrite the fractions with the negative signs in the numerators.

Example

Subtract $\frac{5}{12}$ from $-\frac{5}{6}$.

Solution

$-\frac{5}{6} - \frac{5}{12} = \frac{-5}{6} - \frac{5}{12}$ Rewrite the first fraction with negative sign in the numerator

$= \frac{-10}{12} - \frac{5}{12}$ Rewrite the fractions with a common denominator

$= \frac{-10 - 5}{12}$

$= \frac{-15}{12}$ Subtract

$= -\frac{5}{4} = -1\frac{1}{4}$

Try it

2. Subtract $\frac{6}{7}$ from $-\frac{3}{14}$.

Reflect on it
- To add a fraction with a negative sign, rewrite the fraction with the negative sign in the numerator.
- To subtract fractions with negative signs, first rewrite the fractions with the negative signs in the numerators.

Quiz Yourself 2.3B

1. Add: $-\dfrac{5}{8}+\dfrac{1}{3}$

2. Add: $-\dfrac{1}{3}+\dfrac{5}{12}$

3. Find the sum of $-\dfrac{1}{3}, \dfrac{1}{2}$, and $-\dfrac{2}{5}$.

4. Subtract: $-\dfrac{5}{6}-\dfrac{2}{3}$

5. Subtract: $-\dfrac{7}{12}-\left(-\dfrac{5}{6}\right)$

6. Evaluate $a-b$ for $a=\dfrac{7}{10}$ and $b=-\dfrac{5}{8}$.

Practice Sheet 2.3B

Add or subtract.

1. $-\frac{7}{12}+\frac{2}{3}+\left(-\frac{4}{5}\right)$ 2. $-\frac{5}{8}+\frac{3}{4}+\frac{1}{2}$ 3. $2\frac{1}{6}+\left(-3\frac{1}{2}\right)$

4. $\frac{2}{3}+\left(-\frac{5}{6}\right)+\frac{1}{4}$ 5. $6+9\frac{3}{5}$ 6. $1\frac{2}{3}+2\frac{5}{6}+4\frac{7}{9}$

7. $\frac{5}{n}-\frac{10}{n}$ 8. $-\frac{5}{6}-\frac{1}{9}$ 9. $4\frac{11}{18}-2\frac{5}{18}$

10. $10-4\frac{8}{9}$ 11. $11\frac{1}{6}-8\frac{5}{6}$ 12. $16\frac{1}{3}-\left(-11\frac{5}{12}\right)$

13. $-\frac{3}{10}-\left(-\frac{5}{6}\right)$ 14. $10\frac{2}{5}-8\frac{7}{10}$ 15. $5\frac{5}{6}-4\frac{7}{8}$

16. Find the sum of $7\frac{11}{15}$, $2\frac{7}{10}$, and $5\frac{2}{5}$.   17. What is $-\frac{2}{3}$ more than $-\frac{5}{6}$?

18. What is $-\frac{2}{3}$ less than $-\frac{7}{8}$.   19. Find 9 minus $5\frac{3}{20}$.

Evaluate the variable expression $x + y$ for the given values x and y.

20. $x=\frac{3}{5}$, $y=\frac{4}{5}$ 21. $x=-\frac{3}{8}$, $y=\frac{2}{9}$ 22. $x=-\frac{5}{8}$, $y=-\frac{1}{6}$

Evaluate the variable expression $x - y$ for the given values x and y.

23. $x=\frac{5}{6}$, $y=\frac{1}{6}$ 24. $x=-\frac{11}{12}$, $y=\frac{5}{12}$ 25. $x=6\frac{4}{9}$, $y=-1\frac{1}{6}$

1. _____
2. _____
3. _____
4. _____
5. _____
6. _____
7. _____
8. _____
9. _____
10. _____
11. _____
12. _____
13. _____
14. _____
15. _____
16. _____
17. _____
18. _____
19. _____
20. _____
21. _____
22. _____
23. _____
24. _____
25. _____

Answers

Try it 2.3B

1. $-\dfrac{1}{3}$

2. $-1\dfrac{1}{14}$.

Quiz 2.3B

1. $-\dfrac{7}{24}$

2. $\dfrac{1}{12}$

3. $-\dfrac{7}{30}$

4. $-\dfrac{3}{2}=-1\dfrac{1}{2}$

5. $\dfrac{1}{4}$

6. $\dfrac{53}{40}=1\dfrac{13}{40}$

Solutions to Practice Sheet 2.3B

1. $-\dfrac{7}{12}+\dfrac{2}{3}+\left(-\dfrac{4}{5}\right)=\dfrac{-35}{60}+\dfrac{40}{60}+\dfrac{-48}{60}$
   $$=\dfrac{-35+40+(-48)}{60}$$
   $$=-\dfrac{43}{60}$$

2. $-\dfrac{5}{8}+\dfrac{3}{4}+\dfrac{1}{2}=\dfrac{-5}{8}+\dfrac{6}{8}+\dfrac{4}{8}$
   $$=\dfrac{-5+6+4}{8}$$
   $$=\dfrac{5}{8}$$

3. $2\dfrac{1}{6}+\left(-3\dfrac{1}{2}\right)=2\dfrac{1}{6}+\left(-3\dfrac{3}{6}\right)=-1\dfrac{2}{6}=-1\dfrac{1}{3}$

4. $\dfrac{2}{3}+\left(-\dfrac{5}{6}\right)+\dfrac{1}{4}=\dfrac{8}{12}+\dfrac{-10}{12}+\dfrac{3}{12}$
   $$=\dfrac{8+(-10)+3}{12}$$
   $$=\dfrac{1}{12}$$

5. $6+9\dfrac{3}{5}=15\dfrac{3}{5}$

6. $1\dfrac{2}{3}+2\dfrac{5}{6}+4\dfrac{7}{9}=1\dfrac{12}{18}+2\dfrac{15}{18}+4\dfrac{14}{18}$
   $$=7\dfrac{41}{18}=9\dfrac{5}{18}$$

7. $\dfrac{5}{n}-\dfrac{10}{n}=\dfrac{5+(-10)}{n}=-\dfrac{5}{n}$

8. $-\dfrac{5}{6}-\dfrac{1}{9}=\dfrac{-15}{18}+\dfrac{-2}{18}=\dfrac{-15+(-2)}{18}=-\dfrac{17}{18}$

9. $4\dfrac{11}{18}-2\dfrac{5}{18}=2\dfrac{6}{18}=2\dfrac{1}{3}$

10. $10-4\dfrac{8}{9}=9\dfrac{9}{9}+\left(-4\dfrac{8}{9}\right)=5\dfrac{1}{9}$

11. $\begin{array}{rcl}11\dfrac{1}{6}&=&10\dfrac{7}{6}\\[2mm]-\;\;8\dfrac{5}{6}&=&8\dfrac{5}{6}\\\hline&&2\dfrac{2}{6}=2\dfrac{1}{3}\end{array}$

12. $16\dfrac{1}{3}-\left(-11\dfrac{5}{12}\right)=16\dfrac{4}{12}+11\dfrac{5}{12}$
    $$=27\dfrac{9}{12}=27\dfrac{3}{4}$$

13. $-\dfrac{3}{10}-\left(-\dfrac{5}{6}\right)=\dfrac{-3}{10}+\dfrac{5}{6}$
    $$=\dfrac{-9}{30}+\dfrac{25}{30}$$
    $$=\dfrac{16}{30}=\dfrac{8}{15}$$

14. $\begin{array}{rcl}10\dfrac{2}{5}&=&9\dfrac{14}{10}\\[2mm]-\;\;8\dfrac{7}{10}&=&8\dfrac{7}{10}\\\hline&&1\dfrac{7}{10}\end{array}$

15. $\begin{array}{rcl}5\dfrac{5}{6}&=&4\dfrac{44}{24}\\[2mm]-\;\;4\dfrac{7}{8}&=&4\dfrac{21}{24}\\\hline&&\dfrac{23}{24}\end{array}$

16. $7\dfrac{11}{15}+2\dfrac{7}{10}+5\dfrac{2}{5}=7\dfrac{22}{30}+2\dfrac{21}{30}+5\dfrac{12}{30}$
    $$=14\dfrac{55}{30}$$
    $$=15\dfrac{25}{30}$$
    $$=15\dfrac{5}{6}$$

17. $-\dfrac{5}{6}+\left(-\dfrac{2}{3}\right)=\dfrac{-5}{6}+\dfrac{-4}{6}=-\dfrac{9}{6}=-\dfrac{3}{2}$

18. $-\dfrac{7}{8}-\left(-\dfrac{2}{3}\right)=\dfrac{-7}{8}+\dfrac{2}{3}=\dfrac{-21}{24}+\dfrac{16}{24}=-\dfrac{5}{24}$

19. $9 - 5\frac{3}{20} = 8\frac{20}{20} - 5\frac{3}{20} = 3\frac{17}{20}$

20. $x + y$

$$\frac{3}{5} + \frac{4}{5} = \frac{7}{5}$$

21. $x + y$

$$-\frac{3}{8} + \frac{2}{9} = \frac{-27}{72} + \frac{16}{72} = -\frac{11}{72}$$

22. $x + y$

$$-\frac{5}{8} + \left(-\frac{1}{6}\right) = \frac{-5}{8} + \frac{-1}{6} = \frac{-15}{24} + \frac{-4}{24} = -\frac{19}{24}$$

23. $x - y$

$$\frac{5}{6} - \frac{1}{6} = \frac{4}{6} = \frac{2}{3}$$

24. $x - y$

$$-\frac{11}{12} - \frac{5}{12} = \frac{-11 + (-5)}{12}$$
$$= -\frac{16}{12}$$
$$= -1\frac{4}{12}$$
$$= -1\frac{1}{3}$$

25. $x - y$

$$6\frac{4}{9} - \left(-1\frac{1}{6}\right) = 6\frac{4}{9} + 1\frac{1}{6}$$
$$= 6\frac{8}{18} + 1\frac{3}{18}$$
$$= 7\frac{11}{18}$$

Objective 2.4A – Round a decimal to a given place value

Key Words and Concepts

The number 72.85 is in **decimal notation**.

In decimal notation, the part of the number that appears to the _____ of the decimal point
is the **whole-number part**.   (left/right)

Here, 72 is the whole-number part.

The part of the number that appears to the _____ of the decimal point is the **decimal part**.
                                              (left/right)

Here, 85 is the decimal part. The **decimal** _____ separates the whole-number part from the decimal part.

The decimal part of the number represents a number less than _____.

The place-value chart is extended to the right to show the place values of digits to the right of the decimal point.

Thousands	Hundreds	Tens	Ones		Tenths	Hundredths	Thousandths	Ten-thousandths	Hundred-thousandths
8	2	9	7	.	1	4	5	6	3

The digit 4 in the number 8297.14563 is in the hundredths place. The digit 6 is in the ten-thousandths place.
To write a decimal in words, write the decimal part of the number as though it were a whole number, and then name
the place value of the *last digit*.

Note   The decimal point is read as "and".

The number in the table above is
    eight thousand two hundred ninety-seven and fourteen thousand five hundred sixty-three hundred-thousandths.

Note   32.8045 is read as thirty-two and eight thousand forty-five ten-thousandths.

Example   **Try it**
Name the place value of the digit 4 in the number 32.8045.   1. Name the place value of the digit 0
                                                                   in the number 32.8045.
    Solution
        The digit 4 is in the thousandths place.

In general, rounding decimals is similar to rounding whole numbers except that the digits to the right of the given
place value are dropped instead of being replaced by zeros.

If the digit to the right of the given place value is less than 5, that digit and all digits to the right are dropped.

Example

Round 44.38128 to the nearest thousandth.

Solution

The given place value is the thousandths,

44.38$\boxed{1}$28

and the digit 1 is in the thousandths place.

The digit to the right,

44.38$\boxed{1}$$\boxed{2}$8

2, is less than 5.

The 2 and 8 are dropped.

Then, 44.38128 rounded to the nearest thousandth is 44.381.

Try it

2: Round 38.4588 to the nearest hundredth.

Example

Round 44.38128 to the nearest ten-thousandth.

Solution

The given place value is the ten-thousandths,

44.381$\boxed{2}$8

and the digit 2 is in the ten-thousandths place.

The digit to the right,

44.381$\boxed{2}$$\boxed{8}$

8, is greater than 5.

The 8 is dropped and the 2 is changed to 3.

Then, 44.38128 rounded to the nearest ten-thousandth is 44.3813.

Try it

3. Round 38.4588 to the nearest thousandth.

Reflect on it

- For a decimal the whole-number part is to the left of the decimal point and the decimal part is to the right of the decimal point.
- A decimal point separates the whole number part from the decimal part.
- The decimal part of the number represents a part less than one.

Quiz Yourself 2.4A

1. Name the place value of the digit 3 in the number 84.50397.

2. Round 32.50481 to the nearest hundredth.

3. Round 32.50481 to the nearest thousandth.

4. Round 32.50481 to the nearest ten-thousandth.

5. Round 928.6125 to the nearest tenth.

6. Round 928.6125 to the nearest whole number.

Practice Sheet 2.4A

Name the place value of the digit 8 in each number.

1. 0.81

2. 0.0086

3. 26.389

4. 514.3118

1. _____

2. _____

3. _____

4. _____

Round each decimal to the given place value.

5. 0.064 Tenths

6. 9.138 Tenths

7. 26.349 Tenths

8. 96.4501 Tenths

9. 65.34498 Hundredths

10. 13.01264 Hundredths

11. 517.677 Hundredths

12. 792.246 Hundredths

13. 2.09181 Thousandths

14. 6.27958 Thousandths

15. 79.4625 Thousandths

16. 51.00439 Thousandths

17. 0.04195 Ten-thousandths

18. 0.003642 Ten-thousandths

19. 7.880102 Hundred-thousandths

20. 11.732405 Hundred-thousandths

21. 1.49256 Nearest whole number

22. 3.60021 Nearest whole number

23. 70.50648 Nearest whole number

5. _____

6. _____

7. _____

8. _____

9. _____

10. _____

11. _____

12. _____

13. _____

14. _____

15. _____

16. _____

17. _____

18. _____

19. _____

20. _____

21. _____

22. _____

23. _____

Answers

Try it 2.4A

1. The digit 0 is in the hundredths place.
2. 38.46
3. 38.459

Quiz 2.4A

1. The digit 3 is in the thousandths place.
2. 32.50
3. 32.505
4. 32.5048
5. 928.6
6. 929

Solutions to Practice Sheet 2.4A

1. tenths

2. thousandths

3. hundredths

4. ten-thousandths

5. ┌──Given place value
   0.064
   └─6 > 5
   0.064 rounded to the nearest tenth is 0.1.

6. ┌── Given place value
   9.138
   └─ 3 < 5
   9.138 rounded to the nearest tenth is 9.1.

7. ┌── Given place value
   26.349
   └─4 < 5
   26.349 rounded to the nearest tenth is 26.3.

8. ┌── Given place value
   96.4501
   └─5 = 5
   96.4501 rounded to the nearest tenth is 96.5.

9. ┌── Given place value
   65.34498
   └─4 < 5
   65.34498 rounded to the nearest hundredth is 65.34.

10. ┌── Given place value
    13.01264
    └─2 < 5
    13.01264 rounded to the nearest hundredth is 13.01.

11. ┌──Given place value
    517.677
    └─7 > 5
    517.677 rounded to the nearest hundredth is 517.68.

12. ┌──Given place value
    792.246
    └─6 > 5
    792.246 rounded to the nearest hundredth is 792.25.

13. ┌── Given place value
    2.09181
    └─8 > 5
    2.09181 rounded to the nearest thousandth is 2.092.

14. Given place value
    6.27958
    5 = 5
    6.27958 rounded to the nearest thousandth is 6.280.

15. ┌──Given place value
    79.4625
    └─5 = 5
    79.4625 rounded to the nearest thousandth is 79.463.

16. ┌──Given place value
    51.00439
    └─3 < 5
    51.00439 rounded to the nearest thousandth is 51.004.

17. ┌──Given place value
    0.04195
    └─5 = 5
    0.04195 rounded to the nearest ten-thousandth is 0.0420.

18. ┌──Given place value
    0.003642
    └─4 < 5
    0.003642 rounded to the nearest ten-thousandth is 0.0036.

19.   ┌──Given place value
       7.880102
          └─2 < 5

7.880102 rounded to the nearest
hundred-thousandth is 7.88010.

20.   ┌──Given place value
       11.732405
          └─5 = 5

11.732405 rounded to the nearest
hundred-thousandth is 11.73241.

21. ┌── Given place value
     1.49256
        └─4 < 5

1.49256 rounded to the nearest whole number
is 1.

22. ┌── Given place value
     3.60021
        └─6 > 5

3.60021 rounded to the nearest whole number
is 4.

23. ┌── Given place value
     70.50648
        └─5 = 5

70.50648 rounded to the nearest whole
number is 71.

Objective 2.4B – Compare decimals

Key Words and Concepts

There is a relationship between numbers written in decimal notation and fractions.
For example,

59 hundredths	127 hundred-thousandths
$0.59 = \dfrac{59}{100}$	$0.00127 = \dfrac{127}{100,000}$

This relationship can be used to compare decimals.

To compare decimals:
- Write the numbers as fractions.
- Write the fractions with a common denominator.
- Compare the fractions.

Example

Place the correct symbol, < or >, between the numbers.

$\quad$ 0.0257 $\qquad$ 0.027

Solution

$\quad$ 0.0257 $\quad$ 0.027

$\quad \dfrac{257}{10,000} \quad \dfrac{27}{1000}$ $\qquad$ Write as fractions.

$\quad \dfrac{257}{10,000} \quad \dfrac{270}{10,000}$ $\qquad$ Write fractions with a common denominator

$\quad \dfrac{257}{10,000} < \dfrac{270}{10,000}$ $\qquad$ Compare fractions.

Therefore, 0.0257 < 0.027.

Try it

1. Place the correct symbol, < or >, between the numbers.

$\quad$ 0.00329 $\qquad$ 0.0039

Reflect on it
- What other ways can you compare two decimals?

Quiz Yourself 2.4B

1. Place the correct symbol, < or >, between the numbers.

$\quad$ 0.23 $\qquad$ 0.213

2. Place the correct symbol, < or >, between the numbers.
    0.738   0.72

3. Place the correct symbol, < or >, between the numbers.
    0.4992   0.501

4. Place the correct symbol, < or >, between the numbers.
    0.012   0.01

5. Place the correct symbol, < or >, between the numbers.
    0.08483   0.0942

Practice Sheet 2.4B

Place the correct symbol, < or >, between the two numbers.

1. 0.23   0.3   2. 0.45   0.5   1. _____

                                                  2. _____

3. 4.54   4.45   4. 7.10   7.01   3. _____

                                                  4. _____

5. 9.143   9.134   6. 0.091   0.101   5. _____

                                                  6. _____

7. 0.4103 0.413   8. 0.25   0.256   7. _____

                                                  8. _____

9. 0.63   0.063   10. 0.3   1.003   9. _____

                                                  10. _____

11. 0.7   0.079   12. 0.86   0.859   11. _____

                                                  12. _____

13. 3.025   3.25   14. 0.54   0.0054   13. _____

                                                  14. _____

15. 2.907   2.097   16. 0.8555 0.86   15. _____

                                                  16. _____

Write the given numbers in order from smallest to largest.

17. 0.0037, 0.037, 0.00037, 0.37   18. 0.851, 0.0086, 0.086, 0.86   17. _____

                                                                      18. _____

19. 0.49, 0.05, 0.5, 0.049   20. 0.11, 0.0001, 0.012, 0.21   19. _____

                                                                      20. _____

Answers

Try it 2.4B
   1. <

Quiz 2.4B
   1. >
   2. >
   3. <
   4. >
   5. <

Solutions to Practice Sheet 2.4B

   1. 0.23 < 0.3
   2. 0.45 < 0.5
   3. 4.54 > 4.45
   4. 7.10 > 7.01
   5. 9.143 > 9.134
   6. 0.091 < 0.101
   7. 0.4103 < 0.413
   8. 0.25 < 0.256
   9. 0.63 > 0.063
   10. 0.3 < 1.003
   11. 0.7 > 0.079
   12. 0.86 > 0.859
   13. 3.025 < 3.25
   14. 0.54 > 0.0054
   15. 2.907 > 2.097
   16. 0.8555 < 0.86
   17. 0.00037, 0.0037, 0.037, 0.37
   18. 0.0086, 0.086, 0.851, 0.86
   19. 0.049, 0.05, 0.49, 0.5
   20. 0.0001, 0.012, 0.11, 0.21

Objective 2.5A – Add and subtract decimals

Key Words and Concepts

To add decimals,

write the numbers so that the decimal points are on a vertical line.

Add as you would with whole numbers.

Then write the decimal point in the sum directly below the decimal points in the addends.

Example

Find the sum of 8.238, 19, and 7.8974.

Solution
Arrange the numbers vertically, placing the decimal point on a vertical line. Add.

```
  2 1   1 1
    8 . 2 3 8
  1 9 .
+   7 . 8 9 7 4
  ─────────────
  3 5 . 1 3 5 4
```

The sum is 35.1354.

Try it

1. Find the sum of 5.642, 17, and 2.5921.

To subtract decimals,

write the numbers so that the decimal points are on a vertical line.

Subtract as you would with whole numbers.

Then write the decimal point in the difference directly below the decimal point in the subtrahend.

Note   Insert zeros in the minuend or subtrahend, if necessary, so that each has the same number of decimal places.

In this subtraction problem, identify the minuend, subtrahend, and difference. $15 - 8 = 7$

     minuend: _____   subtrahend: _____   difference: _____

Example

Subtract: 8.35 – 2.9187

Solution
Arrange the numbers vertically, placing the decimal point on a vertical line. Subtract.

```
    7    13 4 9 10
    8 . 3 5 0 0
  − 2 . 9 1 8 7
  ─────────────
    5 . 4 3 1 3
```

The difference is 5.4313.

Try it

2. Subtract: 7.42 – 4.5371

Note   The sign rules for adding and subtracting decimals are
                 the same rules used to add and subtract integers.

Reflect on it
• When adding decimals, first arrange the numbers vertically, placing the decimal points on a vertical line.
• When subtracting decimals, insert zeros in the minuend, if necessary, so that it has the same number of decimal places as the subtrahend.

Quiz Yourself 2.5A
1. Add: $9.87 + 29.992$

2. Subtract: $7.89 - 5.91$

3. Find the difference of 9.021 and 4.88.

4. What is 8.97 less than 2.1?

Practice Sheet 2.5A

Add or subtract.

1.	$4.825 + 31.7894 + 168.67$	**2.**	$25.25 + 7.4418 + 18.5$	**1.** _____
				2. _____
3.	$46.287 - 13.91$	**4.**	$23.031 - 17.61$	**3.** _____
				4. _____
5.	$6.841 + 54 + 59.3254$	**6.**	$85.0013 + 1.407 + 3.1114$	**5.** _____
				6. _____
7.	$145.03 - 8.2174$	**8.**	$650 - 56.413$	**7.** _____
				8. _____
9.	$6.421 + 52.118 + 3 + 0.0098$	**10.**	$4.46 + 2.3845 + 2.5 + 0.0231$	**9.** _____
				10. _____
11.	$14.1 - 11.7809$	**12.**	$43.001 - 19.875$	**11.** _____
				12. _____
13.	$0.0014 + 83.9 + 46 + 148.0908$	**14.**	$75.514 + 0.199 + 29 + 8.356$	**13.** _____
				14. _____
15.	$143.24 - 80.794$	**16.**	$9.08 - 6.324$	**15.** _____
				16. _____
17.	$2.7156 + 45.08 + 6.0406$	**18.**	$5.52 + 94.099 + 7.2148$	**17.** _____
				18. _____
19.	$11 + 77.29 + 5.0531$	**20.**	$0.452 - 0.39$	**19.** _____
				20. _____
21.	$0.847 - 0.25$	**22.**	$9.406 - 6.315$	**21.** _____
				22. _____

Answers

Try it 2.5A
1. 25.2341
2. 2.8829

Quiz 2.5A
1. 39.862
2. 1.98
3. 4.141
4. −6.87

Solutions to Practice Sheet 2.5A

1. $4.825 + 31.7894 + 168.67$
   $= 36.6144 + 168.67$
   $= 205.2844$

2. $25.25 + 7.4418 + 18.5$
   $= 32.6918 + 18.5$
   $= 51.1918$

3. $46.287 - 13.91 = 32.377$

4. $23.031 - 17.61 - 5.421$

5. $6.841 + 54 + 59.3254$
   $= 60.841 + 59.3254$
   $= 120.1664$

6. $85.0013 + 1.407 + 3.1114$
   $= 86.4083 + 3.1114$
   $= 89.5197$

7. $145.03 - 8.2174 = 136.8126$

8. $650 - 56.413 = 593.587$

9. $6.421 + 52.118 + 3 + 0.0098$
   $= 58.539 + 3 + 0.0098$
   $= 61.539 + 0.0098$
   $= 61.5488$

10. $4.46 + 2.3845 + 2.5 + 0.0231$
    $= 6.8445 + 2.5 + 0.0231$
    $= 9.3445 + 0.0231$
    $= 9.3676$

11. $14.1 - 11.7809 = 2.3191$

12. $43.001 - 19.875 = 23.126$

13. $0.0014 + 83.9 + 46 + 148.0908$
    $= 83.9014 + 46 + 148.0908$
    $= 129.9014 + 148.0908$
    $= 277.9922$

14. $75.514 + 0.199 + 29 + 8.356$
    $= 75.713 + 29 + 8.356$
    $= 104.713 + 8.356$
    $= 113.069$

15. $143.24 - 80.794 = 62.446$

16. $9.08 - 6.324 = 2.756$

17. $2.7156 + 45.08 + 6.0406$
    $= 47.7956 + 6.0406$
    $= 53.8362$

18. $5.52 + 94.099 + 7.2148$
    $= 99.619 + 7.2148$
    $= 106.8338$

19. $11 + 77.29 + 5.0531$
    $= 88.29 + 5.0531$
    $= 93.3431$

20. $0.452 - 0.39 = 0.062$

21. $0.847 - 0.25 = 0.597$

22. $9.406 - 6.315 = 3.091$

Objective 2.5B – Multiply decimals

Key Words and Concepts

To multiply decimals,

multiply the numbers as you would whole numbers.

Then write the decimal point in the product so that the number of decimal places in the product is the sum of the numbers of decimal places in the factors.

The product of two numbers with the same sign is _____.
                                                                              (positive/negative)

The product of two numbers with _____ sign(s) is negative.
                                            (same/different)

Note   The sign rules for multiplying decimals are the same rules used to multiply integers.

Example	**Try it**
Multiply (0.361)(20.7).	1. Multiply (0.482)(30.8).

Solution

```
       0.  3   6   1     3 decimal places
   ×       2   0.  7     1 decimal place
   ─────────────────
       2   5   2   7
   7   2   2
   ─────────────────
   7.  4   7   2   7     4 decimal places
```

The product is 7.4727.

To divide decimals,

move the decimal point in the divisor to the right so that the divisor is a whole number.

Move the decimal point in the dividend the same number of places to the right.

Place the decimal point in the quotient directly above the decimal point in the dividend.

Then divide as you would with whole numbers.

The quotient of two numbers with _____ sign(s) is positive.
                                            (same/different)

The quotient of two numbers with different signs is _____.
                                                                       (positive/negative)

Example
Divide $41.756 \div 5.72$.

> Solution
> Move the decimal point 2 places to the right in the divisor.
> Move the decimal point 2 places to the right in the dividend.
> Place the decimal point in the quotient.
> Then divide as shown below.

$$
\begin{array}{r}
7.3 \\
572\overline{)4175.6} \\
\underline{4004} \\
1716 \\
\underline{1716} \\
0
\end{array}
$$

> The quotient is 7.3.

Try it
2. Divide $19.154 \div 0.61$.

In division of decimals, rather than writing the quotient with a remainder, we usually round the quotient to a specified place value. The symbol $\approx$ is read "is approximately equal to"; it is used to indicate that the quotient is an approximate value after being rounded.

Example
Divide and round to the nearest tenth: $0.47 \div 0.6$.

> Solution
> Move the decimal point 1 place to the right in the divisor.
> Move the decimal point 1 place to the right in the dividend.
> Then divide as shown below.

$$
\begin{array}{r}
0.78 \\
6\overline{)4.70} \\
\underline{42} \\
50 \\
\underline{48} \\
2
\end{array}
$$

> To round the quotient to the nearest tenth, the division must be
> carried to the hundredths place. Therefore, a zero must be inserted
> in the dividend so that the quotient has a digit in the hundredths
> place.
> $$0.47 \div 0.6 \approx 0.8$$

Try it
3. Divide and round to the nearest tenth: $6.25 \div 0.4$.

The sign rules for dividing decimals are the same rules used to divide integers.

Reflect on it
- The product (or quotient) of two numbers with the same sign is positive.
- The product (or quotient) of two numbers with different signs is negative.
- How is the placement of decimal point different when multiplying decimals and dividing decimals?

Quiz Yourself 2.5B

1. Multiply: $(9.85)(0.21)$

2. Find the product of -5.43 and 8.6.

3. Multiply -2.9 by 48.2

4. Find the product of -2.1, 3, and 0.4.

5. Divide: $14.088 \div 2.4$

6. Find the quotient of -5.88 and 0.6.

7. Divide -128.92 by -4.4.

142

Practice Sheet 2.5B

Multiply.

1. $(0.5)(0.7)$ 2. $(6.4)(0.3)$ 1. _____

                                                                             2. _____

3. What is 5.6 times 9?   4. Find the product of 0.28 and 0.6.   3. _____

                                                                             4. _____

5. $(5.1)(4.5)$ 6. $(0.96)(3.7)$ 5. _____

                                                                             6. _____

7. $(2.64)(0.03)$ 8. $5.83(0.008)$ 7. _____

                                                                             8. _____

9. $(0.67)(0.41)$ 10. $(52.9)(0.2)$ 9. _____

                                                                             10. _____

11. $(24.8)(0.0019)$ 12. 5.92×0.8 11. _____

                                                                             12. _____

13. 0.76×0.6 14. 3.9×0.44 13. _____

                                                                             14. _____

Divide.

15. $5.64 \div 6$ 16. $2.24 \div 0.7$ 17. $40 \div 0.8$ 15. _____

                                                                                       16. _____

                                                                                       17. _____

18. $25.95 \div 0.3$ 19. $6.515 \div 5$ 20. $0.899 \div 2.9$ 18. _____

                                                                                       19. _____

                                                                                       20. _____

21. $1.288 \div 0.46$ 22. $42.3 \div 0.09$ 23. $0.1116 \div 0.012$ 21. _____

                                                                                       22. _____

                                                                                       23. _____

Divide. Round to the nearest tenth.

24. $73.85 \div 9.6$ 25. $0.473 \div 0.54$ 26. $1.265 \div 0.043$ 24. _____

                                                                                       25. _____

                                                                                       26. _____

Answers

Try it 2.5B
1. 14.8456
2. 31.4
3. 15.6

Quiz 2.5B
1. 2.0685
2. −46.698
3. −139.78
4. −2.52
5. 5.87
6. −9.8
7. 29.3

20. $0.899 \div 2.9 = 0.31$
21. $1.288 \div 0.46 = 2.8$
22. $42.3 \div 0.09 = 470$
23. $0.1116 \div 0.012 = 9.3$
24. $73.85 \div 9.6 \approx 7.7$
25. $0.473 \div 0.54 \approx 0.9$
26. $1.265 \div 0.043 \approx 29.4$

Solutions to Practice Sheet 2.5B

1. $(0.5)(0.7) = 0.35$

2. $(6.4)(0.3) = 1.92$

3. $5.6(9) = 50.4$

4. $0.28(0.6) = 0.168$

5. $(5.1)(4.5) = 22.95$

6. $(0.96)(3.7) = 3.552$

7. $(2.64)(0.03) = 0.0792$

8. $5.83(0.008) = 0.04664$

9. $(0.67)(0.41) = 0.2747$

10. $(52.9)(0.2) = 10.58$

11. $(24.8)(0.0019) = 0.04712$

12. $5.92 \times 0.8 = 4.736$

13. $0.76 \times 0.6 = 0.456$

14. $3.9 \times 0.44 = 1.716$

15. $5.64 \div 6 = 0.94$

16. $2.24 \div 0.7 = 3.2$

17. $40 \div 0.8 = 50$

18. $25.95 \div 0.3 = 86.5$

19. $6.515 \div 5 = 1.303$

Objective 2.6A – Convert fractions to decimals

Key Words and Concepts

To convert a fraction to a decimal, divide the numerator by the denominator.

What is the difference between a terminating decimal and a repeating decimal? _____

_____

_____

It is common practice to write a _____ over the repeating digits of a decimal.

Example

Convert $\frac{3}{16}$ to a decimal.

Solution

$$
\begin{array}{r}
0.1875 \\
16\overline{)3.0000}
\end{array}
$$

16	Multiply 16×1
140	Subtract 30−16
128	Multiply 16×8
120	Subtract 140−128
112	Multiply 16×7
80	Subtract 120−112
80	Multiply 16×5
0	Subtract 80−80

Therefore, $\frac{3}{16} = 0.1875$.

Try it

1. Convert $\frac{2}{25}$ to a decimal.

Example

Convert $\frac{7}{12}$ to a decimal.

Solution

$$
\begin{array}{r}
0.5833 \\
12\overline{)7.0000}
\end{array}
$$

60	Multiply 12×5
100	Subtract 70−60
96	Multiply 12×8
40	Subtract 110−96
36	Multiply 12×3
40	Subtract 40−36
36	Multiply 12×3
4	Subtract 40−36

Notice that the remainder of 4 repeats.

Therefore, $\frac{7}{12} = 0.58\overline{3}$.

Try it

2. Convert $\frac{8}{11}$ to a decimal.

Reflect on it
• If the decimal eventually has a remainder of 0, then the decimal is a terminating decimal.
• If the remainder is never 0, then the decimal is a repeating decimal.

Quiz Yourself 2.6A

1. Convert $\frac{5}{6}$ to a decimal. Place a bar over any repeating digits.

2. Convert $\frac{7}{16}$ to a decimal. Place a bar over any repeating digits.

3. Convert $\frac{3}{37}$ to a decimal. Place a bar over any repeating digits.

4. Convert $\frac{7}{13}$ to a decimal. Place a bar over any repeating digits.

5. Convert $\frac{13}{7}$ to a decimal. Place a bar over any repeating digits.

Practice Sheet 2.6A

Convert the fraction to a decimal.

1. $\dfrac{5}{7}$

2. $\dfrac{1}{15}$

3. $\dfrac{5}{12}$

4. $\dfrac{8}{9}$

5. $\dfrac{3}{14}$

6. $\dfrac{6}{11}$

7. $2\dfrac{10}{17}$

8. $\dfrac{7}{16}$

9. $5\dfrac{2}{3}$

10. $\dfrac{9}{14}$

11. $\dfrac{6}{7}$

12. $\dfrac{5}{4}$

13. $8\dfrac{3}{22}$

14. $15\dfrac{1}{2}$

15. $\dfrac{53}{100}$

16. $\dfrac{35}{8}$

17. $41\dfrac{3}{10}$

18. $1\dfrac{2}{23}$

19. $6\dfrac{1}{3}$

20. $\dfrac{5}{9}$

21. $\dfrac{39}{11}$

22. $27\dfrac{1}{8}$

23. $\dfrac{46}{7}$

24. $\dfrac{13}{1000}$

25. $\dfrac{17}{18}$

26. $6\dfrac{5}{9}$

27. $10\dfrac{12}{21}$

1. _____
2. _____
3. _____
4. _____
5. _____
6. _____
7. _____
8. _____
9. _____
10. _____
11. _____
12. _____
13. _____
14. _____
15. _____
16. _____
17. _____
18. _____
19. _____
20. _____
21. _____
22. _____
23. _____
24. _____
25. _____
26. _____
27. _____

Answers

Try it 2.6A
1. 0.08
2. $0.\overline{72}$

Quiz 2.6A
1. $0.8\overline{3}$
2. 0.4375
3. $0.0\overline{81}$
4. $0.\overline{538461}$
5. $1.\overline{857142}$

Solutions to Practice Sheet 2.6A

1.
$$
\begin{array}{r}
0.7142 \\
7\overline{)5.0000} \\
\underline{49} \\
10 \\
\underline{7} \\
30 \\
\underline{28} \\
20 \\
\underline{14} \\
6
\end{array}
$$

$\frac{5}{7} \approx 0.714$

2.
$$
\begin{array}{r}
0.066 \\
15\overline{)1.000} \\
\underline{90} \\
100 \\
\underline{90} \\
10
\end{array}
$$

$\frac{1}{15} = 0.0666... = 0.0\overline{6}$

3.
$$
\begin{array}{r}
0.416 \\
12\overline{)5.000} \\
\underline{48} \\
20 \\
\underline{12} \\
80 \\
\underline{72} \\
80
\end{array}
$$

$\frac{5}{12} = 0.4166... = 0.41\overline{6}$

4.
$$
\begin{array}{r}
0.88 \\
9\overline{)8.00} \\
\underline{72} \\
80 \\
\underline{72} \\
8
\end{array}
$$

$\frac{8}{9} = 0.888... = 0.\overline{8}$

5.
$$
\begin{array}{r}
0.2142 \\
14\overline{)3.0000} \\
\underline{28} \\
20 \\
\underline{14} \\
60 \\
\underline{56} \\
40 \\
\underline{28} \\
12
\end{array}
$$

$\frac{3}{14} \approx 0.214$

6.
$$
\begin{array}{r}
0.5454 \\
11\overline{)6.0000} \\
\underline{55} \\
50 \\
\underline{44} \\
60 \\
\underline{55} \\
50 \\
\underline{44} \\
6
\end{array}
$$

$\frac{6}{11} = 0.5454... = 0.\overline{54}$

7.
$$
\begin{array}{r}
0.5882 \\
17\overline{)10.0000} \\
\underline{85} \\
150 \\
\underline{136} \\
140 \\
\underline{136} \\
40 \\
\underline{34} \\
6
\end{array}
$$

$\frac{10}{17} \approx 0.588$

$2\frac{10}{17} \approx 2.588$

8.
$$16)\overline{7.0000}$$ (quotient 0.4375)
64
60
48
120
112
80
80
0

$$\frac{7}{16} \approx 0.438$$

9.
$$3)\overline{2.00}$$ (quotient 0.66)
18
20
18
2

$$\frac{2}{3} = 0.666... = 0.\overline{6}$$
$$5\frac{2}{3} = 5.666... = 5.\overline{6}$$

10.
$$14)\overline{9.0000}$$ (quotient 0.6428)
84
60
56
40
28
120
112
8

$$\frac{9}{14} \approx 0.643$$

11.
$$7)\overline{6.0000}$$ (quotient 0.8571)
56
40
35
50
49
10
7
3

$$\frac{6}{7} \approx 0.857$$

12.
$$4)\overline{5.00}$$ (quotient 1.25)
4
10
8
20
20
0

$$\frac{5}{4} = 1.25$$

13.
$$22)\overline{3.00000}$$ (quotient 0.13636)
22
80
66
140
132
80
66
14

$$\frac{3}{22} = 0.13636... = 0.1\overline{36}$$
$$8\frac{3}{22} = 8.13636... = 8.1\overline{36}$$

14.
$$2)\overline{1.0}$$ (quotient 0.5)
10
0

$$\frac{1}{2} = 0.5$$
$$15\frac{1}{2} = 15.5$$

15.
$$100)\overline{53.00}$$ (quotient 0.53)
500
300
300
0

$$\frac{53}{100} = 0.53$$

16.
$$\begin{array}{r} 4.375 \\ 8\overline{)35.000} \\ \underline{32} \\ 30 \\ \underline{24} \\ 60 \\ \underline{56} \\ 40 \\ \underline{40} \\ 0 \end{array}$$

$$\frac{35}{8} = 4.375$$

17.
$$\begin{array}{r} 0.3 \\ 10\overline{)3.0} \\ \underline{30} \\ 0 \end{array}$$

$$\frac{3}{10} = 0.3$$

$$41\frac{3}{10} = 41.3$$

18.
$$\begin{array}{r} 0.0869 \\ 23\overline{)2.0000} \\ \underline{184} \\ 160 \\ \underline{138} \\ 220 \\ \underline{207} \\ 13 \end{array}$$

$$\frac{2}{23} \approx 0.087$$

$$1\frac{2}{23} \approx 1.087$$

19.
$$\begin{array}{r} 0.33 \\ 3\overline{)1.00} \\ \underline{9} \\ 10 \\ \underline{9} \\ 1 \end{array}$$

$$\frac{1}{3} = 0.333... = 0.\overline{3}$$

$$6\frac{1}{3} = 6.333... = 6.\overline{3}$$

20.
$$\begin{array}{r} 0.55 \\ 9\overline{)5.00} \\ \underline{45} \\ 50 \\ \underline{45} \\ 5 \end{array}$$

$$\frac{5}{9} = 0.555... = 0.\overline{5}$$

21.
$$\begin{array}{r} 3.5454 \\ 11\overline{)39.0000} \\ \underline{33} \\ 60 \\ \underline{55} \\ 50 \\ \underline{44} \\ 60 \\ \underline{55} \\ 50 \end{array}$$

$$\frac{39}{11} = 3.5454... = 3.\overline{54}$$

22.
$$\begin{array}{r} 0.125 \\ 8\overline{)1.000} \\ \underline{8} \\ 20 \\ \underline{16} \\ 40 \\ \underline{40} \\ 0 \end{array}$$

$$\frac{1}{8} = 0.125$$

$$27\frac{1}{8} = 27.125$$

23.
$$\begin{array}{r} 6.5714 \\ 7\overline{)46.0000} \\ \underline{42} \\ 40 \\ \underline{35} \\ 50 \\ \underline{49} \\ 10 \\ \underline{7} \\ 3 \end{array}$$

$$\frac{46}{7} \approx 6.571$$

24.
$$\begin{array}{r} 0.013 \\ 1000\overline{)13.000} \\ \underline{1000} \\ 3000 \\ \underline{3000} \\ 0 \end{array}$$

$$\frac{13}{1000} = 0.013$$

25.

$$
\begin{array}{r}
0.944 \\
18\overline{)17.000} \\
\underline{162} \\
80 \\
\underline{72} \\
80 \\
\underline{72} \\
8
\end{array}
$$

$$\frac{17}{18} = 0.9444\ldots = 0.9\overline{4}$$

26.

$$
\begin{array}{r}
0.55 \\
9\overline{)5.00} \\
\underline{45} \\
50 \\
\underline{45} \\
5
\end{array}
$$

$$\frac{5}{9} = 0.555\ldots = 0.\overline{5}$$

$$6\frac{5}{9} = 6.555\ldots = 6.\overline{5}$$

27.

$$
\begin{array}{r}
0.5714 \\
21\overline{)12.0000} \\
\underline{105} \\
150 \\
\underline{147} \\
30 \\
\underline{21} \\
90 \\
\underline{84} \\
6
\end{array}
$$

$$\frac{12}{21} \approx 0.571$$

$$10\frac{12}{21} \approx 10.571$$

Objective 2.6B – Convert decimals to fractions

Key Words and Concepts

To convert a decimal to a fraction, remove the decimal point and place the decimal part over a denominator equal to the place value of the last digit in the decimal. Then write the fraction in simplest form.

Example
Convert 0.368 to a fraction.

Solution

$$0.368 = \frac{368}{1000}$$ Write 0.368 as 368 over 1000.

$$= \frac{\cancel{2}\cdot\cancel{2}\cdot\cancel{2}\cdot 2\cdot 23}{\cancel{2}\cdot\cancel{2}\cdot\cancel{2}\cdot 5\cdot 5\cdot 5}$$ Factor 368 and 1000.

$$= \frac{46}{125}$$ Simplify.

Try it
1. Convert 0.425 to a fraction.

Example
Convert 0.91 to a fraction.

Solution

$$0.91 = \frac{91}{100}$$ Write 0.91 as 91 over 100.

$$= \frac{7\cdot 13}{100}$$ Factor 91.

$$= \frac{91}{100}$$ The fraction is already in simplest form.

Try it
2. Convert 0.37 to a fraction.

Example
Convert 3.25 to a mixed number.

Solution

$$3.25 = 3\frac{25}{100}$$ Write 0.25 as 25 over 100.

$$= 3\frac{1}{4}$$ Simplify.

Try it
3. Convert 2.375 to a mixed number.

Quiz Yourself 2.6B

1. Convert 0.8 to a fraction.

2. Convert 0.72 to a fraction.

3. Convert 0.735 to a fraction.

4. Convert 2.25 to a mixed number.

5. Convert 0.096 to a fraction.

6. Convert 0.00032 to a fraction.

Practice Sheet 2.6B

Convert the decimal to a fraction.

1. 0.7	**2.** 0.5	**3.** 0.46	**1.** _____	
			2. _____	
			3. _____	
4. 0.74	**5.** 0.375	**6.** 0.205	**4.** _____	
			5. _____	
			6. _____	
7. 2.55	**8.** 6.75	**9.** 18.4	**7.** _____	
			8. _____	
			9. _____	
10. 12.3	**11.** 9.2	**12.** 14.5	**10.** _____	
			11. _____	
			12. _____	
13. 4.138	**14.** 6.064	**15.** 3.35	**13.** _____	
			14. _____	
			15. _____	
16. 9.93	**17.** 0.16	**18.** 0.93	**16.** _____	
			17. _____	
			18. _____	
19. 0.11	**20.** 4.81	**21.** 0.055	**19.** _____	
			20. _____	
			21. _____	
22. 0.015	**23.** 23.62	**24.** 0.44	**22.** _____	
			23. _____	
			24. _____	
25. 0.83	**26.** 0.25	**27.** 0.77	**25.** _____	
			26. _____	
			27. _____	

Answers

Try it 2.6B

1. $\dfrac{17}{40}$

2. $\dfrac{37}{100}$

3. $2\dfrac{3}{8}$

Quiz 2.6B

1. $\dfrac{4}{5}$

2. $\dfrac{18}{25}$

3. $\dfrac{147}{200}$

4. $2\dfrac{1}{4}$

5. $\dfrac{12}{125}$

6. $\dfrac{1}{3125}$

Solutions to Practice Sheet 2.6B

1. $0.7 = \dfrac{7}{10}$

2. $0.5 = \dfrac{5}{10} = \dfrac{1}{2}$

3. $0.46 = \dfrac{46}{100} = \dfrac{23}{50}$

4. $0.74 = \dfrac{74}{100} = \dfrac{37}{50}$

5. $0.375 = \dfrac{375}{1000} = \dfrac{3 \cdot \cancel{5} \cdot \cancel{5} \cdot \cancel{5}}{2 \cdot 2 \cdot 2 \cdot \cancel{5} \cdot \cancel{5} \cdot \cancel{5}} = \dfrac{3}{8}$

6. $0.205 = \dfrac{205}{1000} = \dfrac{41}{200}$

7. $2.55 = 2\dfrac{55}{100} = 2\dfrac{11}{20}$

8. $6.75 = 6\dfrac{75}{100} = 6\dfrac{3}{4}$

9. $18.4 = 18\dfrac{4}{10} = 18\dfrac{2}{5}$

10. $12.3 = 12\dfrac{3}{10}$

11. $9.2 = 9\dfrac{2}{10} = 9\dfrac{1}{5}$

12. $14.5 = 14\dfrac{5}{10} = 14\dfrac{1}{2}$

13. $4.138 = 4\dfrac{138}{1000} = 4\dfrac{69}{500}$

14. $6.064 = 6\dfrac{64}{1000} = 6\dfrac{8}{125}$

15. $3.35 = 3\dfrac{35}{100} = 3\dfrac{7}{20}$

16. $9.93 = 9\dfrac{93}{100}$

17. $0.16 = \dfrac{16}{100} = \dfrac{4}{25}$

18. $0.93 = \dfrac{93}{100}$

19. $0.11 = \dfrac{11}{100}$

20. $4.81 = 4\dfrac{81}{100}$

21. $0.055 = \dfrac{55}{1000} = \dfrac{11}{200}$

22. $0.015 = \dfrac{15}{1000} = \dfrac{3}{200}$

23. $23.62 = 23\dfrac{62}{100} = 23\dfrac{31}{50}$

24. $0.44 = \dfrac{44}{100} = \dfrac{11}{25}$

25. $0.83 = \dfrac{83}{100}$

26. $0.25 = \dfrac{25}{100} = \dfrac{1}{4}$

27. $0.77 = \dfrac{77}{100}$

Objective 2.6C – Compare fractions and decimals

Key Words and Concepts

One way to determine the order relation between a decimal and a fraction is to write the decimal as a fraction and then compare the fractions.

Example

Place the correct symbol, $<$ or $>$, between the two numbers.

$$0.3 \quad \frac{1}{4}$$

Solution Write 0.3 as a fraction and compare.

$$0.3 \quad \frac{1}{4}$$

$$\frac{3}{10} \quad \frac{1}{4}$$

$$\frac{6}{20} \quad \frac{5}{20}$$

$$\frac{6}{20} > \frac{5}{20}$$

Since $\frac{6}{20} > \frac{5}{20}$ then $0.3 > \frac{1}{4}$.

Try it

1. Place the correct symbol, $<$ or $>$, between the two numbers.

$$0.7 \quad \frac{4}{5}$$

Quiz Yourself 2.6C

1. Place the correct symbol, $<$ or $>$, between the two numbers.

$$0.34 \quad \frac{7}{20}$$

2. Place the correct symbol, $<$ or $>$, between the two numbers.

$$\frac{11}{12} \quad 0.85$$

3. Place the correct symbol, $<$ or $>$, between the two numbers.

$$0.75 \quad \frac{29}{40}$$

Practice Sheet 2.6C

Place the correct symbol, < or >, between the numbers.

1. 0.23   0.3   2. 0.45   0.5   3. 4.54   4.45

4. 7.10   7.01   5. 9.143   9.134   6. 0.091   0.101

7. 0.399 $\frac{2}{5}$ 8. 0.433 $\frac{7}{16}$ 9. $\frac{5}{9}$ 0.54

10. 0.58 $\frac{7}{12}$ 11. $\frac{5}{7}$ 0.72   12. 0.26 $\frac{4}{15}$

13. $\frac{1}{3}$ 0.32   14. $\frac{13}{16}$ 0.82   15. 0.626 $\frac{5}{8}$

16. 0.4103   0.413   17. 0.25   0.256   18. 0.63   0.063

19. 0.3   1.003   20. 0.7   0.079   21. 0.86   0.859

22. $\frac{1}{8}$ 0.124   23. $\frac{11}{15}$ 0.734   24. 0.589 $\frac{3}{5}$

25. 0.708 $\frac{17}{24}$ 26. 0.167 $\frac{1}{6}$ 27. $\frac{4}{7}$ 0.572

1. _____
2. _____
3. _____
4. _____
5. _____
6. _____
7. _____
8. _____
9. _____
10. _____
11. _____
12. _____
13. _____
14. _____
15. _____
16. _____
17. _____
18. _____
19. _____
20. _____
21. _____
22. _____
23. _____
24. _____
25. _____
26. _____
27. _____

Answers

Try it 2.6C
   1. <

Quiz 2.6C
   1. <
   2. >
   3. >

Solutions to Practice Sheet 2.6C

1. $0.23 < 0.3$

2. $0.45 < 0.5$

3. $4.54 > 4.45$

4. $7.10 > 7.01$

5. $9.143 > 9.134$

6. $0.091 < 0.101$

7. $0.399 \quad \dfrac{2}{5}$

$\dfrac{399}{1000} \quad \dfrac{2}{5}$

$\dfrac{399}{1000} \quad \dfrac{400}{1000}$

$\dfrac{399}{1000} < \dfrac{400}{1000}$

$0.399 < \dfrac{2}{5}$

8. $0.433 \quad \dfrac{7}{16}$

$\dfrac{433}{1000} \quad \dfrac{7}{16}$

$\dfrac{866}{2000} \quad \dfrac{875}{2000}$

$\dfrac{866}{2000} < \dfrac{875}{2000}$

$0.433 < \dfrac{7}{16}$

9. $\dfrac{5}{9} \quad 0.54$

$\dfrac{5}{9} \quad \dfrac{54}{100}$

$\dfrac{500}{900} \quad \dfrac{486}{900}$

$\dfrac{500}{900} > \dfrac{486}{900}$

$\dfrac{5}{9} > 0.54$

10. $0.58 \quad \dfrac{7}{12}$

$\dfrac{58}{100} \quad \dfrac{7}{12}$

$\dfrac{174}{300} \quad \dfrac{175}{300}$

$\dfrac{174}{300} < \dfrac{175}{300}$

$0.58 < \dfrac{7}{12}$

11. $\dfrac{5}{7} \quad 0.72$

$\dfrac{5}{7} \quad \dfrac{72}{100}$

$\dfrac{5}{7} \quad \dfrac{18}{25}$

$\dfrac{125}{175} \quad \dfrac{126}{175}$

$\dfrac{125}{175} < \dfrac{126}{175}$

$\dfrac{5}{7} < 0.72$

12. $0.26 \quad \dfrac{4}{15}$

$\dfrac{26}{100} \quad \dfrac{4}{15}$

$\dfrac{78}{300} \quad \dfrac{80}{300}$

$\dfrac{78}{300} < \dfrac{80}{300}$

$0.26 < \dfrac{4}{15}$

13. $\dfrac{1}{3} \quad 0.32$

$\dfrac{1}{3} \quad \dfrac{32}{100}$

$\dfrac{100}{300} \quad \dfrac{96}{300}$

$\dfrac{100}{300} > \dfrac{96}{300}$

$\dfrac{1}{3} > 0.32$

14. $\dfrac{13}{16} \quad 0.82$

$\dfrac{13}{16} \quad \dfrac{82}{100}$

$\dfrac{325}{400} \quad \dfrac{328}{400}$

$\dfrac{325}{400} < \dfrac{328}{400}$

$\dfrac{13}{16} < 0.82$

15. $0.626 \quad \dfrac{5}{8}$

$\dfrac{626}{1000} \quad \dfrac{5}{8}$

$\dfrac{626}{1000} \quad \dfrac{625}{1000}$

$\dfrac{626}{1000} > \dfrac{625}{1000}$

$0.626 > \dfrac{5}{8}$

16. $0.4103 < 0.413$

17. $0.25 < 0.256$

18. $0.63 > 0.063$

19. $0.3 < 1.003$

20. $0.7 > 0.079$

21. $0.86 > 0.859$

22. $\dfrac{1}{8} \quad 0.124$

$\dfrac{1}{8} \quad \dfrac{124}{1000}$

$\dfrac{125}{1000} \quad \dfrac{124}{1000}$

$\dfrac{125}{1000} > \dfrac{124}{1000}$

$\dfrac{1}{8} > 0.124$

23. $\dfrac{11}{15} \quad 0.734$

$\dfrac{11}{15} \quad \dfrac{734}{1000}$

$\dfrac{11}{15} \quad \dfrac{367}{500}$

$\dfrac{1100}{1500} \quad \dfrac{1101}{1500}$

$\dfrac{1100}{1500} < \dfrac{1101}{1500}$

$\dfrac{11}{15} < 0.734$

24. $0.589 \quad \dfrac{3}{5}$

$\dfrac{589}{1000} \quad \dfrac{3}{5}$

$\dfrac{589}{1000} \quad \dfrac{600}{1000}$

$\dfrac{589}{1000} < \dfrac{600}{1000}$

$0.589 < \dfrac{3}{5}$

25. $0.708 \quad \dfrac{17}{24}$

$\dfrac{708}{1000} \quad \dfrac{17}{24}$

$\dfrac{2124}{3000} \quad \dfrac{2125}{3000}$

$\dfrac{2124}{3000} < \dfrac{2125}{3000}$

$0.708 < \dfrac{17}{24}$

26. $0.167 \quad \dfrac{1}{6}$

$\dfrac{167}{1000} \quad \dfrac{1}{6}$

$\dfrac{501}{3000} \quad \dfrac{500}{3000}$

$\dfrac{501}{3000} > \dfrac{500}{3000}$

$0.167 > \dfrac{1}{6}$

27. $\dfrac{4}{7} \quad 0.572$

$\dfrac{4}{7} \quad \dfrac{572}{1000}$

$\dfrac{4}{7} \quad \dfrac{143}{250}$

$\dfrac{1000}{1750} \quad \dfrac{1001}{1750}$

$\dfrac{1000}{1750} < \dfrac{1001}{1750}$

$\dfrac{4}{7} < 0.572$

Objective 2.7A – Find the square root of a perfect square

Key Words and Concepts

Recall that the square of a number is equal to the number multiplied by itself. The square of an integer is called a _____ **square**.

Why is 25 an example of a perfect square? _____

_____

Write the perfect squares from 1 to 100. _____

_____

Give the definition of a square root. _____

_____

_____

The symbol $\sqrt{}$ is used to indicate the _____ square root of a number.
                                                    (positive/negative)

Note $\sqrt{25} = 5$ and $-\sqrt{25} = -5$

The square root symbol, $\sqrt{}$, is also called a **radical**. The number under the radical is called a **radicand**. For example, in the radical expression $\sqrt{25}$, the radicand is 25.

Example	**Try it**
Simplify $\sqrt{144}$.	1. Simplify $\sqrt{81}$.

   Solution Look at the radicand, 144.
      Is 144 a perfect square?
      Since $12^2 = 144$, then $\sqrt{144} = 12$.

Example	**Try it**
Simplify $6 + 5\sqrt{16}$.	2. Simplify $3 + 7\sqrt{4}$.

   Solution

   $\begin{aligned} 6 + 5\sqrt{16} \quad & \text{Look at the radicand, 16, a perfect square.} \\ = 6 + 5 \cdot 4 \quad & \text{Simplify } \sqrt{16} = 4 \\ = 6 + 20 \quad & \text{Multiply } 5 \cdot 4 = 20 \\ = 26 \quad & \text{Add } 6 + 20 \end{aligned}$

Example

Simplify $\sqrt{25} - 2\sqrt{4}$.

Solution

$\sqrt{25} - 2\sqrt{4}$ Each radicand, 25 and 4, is a perfect square.

$= 5 - 2 \cdot 2$ Simplify $\sqrt{25} = 5$ and $\sqrt{4} = 2$

$= 5 - 4$ Multiply $2 \cdot 2 = 4$

$= 1$ Subtract $5 - 4$

Try it

3. Simplify $\sqrt{64} - 2\sqrt{9}$.

Example

Evaluate $3\sqrt{xy}$ for $x = 3$ and $y = 12$.

Solution $3\sqrt{xy}$ Start with the given expression.

$3\sqrt{3 \cdot 12}$ Substitute 3 for x and 12 for y

$= 3\sqrt{36}$ Multiply $3 \cdot 12 = 36$

$= 3 \cdot 6$ Simplify $\sqrt{36} = 6$

$= 18$ Multiply $3 \cdot 6$

Try it

4. Evaluate $2\sqrt{xy}$ for $x = 4$ and $y = 16$.

Reflect on it

• A perfect square is the square of an integer.
• A square root of a positive number x is a number whose square is x.
• The symbol $\sqrt{}$ is used to indicate the positive square root of a number.

Quiz Yourself 2.7A

1. Simplify: $\sqrt{64}$

2. Simplify: $-\sqrt{49}$

3. Simplify: $\sqrt{121} - \sqrt{16}$

4. Simplify: $3 + 2\sqrt{49}$

5. Evaluate $6\sqrt{a+b}$ for $a = 59$ and $b = 5$.

6. What is 9 decreased by the square root of 49?

Practice Sheet 2.7A

Simplify.

1. $\sqrt{49}$ 2. $-\sqrt{16}$

3. $\sqrt{144}$ 4. $-\sqrt{121}$

5. $\sqrt{9+40}$ 6. $\sqrt{69+12}$

7. $\sqrt{64}+\sqrt{25}$ 8. $\sqrt{169}-\sqrt{64}$

9. $-4\sqrt{81}$ 10. $6\sqrt{25}-11$

11. $15\sqrt{4}-\sqrt{49}$ 12. $\dfrac{\sqrt{25}}{\sqrt{64}}$

13. $\dfrac{\sqrt{4}}{\sqrt{49}}$ 14. $\dfrac{1}{\sqrt{49}}-\dfrac{1}{\sqrt{81}}$

15. $\dfrac{1}{\sqrt{16}}+\dfrac{1}{\sqrt{25}}$ 16. $\sqrt{25}+\sqrt{1}$

Evaluate the expression for the given values of the variables.

17. $-5\sqrt{xy}$
    for $x=2$ and $y=8$

18. $9\sqrt{x+y}$
    for $x=54$ and $y=10$

19. $\sqrt{b^2-4ac}$
    for $a=2$, $b=7$
    and $c=-4$

1. _____

2. _____

3. _____

4. _____

5. _____

6. _____

7. _____

8. _____

9. _____

10. _____

11. _____

12. _____

13. _____

14. _____

15. _____

16. _____

17. _____

18. _____

19. _____

Answers

Try it 2.7A
1. 9
2. 17
3. 2
4. 16

Quiz 2.7A
1. 8
2. −7
3. 7
4. 17
5. 48
6. 2

Solutions to Practice Sheet 2.7A

1. Since $7^2 = 49$, $\sqrt{49} = 7$.

2. Since $4^2 = 16$, $-\sqrt{16} = -4$.

3. Since $12^2 = 144$, $\sqrt{144} = 12$.

4. Since $11^2 = 121$, $-\sqrt{121} = -11$.

5. $\sqrt{40+9} = \sqrt{49} = 7$

6. $\sqrt{69+12} = \sqrt{81} = 9$

7. $\sqrt{64} + \sqrt{25} = 8+5 = 13$

8. $\sqrt{169} - \sqrt{64} = 13-8 = 5$

9. $-4\sqrt{81} = -4(9) = -36$

10. $6\sqrt{25} - 11 = 6(5) - 11$
$$= 30 - 11$$
$$= 19$$

11. $15\sqrt{4} - \sqrt{49} = 15(2) - 7$
$$= 30 - 7$$
$$= 23$$

12. $\dfrac{\sqrt{25}}{\sqrt{64}} = \dfrac{5}{8}$

13. $\dfrac{\sqrt{4}}{\sqrt{49}} = \dfrac{2}{7}$

14. $\dfrac{1}{\sqrt{49}} - \dfrac{1}{\sqrt{81}} = \dfrac{1}{7} - \dfrac{1}{9}$
$$= \dfrac{9}{63} - \dfrac{7}{63}$$
$$= \dfrac{2}{63}$$

15. $\dfrac{1}{\sqrt{16}} + \dfrac{1}{\sqrt{25}} = \dfrac{1}{4} + \dfrac{1}{5}$
$$= \dfrac{5}{20} + \dfrac{4}{20}$$
$$= \dfrac{9}{20}$$

16. $\sqrt{25} + \sqrt{1} = 5+1 = 6$

17. $-5\sqrt{xy}$
$$-5\sqrt{(2)(8)} = -5\sqrt{16} = -5(4) = -20$$

18. $9\sqrt{x+y}$
$$9\sqrt{54+10} = 9\sqrt{64} = 9(8) = 72$$

19. $\sqrt{b^2 - 4ac}$
$$\sqrt{7^2 - 4(2)(-4)} = \sqrt{49+32} = \sqrt{81} = 9$$

Objective 2.7B – Approximate the square root of a natural number

Key Words and Concepts

If the radicand is not a perfect square, the square root can only be approximated. Recall that a rational number has a decimal representation that terminates or repeats.

State the definition of an irrational number. _____

_____

_____

The rational numbers and the irrational numbers taken together are called the _____ **numbers**.

Example

Approximate $\sqrt{23}$ to the nearest ten-thousandth.

Solution
23 is not a perfect square.
Using a calculator, $\sqrt{23} \approx 4.7958$.

Try it

1. Approximate $\sqrt{18}$ to the nearest ten-thousandth.

Example

Between what two whole number is the value $\sqrt{72}$?

Solution
Since the number 72 is between the perfect squares 64 and 81, the value of $\sqrt{72}$ is between $\sqrt{64}$ and $\sqrt{81}$.
Because $\sqrt{64} = 8$ and $\sqrt{81} = 9$, the value of $\sqrt{72}$ is between the whole numbers 8 and 9. This can be written
$$8 < \sqrt{72} < 9.$$

Try it

2. Between what two whole number is the value $\sqrt{55}$?

Reflect on it

- A square root of a non-perfect square can only be approximated.
- An irrational number is a number whose decimal representation never terminates or repeats.
- The real numbers are all the rational numbers together with all the irrational numbers.

Quiz Yourself 2.7B

1. Approximate $\sqrt{14}$ to the nearest ten-thousandth.

2. Approximate $3\sqrt{99}$ to the nearest ten-thousandth.

3. Approximate $-25\sqrt{17}$ to the nearest ten-thousandth.

4. Approximate $8\sqrt{41}$ to the nearest ten-thousandth.

5. Between what two whole numbers is the value of $\sqrt{95}$?

6. Between what two whole numbers is the value of $\sqrt{150}$?

Practice Sheet 2.7B

Approximate to the nearest ten-thousandth.

1. $\sqrt{17}$

2. $7\sqrt{18}$

3. $-9\sqrt{29}$

4. $-13\sqrt{42}$

5. $4\sqrt{26}$

6. $-2\sqrt{31}$

7. $5\sqrt{33}$

8. $11\sqrt{15}$

9. $-21\sqrt{6}$

1. _____

2. _____

3. _____

4. _____

5. _____

6. _____

7. _____

8. _____

9. _____

Between what two whole numbers is the value of the radical expression?

10. $\sqrt{145}$

11. $\sqrt{75}$

12. $\sqrt{111}$

10. _____

11. _____

12. _____

Approximate to the nearest hundredth.

13. $\sqrt{24}$

14. $\sqrt{60}$

15. $\sqrt{125}$

16. $\sqrt{160}$

17. $\sqrt{280}$

18. $\sqrt{300}$

19. $\sqrt{252}$

20. $\sqrt{50}$

21. $\sqrt{126}$

13. _____

14. _____

15. _____

16. _____

17. _____

18. _____

19. _____

20. _____

21. _____

Answers

Try it: 2.7B
1. 4.2426
2. $7 < \sqrt{55} < 8.$

Quiz 2.7B
1. 3.7417
2. 29.8496
3. −103.0776
4. 51.2250
5. $9 < \sqrt{95} < 10$
6. $12 < \sqrt{150} < 13$

15. $\sqrt{125} \approx 11.18$

16. $\sqrt{160} \approx 12.65$

17. $\sqrt{280} \approx 16.73$

18. $\sqrt{300} \approx 17.32$

19. $\sqrt{252} \approx 15.87$

20. $\sqrt{50} \approx 7.07$

21. $\sqrt{126} \approx 11.22$

Solutions to Practice Sheet 2.7B

1. $\sqrt{17} \approx 4.1231$

2. $7\sqrt{18} \approx 29.6985$

3. $-9\sqrt{29} \approx -48.4665$

4. $-13\sqrt{42} \approx -84.2496$

5. $4\sqrt{26} \approx 20.3961$

6. $-2\sqrt{31} \approx -11.1355$

7. $5\sqrt{33} \approx 28.7228$

8. $11\sqrt{15} \approx 42.6028$

9. $-21\sqrt{6} \approx -51.4393$

10. 145 is between the perfect squares
    144 and 169
    $\sqrt{144} = 12$ and $\sqrt{169} = 13$
    $12 < \sqrt{145} < 13$

11. 75 is between the perfect squares 64 and 81
    $\sqrt{64} = 8$ and $\sqrt{81} = 9$
    $8 < \sqrt{75} < 9$

12. 111 is between the perfect squares
    100 and 121
    $\sqrt{100} = 10$ and $\sqrt{121} = 11$
    $10 < \sqrt{111} < 11$

13. $\sqrt{24} \approx 4.90$

14. $\sqrt{60} \approx 7.75$

Objective 2.7C – Solve application problems

Example

The formula $t = \sqrt{\dfrac{d}{16}}$ gives the time t, in seconds, it takes for an object to fall a distance of d feet assuming no air resistance. Find the time that it would take for an object to fall 100 feet. Round to the nearest tenth of a second.

Solution:

$$t = \sqrt{\dfrac{d}{16}} \qquad \text{Use the given formula}$$

$$= \sqrt{\dfrac{100}{16}} \qquad \text{Substitute 100 for } d$$

$$\approx 2.5 \text{ seconds} \quad \text{Simplify.}$$

Try it

1. Use the formula $t = \sqrt{\dfrac{d}{16}}$ as described in the Example. Find the time that it would take for an object to fall 160 feet. Round to the nearest tenth of a second.

Quiz Yourself 2.7C

In Exercises 1 and 2, use the formula $t = \sqrt{\dfrac{d}{16}}$ as described above.

1. If an object is dropped from a plane, how long will it take for the object to fall 400 feet?

2. If an object is dropped from a plane, how long will it take for the object to fall 80 feet? Round to the nearest tenth of a second.

3. If an object is dropped from a plane, how long will it take for the object to fall 85 feet? Round to the nearest tenth of a second.

Practice Sheet 2.7C

Solve.

1. Use the formula $v = 3\sqrt{d}$, where v is the velocity in feet per second of a tsunami as it approaches land and d is the depth in feet of the water. Find the velocity of the tsunami when the depth of the water is 121 ft.

2. Use the formula $v = 3\sqrt{d}$, where v is the velocity in feet per second of a tsunami as it approaches land and d is the depth in feet of the water. Find the velocity of the tsunami when the depth of the water is 196 ft.

1. _____

2. _____

3. Use the formula $v = 3\sqrt{d}$, where v is the velocity in feet per second of a tsunami as it approaches land and d is the depth in feet of the water. Find the velocity of the tsunami when the depth of the water is 81 ft.

4. Use the formula $t = \sqrt{\dfrac{d}{16}}$, where t is the time in seconds that an object falls and d is the distance in feet that the object falls. If an object is dropped from a plane, how long will it take for the object to fall 2304 ft?

3. _____

4. _____

5. Use the formula $t = \sqrt{\dfrac{d}{16}}$, where t is the time in seconds that an object falls and d is the distance in feet that the object falls. If an object is dropped from a plane, how long will it take for the object to fall 400 ft?

6. Use the formula $t = \sqrt{\dfrac{d}{16}}$, where t is the time in seconds that an object falls and d is the distance in feet that the object falls. If an object is dropped from a plane, how long will it take for the object to fall 256 ft?

5. _____

6. _____

Answers

Try it: 2.7C
1. 3.2 s

Quiz 2.7C
1. 5 s
2. 2.2 s
3. 2.3 s

Solutions to Practice Sheet 2.7C

1. **STRATEGY**

   To find the velocity, substitute 121 for d in the given formula and solve for v.

   SOLUTION

   $v = 3\sqrt{d}$

   $v = 3\sqrt{121} = 3(11) = 33$

   The velocity is 33 feet per second.

2. **STRATEGY**

   To find the velocity, substitute 196 for d in the given formula and solve for v.

   SOLUTION

   $v = 3\sqrt{d}$

   $v = 3\sqrt{196} = 3(14) = 42$

   The velocity is 42 feet per second.

3. **STRATEGY**

   To find the velocity, substitute 81 for d in the given formula and solve for v.

   SOLUTION

   $v = 3\sqrt{d}$

   $v = 3\sqrt{81} = 3(9) = 27$

   The velocity is 27 feet per second.

4. **STRATEGY**

   To find the time, substitute 2304 for d in the given formula and solve for t.

   SOLUTION

   $t = \sqrt{\dfrac{d}{16}}$

   $t = \sqrt{\dfrac{2304}{16}} = \sqrt{144} = 12$

   It will take 12 s for the object to fall 2304 ft.

5. **STRATEGY**

   To find the time, substitute 400 for d in the given formula and solve for t.

   SOLUTION

   $t = \sqrt{\dfrac{d}{16}}$

   $t = \sqrt{\dfrac{400}{16}} = \sqrt{25} = 5$

   It will take 5 s for the object to fall 400 ft.

6. **STRATEGY**

   To find the time, substitute 256 for d in the given formula and solve for t.

   SOLUTION

   $t = \sqrt{\dfrac{d}{16}}$

   $t = \sqrt{\dfrac{256}{16}} = \sqrt{16} = 4$

   It will take 4 s for the object to fall 256 ft.

Application and Activities

Discussion and reflection questions:

1. Have you ever built something with wood? If so, did you have to take measurements to cut the wood?

2. If you are measuring something, which would you think would be more useful to measure part of an inch; fractions or decimals?

3. If you are building a box to fit a speaker in, and the paperwork stated the speaker was 1.125 feet wide, would you be able to measure the wood to the nearest thousandths place? How would you round the length to most accurately cut the wood?

Group Activity (2-3 people):

Objectives for the lesson:

In this lesson you will use and enhance your understanding of the following:

- Finding the least common multiple and other common multiples of a pair of numbers
- Rounding to help estimate quantities needed to build something
- Performing operations on decimals
- Using a formula that contains a square root
- Creating a common denominator to compare fractions
- Performing operations on fractions
- Converting a fraction to decimal form

Jon and Jan want to build an entertainment center for their family room that will hold a TV, books, videos, and has room for storage. They drew up the following design, based on the dimensions of the wall the center will be placed against.

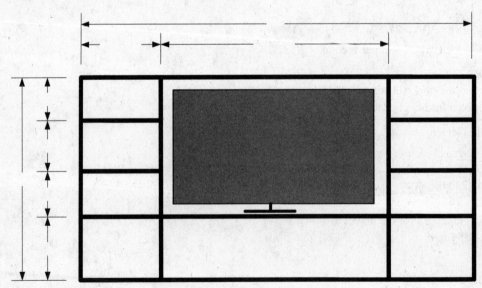

Module 2: Problem Solving with Fractions and Decimals

1. Jon and Jan are going to build the entertainment center themselves. Jon can only work on the project every 4 days and Jan can only work every 3 days. If the first time they can work together is January 1st, what is the next date when they can work together? They believe they can complete the project if they get together 4 times. Find the next 2 dates that they can get together to work. Show your work or explain your thought process on how you arrived at the dates.

2. Jon and Jan want to build the entertainment center using as few pieces of wood as possible. Their plan is shown below. (A) They will use one solid piece of wood for the top and one for the bottom of the center. (B) They will use 4 solid pieces of wood for the vertical parts. (C) They will use 6 pieces of wood for the book shelves. (D) They will use 1 piece of wood where the TV and stand will sit. Estimate by rounding to the nearest whole number, how much wood Jon and Jan will need to build the center. Be sure to show all the work you did to arrive at your estimate.

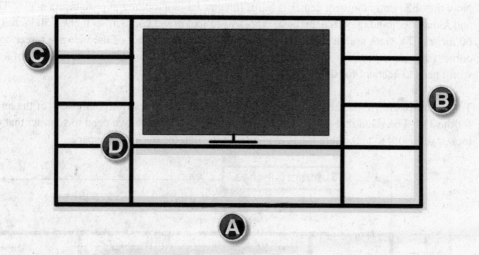

3. The wood that Jon and Jan want to use comes in lengths of 8 feet, 12 feet and 16 feet. Discuss the best ways to go about purchasing the least number of boards. Based on your estimates for exercise 2, decide how many boards, and what sizes they should purchase. Justify your choices.

4. Jon and Jan researched the prices of the boards they wanted at their local hardware store. Boards that are 8 ft. long cost $11.42 each, boards that are 12 ft. long cost $17.12 each, and boards that are 16 ft. long cost $22.82 each. Find how much it will cost for the boards they need.

5. The taxes where they live are 7.2%. Adding in $40 for hardware (screws and brackets and some tools), and the tax, find their total bill. Show how you arrived at the total bill.

6. Now that their entertainment center is built, Jon and Jan are looking at purchasing a TV. They researched and found the brand they like the best. The TVs in that brand come in 45 inches, 50 inches, 55 inches, and 60 inches. TV sizes are measured by the diagonal from one corner of the viewing screen to the opposite corner (they do not include the TV frame itself). The TVs they like do not have much of a frame, so we won't need to account for that.

    To make sure the TV has ample room around it, subtract 2 inches from both sides of the length of the space for the TV. The stand that comes with the TV is 3 inches high, so we need to subtract that off, along with 2 inches vertically to make sure it has ample room to fit in the designated space.

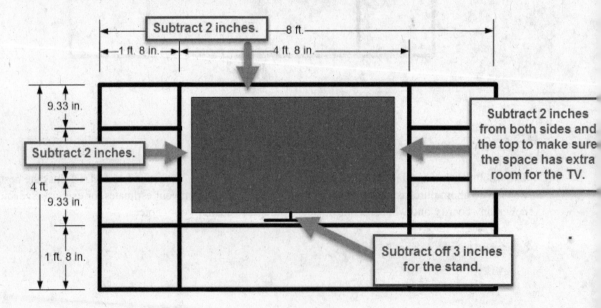

Use the following formula for finding the TV size, using your new length and height. What is the largest of the four TVs that will fit in their new entertainment center? Justify your answer.

$$\text{TV size} = \sqrt{(\text{length})^2 + (\text{height})^2}$$

7. Jon and Jan start arranging their books on the bookshelf. Jon has 5 books that are $\frac{7}{4}$ inches wide. Jan has 7 books that are $\frac{6}{5}$ inches wide. Using fractions, decide whose books take up more space on the bookshelf, and by how much. Show the work you used to arrive at your conclusion.

8. According to the diagram, the bookshelves are each 1ft 8 in. long, or 20 inches. Assume the average book size for Jon's and Jan's books is $\frac{6}{5}$ inches wide. Convert $\frac{6}{5}$ to a decimal and use that value to decide the approximate number of books Jon and Jan can place on each bookshelf. Show all work to justify your answer.

Objective 3.1A – Evaluate a variable expression

Key Words and Concepts

Recall that a letter of the alphabet, called a **variable**, is used to stand for a quantity that is unknown or that can change, or vary.

An expression that contains one or more variables is called a **variable expression**.

The **terms** of a variable expression are the addends of the expression.

The expression $7x^3 + 3x - 9y - 8xy + 5$ is a variable expression with 5 terms. Identify each of the terms as "variable" or "constant".

$7x^3$ _____ $3x$ _____ $-9y$ _____

$-8xy$ _____ 5 _____

Each variable term is composed of a **numerical coefficient** and a **variable part** (the variable or variables and their exponents).

  For the term $-8xy$, the numerical coefficient is -8 and the variable part is xy.

  A term such as $-xy$ has a numerical coefficient of -1.

Replacing each variable by its value and then simplifying the resulting numerical expression is called **evaluating a variable expression**.

Example

Evaluate $(a - 2b)^2$ for $a = -2$ and $b = -3$.

  Solution
    Start with the variable expression.
    Substitute the given values and simplify.

$$(a - 2b)^2 \qquad a = -2 \text{ and } b = -3$$
$$(-2 - 2(-3))^2 = (-2 + 6)^2$$
$$= 4^2$$
$$= 16$$

Try it

1. Evaluate $(a + 3b)^2$ for $a = -3$ and $b = -2$.

Example

Evaluate $x - \left(2 - \dfrac{y}{z}\right)^3$ for $x = -4$, $y = 12$, and $z = -3$.

Solution

Start with the variable expression.
Substitute the given values and simplify.

$$x - \left(2 - \frac{y}{z}\right)^3 \qquad x = -4, \ y = 12, \text{ and } z = -3$$

$$-4 - \left(2 - \frac{12}{-3}\right)^3 = -4 - (2 + 4)^3$$

$$= -4 - 6^3$$

$$= -4 - 216$$

$$= -220$$

Try it

2. Evaluate $(z - 2y)^2 - \dfrac{x+5}{y}$ for

$x = -8$, $y = 3$, and $z = -4$.

Reflect on it

- A variable is used to stand for a quantity that is unknown or that can change, or vary.
- The terms of a variable expression are the addends of the expression.
- Replacing each variable by its value and then simplifying the resulting numerical expression is called evaluating a variable expression.

Quiz Yourself 3.1A

1. Evaluate $4a - 3b$ for $a = 3$ and $b = -2$.

2. Evaluate $a \div (-3bc)$ for $a = 36$, $b = -4$, and $c = -1$.

3. Evaluate $(2x - y)^2$ for $x = -1$ and $y = 3$.

4. Evaluate $-a - 2b + 3c$ for $a = 3$, $b = 4$, and $c = -5$.

5. Evaluate $\dfrac{1}{3}w - \dfrac{1}{4}(wx - yz)$ for $w = 9$, $x = -1$, $y = 3$, and $z = 5$.

Practice Sheet 3.1A

Evaluate the variable expression when $a = -2$, $b = 3$, and $c = 4$.

1. $-2a + 3b$
2. $a^2 - 2$
3. $5(b - a)$

4. $4a + c^2$
5. $-2ab \div c$
6. $c^2 - ab$

7. $(b + c)^2$
8. $(c - a)^2 + 2b$
9. $\dfrac{c - a}{b}$

10. $b^2 - 2ac$
11. $-\dfrac{3}{4}c^2 + \dfrac{2}{3}b$
12. $\dfrac{a + 2b}{c}$

Evaluate the variable expression when $a = 3$, $b = -1$, $c = 2$ and $d = -4$.

13. $2ad - 4c$
14. $(c - b)^2 - 2d$
15. $2(b - d) + 3a$

16. $\dfrac{d - b}{ac}$
17. $\dfrac{a - d}{b}$
18. $\dfrac{b^2 - 4c}{-2b - a^2}$

19. $\dfrac{1}{4}d^2 + 3ac$
20. $\dfrac{5}{9}a^2 - bcd$
21. $\dfrac{-3cd}{ab^3}$

22. $c - \dfrac{1}{5}(ad - bc)$
23. $(d - a)^2 + (b - c)^2$
24. $\dfrac{ab - cd}{a - bc}$

1.	_____
2.	_____
3.	_____
4.	_____
5.	_____
6.	_____
7.	_____
8.	_____
9.	_____
10.	_____
11.	_____
12.	_____
13.	_____
14.	_____
15.	_____
16.	_____
17.	_____
18.	_____
19.	_____
20.	_____
21.	_____
22.	_____
23.	_____
24.	_____

Answers

Try it 3.1A
1. 81
2. 101

Quiz 3.1A
1. 18
2. −3
3. 25
4. −26
5. 9

Solutions to Practice Sheet 3.1A

1. $-2a + 3b$; $a = -2$, $b = 3$
$-2(-2) + 3(3) = 4 + 9 = 13$

2. $a^2 - 2$; $a = -2$
$(-2)^2 - 2 = 4 - 2 = 2$

3. $5(b - a)$; $a = -2$, $b = 3$
$5(3 - (-2)) = 5(3 + 2)$
$\qquad\qquad = 5(5) = 25$

4. $4a + c^2$; $a = -2$, $c = 4$
$4(-2) + 4^2 = 4(-2) + 16$
$\qquad\qquad = -8 + 16 = 8$

5. $-2ab \div c$; $a = -2$, $b = 3$, $c = 4$
$-2(-2)(3) \div 4 = 4(3) \div 4$
$\qquad\qquad\qquad = 12 \div 4$
$\qquad\qquad\qquad = 3$

6. $c^2 - ab$; $a = -2$, $b = 3$, $c = 4$
$(4)^2 - (-2)(3) = 16 - (-2)(3)$
$\qquad\qquad\qquad = 16 - (-6)$
$\qquad\qquad\qquad = 16 + 6$
$\qquad\qquad\qquad = 22$

7. $(b + c)^2$; $b = 3$, $c = 4$
$(3 + 4)^2 = 7^2 = 49$

8. $(c - a)^2 + 2b$; $a = -2$, $b = 3$, $c = 4$
$[4 - (-2)]^2 + 2(3) = (4 + 2)^2 + 2(3)$
$\qquad\qquad\qquad\quad = 6^2 + 2(3)$
$\qquad\qquad\qquad\quad = 36 + 2(3)$
$\qquad\qquad\qquad\quad = 36 + 6 = 42$

9. $\dfrac{c - a}{b}$; $a = -2$, $b = 3$, $c = 4$
$\dfrac{4 - (-2)}{3} = \dfrac{4 + 2}{3}$
$\qquad\quad = \dfrac{6}{3} = 2$

10. $b^2 - 2ac$; $a = -2$, $b = 3$, $c = 4$
$3^2 - 2(-2)(4) = 9 - 2(-2)(4)$
$\qquad\qquad\qquad = 9 + 4(4)$
$\qquad\qquad\qquad = 9 + 16 = 25$

11. $-\dfrac{3}{4}c^2 + \dfrac{2}{3}b$; $b = 3$, $c = 4$
$-\dfrac{3}{4}(4)^2 + \dfrac{2}{3}(3) = -\dfrac{3}{4}(16) + \dfrac{2}{3}(3)$
$\qquad\qquad\qquad\qquad = -12 + 2 = -10$

12. $\dfrac{a + 2b}{c}$; $a = -2$, $b = 3$, $c = 4$
$\dfrac{-2 + 2(3)}{4} = \dfrac{-2 + 6}{4} = \dfrac{4}{4} = 1$

13. $2ad - 4c$; $a = 3$, $c = 2$, $d = -4$
$2(3)(-4) - 4(2) = 6(-4) - 8$
$\qquad\qquad\qquad = -24 - 8 = -32$

14. $(c - b)^2 - 2d$; $b = -1$, $c = 2$, $d = -4$
$[2 - (-1)]^2 - 2(-4) = [2 + 1]^2 - 2(-4)$
$\qquad\qquad\qquad\qquad = 3^2 - 2(-4)$
$\qquad\qquad\qquad\qquad = 9 - 2(-4)$
$\qquad\qquad\qquad\qquad = 9 + 8 = 17$

15. $2(b - d) + 3a$; $a = 3$, $b = -1$, $d = -4$
$2[-1 - (-4)] + 3(3) = 2[-1 + 4] + 3(3)$
$\qquad\qquad\qquad\qquad = 2(3) + 3(3)$
$\qquad\qquad\qquad\qquad = 6 + 9 = 15$

16. $\dfrac{d - b}{ac}$; $a = 3$, $b = -1$, $c = 2$, $d = -4$
$\dfrac{-4 - (-1)}{3(2)} = \dfrac{-4 + 1}{3(2)} = \dfrac{-3}{3(2)} = -\dfrac{1}{2}$

17. $\dfrac{a - d}{b}$; $a = 3$, $b = -1$, $d = -4$
$\dfrac{3 - (-4)}{-1} = \dfrac{3 + 4}{-1} = \dfrac{7}{-1} = -7$

18. $\dfrac{b^2 - 4c}{-2b - a^2}$; $a = 3$, $b = -1$, $c = 2$
$\dfrac{(-1)^2 - 4(2)}{-2(-1) - 3^2} = \dfrac{1 - 4(2)}{-2(-1) - 9}$
$\qquad\qquad\qquad = \dfrac{1 - 8}{2 - 9}$
$\qquad\qquad\qquad = \dfrac{-7}{-7} = 1$

19. $\dfrac{1}{4}d^2 - 3ac$; $a = 3$, $c = 2$, $d = -4$
$\dfrac{1}{4}(-4)^2 + 3(3)(2) = \dfrac{1}{4}(16) + 3(3)(2)$
$\qquad\qquad\qquad\qquad = 4 + 9(2)$
$\qquad\qquad\qquad\qquad = 4 + 18 = 22$

20. $\frac{5}{9}a^2 - bcd$; $a = 3, \; b = -1, \; c = 2, \; d = -4$

$$\frac{5}{9}(3)^2 - (-1)(2)(-4) = \frac{5}{9}(9) - (-1)(2)(-4)$$
$$= 5 - (-2)(-4)$$
$$= 5 - 8 = -3$$

21. $\frac{-3cd}{ab^3}$; $a = 3, \; b = -1, \; c = 2, \; d = -4$

$$\frac{-3(2)(-4)}{3(-1)^3} = \frac{-3(2)(-4)}{3(-1)}$$
$$= \frac{-6(-4)}{-3}$$
$$= \frac{24}{-3} = -8$$

22. $c - \frac{1}{5}(ad - bc)$; $a = 3, \; b = -1, \; c = 2, \; d = -4$

$$2 - \frac{1}{5}[3(-4) - (-1)(2)] = 2 - \frac{1}{5}[-12 - (-2)]$$
$$= 2 - \frac{1}{5}[-12 + 2]$$
$$= 2 - \frac{1}{5}(-10)$$
$$= 2 + 2 = 4$$

23. $(d - a)^2 + (b - c)^2$; $a = 3, \; b = -1, \; c = 2, \; d = -4$

$$(-4 - 3)^2 + (-1 - 2)^2 = (-7)^2 + (-3)^2$$
$$= 49 + 9 = 58$$

24. $\frac{ab - cd}{a - bc}$; $a = 3, \; b = -1, \; c = 2, \; d = -4$

$$\frac{3(-1) - 2(-4)}{3 - (-1)(2)} = \frac{-3 + 8}{3 - (-2)}$$
$$= \frac{-3 + 8}{3 + 2}$$
$$= \frac{5}{5} = 1$$

Objective 3.1B – Simplify a variable expression using the Properties of Real Numbers

Key Words and Concepts

Define **like terms**. _____

_____

To simplify a variable expression, we use the Distributive Property to add the numerical coefficients of like variable terms. The variable part remains unchanged. This is called **combining like terms**.

The Distributive Property

If a, b, and c are real numbers, then $a(b+c) = ab + ac$ or $(b+c)a = ba + ca$.

For example, $3(5+8) = 3(5) + 3(8)$
$3(13) = 15 + 24$
$39 = 39$

The following Properties of Addition and Multiplication are used to simplify variable expressions.

The Associative Property of Addition

If a, b, and c are real numbers, then $(a+b)+c = a+(b+c)$.

For example, $(6x + 2x) + 7x = 6x + (2x + 7x)$
$8x + 7x = 6x + 9x$
$15x = 15x$

The Commutative Property of Addition

If a and b are real numbers, then $a + b = b + a$.

For example, $16x + (-29x) = -29x + 16x$
$-13x = -13x$

The Addition Property of Zero

If a is a real number, then $a + 0 = a$ and $0 + a = a$.

For example, $-6x + 0 = -6x$ and $0 + (-6x) = -6x$

The Inverse Property of Addition

If a is a real number, then $a + (-a) = 0$ and $(-a) + a = 0$.

For example, $-4x + 4x = 0$ and $4x + (-4x) = 0$

The Associative Property of Multiplication

If a, b, and c are real numbers, then $(ab)c = a(bc)$.

For example, $7(3x) = (7 \cdot 3)x$
$= 21x$

The Commutative Property of Multiplication
If a and b are real numbers, then $ab = ba$.

For example, $(6x) \cdot 9 = 9 \cdot (6x)$
$$= (9 \cdot 6)x$$
$$= 54x$$

The Multiplication Property of One
If a is a real number, then $a \cdot 1 = a$ and $1 \cdot a = a$.

For example, $(5x) \cdot 1 = 5x$

The Inverse Property of Multiplication
If a is a real number and a is not equal to zero, then $a \cdot \frac{1}{a} = 1$ and $\frac{1}{a} \cdot a = 1$. $\frac{1}{a}$ is called the **reciprocal** or

multiplicative inverse of a.

For example, $3 \cdot \frac{1}{3} = 1$ and $\frac{1}{3} \cdot 3 = 1$

Note   The product of a number and its reciprocal is 1.

The Distributive Property is also used to remove parentheses from a variable expression.

Example

Simplify: $-(4x^2 - 2x - 3)$

Solution
Use the Distributive Property.
$$-(4x^2 - 2x - 3) = -1(4x^2 - 2x - 3)$$
$$= -1(4x^2) - (-1)(2x) - (-1)(3)$$
$$= -4x^2 + 2x + 3$$

Try it

1. Simplify: $-(3x^2 + 6x - 2)$

Example

Simplify: $-4\left(-x - 2y + \frac{3}{2}z\right)$

Solution
Use the Distributive Property.
$$-4\left(-x - 2y + \frac{3}{2}z\right) = -4(-x) - 4(-2y) - 4\left(\frac{3}{2}z\right)$$
$$= 4x + 8y - 6z$$

Try it

2. Simplify: $-5(-2a + 6b - 3c)$

When simplifying variable expressions, use the Distributive Property to remove parentheses and brackets used as grouping symbols.

Example

Simplify: $4(2-3y)-(3-2y)$

    Solution
        Use the Distributive Property.
        Combine like terms.

$$4(2-3y)-(3-2y)=8-12y-3+2y$$
$$=5-10y$$

Try it

3. Simplify: $3(5-6x)-2(4-7x)$

Example

Simplify: $3x-2\left[4x-3(-x-5)\right]$

    Solution
        Sometimes you need to use the distributive property more
        than once.

$$3x-2\left[4x-3(-x-5)\right]=3x-2\left[4x+3x+15\right]$$
$$=3x-2\left[7x+15\right]$$
$$=3x-14x-30$$
$$=-11x-30$$

Try it

4. Simplify: $5x-3\left[6x-2(-4x-3)\right]$

Reflect on it
- Like terms of a variable expression are terms with the same variable part.
- Review all the Properties of Addition and Multiplication.
- For extra practice with the properties, rewrite all the examples of the properties *without* the variable x.

Quiz Yourself 3.1B

1. Simplify: $-2\left(4x^2-3x-1\right)$

2. Simplify: $-\dfrac{3}{4}(8a-4b+c)$

3. Simplify: $-3\left(\dfrac{1}{2}y - \dfrac{1}{3}z + 1\right)$

4. Simplify: $7x - 2(2x - 4)$

5. Simplify: $7 - 2(2x - 4)$

6. Simplify: $3x - \left[4 - 2(x + 5)\right]$

7. Simplify: $5(4x - 2y) - (x - 2y)$

Practice Sheet 3.1B

Simplify.

1. $4x + 7x$

2. $10x + 11x$

3. $7a - 2a$

4. $9y + (-11x) - 6y + 4y$

5. $x^2 - 8x + (-3x^2) + 7x$

6. $-4b + 3a - 9b + 11a$

7. $-3(-4y)$

8. $-7(9x^2)$

9. $-\frac{1}{5}(-5a)$

10. $-3(a + 2)$

11. $-4(3y - 4)$

12. $3(-2x^2 - 12)$

13. $6(x^2 + x - 7)$

14. $-2(y^2 + 3y - 5)$

15. $-7(-2x^2 + 4x + 6)$

16. $5x - 3(2x + 7)$

17. $9 - (10x - 4)$

18. $3(x - 3) - 2(x + 4)$

19. $8(y - 1) + 2(5 - 2y)$

20. $-4[x + 2(6 - x)]$

21. $-2[3x - (x + 8)]$

22. $-6x + 4[x - 6(2 - x)]$

23. $-4x - 3[3x - 3(x + 6)] - 5$

24. $3a - 2[3b - (2b - a)] + 4b$

25. $3a - 3[b - (2b - a)] + 5b$

1. _____

2. _____

3. _____

4. _____

5. _____

6. _____

7. _____

8. _____

9. _____

10. _____

11. _____

12. _____

13. _____

14. _____

15. _____

16. _____

17. _____

18. _____

19. _____

20. _____

21. _____

22. _____

23. _____

24. _____

25. _____

Answers

Try it 3.1B

1. $-3x^2 - 6x + 2$
2. $10a - 30b + 15c$
3. $-4x + 7$
4. $-37x - 18$

Quiz 3.1B

1. $-8x^2 + 6x + 2$
2. $-6a + 3b - \dfrac{3}{4}c$
3. $-\dfrac{3}{2}y + z - 3$
4. $3x + 8$
5. $15 - 4x$
6. $5x + 6$
7. $19x - 8y$

Solutions to Practice Sheet 3.1B

1. $4x + 7x = 11x$
2. $10x + 11x = 21x$
3. $7a - 2a = 5a$
4. $9y + (-11x) - 6y + 4y = -11x + 7y$
5. $x^2 - 8x + (-3x^2) + 7x = -2x^2 - x$
6. $-4b + 3a - 9b + 11a = 14a - 13b$
7. $-3(-4y) = 12y$
8. $-7(9x^2) = -63x^2$
9. $-\dfrac{1}{5}(-5a) = a$
10. $-3(a + 2) = -3a - 6$
11. $-4(3y - 4) = -12y + 16$
12. $3(-2x^2 - 12) = -6x^2 - 36$
13. $6(x^2 + x - 7) = 6x^2 + 6x - 42$
14. $-2(y^2 + 3y - 5) = -2y^2 - 6y + 10$
15. $-7(-2x^2 + 4x + 6) = 14x^2 - 28x - 42$
16. $5x - 3(2x + 7) = 5x - 6x - 21 = -x - 21$
17. $9 - (10x - 4) = 9 - 10x + 4 = -10x + 13$
18. $3(x - 3) - 2(x + 4) = 3x - 9 - 2x - 8$
    $\qquad\qquad\qquad\quad = x - 17$

19. $8(y - 1) + 2(5 - 2y) = 8y - 8 + 10 - 4y$
    $\qquad\qquad\qquad\qquad\quad = 4y + 2$

20. $-4[x + 2(6 - x)] = -4[x + 12 - 2x]$
    $\qquad\qquad\qquad\quad = -4[-x + 12] = 4x - 48$

21. $-2[3x - (x + 8)] = -2[3x - x - 8]$
    $\qquad\qquad\qquad\quad = -2[2x - 8] = -4x + 16$

22. $-6x + 4[x - 6(2 - x)] = -6x + 4[x - 12 + 6x]$
    $\qquad\qquad\qquad\qquad\quad = -6x + 4[7x - 12]$
    $\qquad\qquad\qquad\qquad\quad = -6x + 28x - 48$
    $\qquad\qquad\qquad\qquad\quad = 22x - 48$

23. $-4x - 3[3x - 3(x + 6)] - 5$
    $\quad = -4x - 3[3x - 3x - 18] - 5$
    $\quad = -4x - 3[-18] - 5$
    $\quad = -4x + 54 - 5 = -4x + 49$

24. $3a - 2[3b - (2b - a)] + 4b$
    $\quad = 3a - 2[3b - 2b + a] + 4b$
    $\quad = 3a - 2[b + a] + 4b$
    $\quad = 3a - 2b - 2a + 4b$
    $\quad = a + 2b$

25. $3a - 3[b - (2b - a)] + 5b$
    $\quad = 3a - 3[b - 2b + a] + 5b$
    $\quad = 3a - 3[-b + a] + 5b$
    $\quad = 3a + 3b - 3a + 5b$
    $\quad = 8b$

Objective 3.2A – Solve an equation of the form $x + a = b$ or $ax = b$

Key Words and Concepts

An **equation** expresses the equality of two mathematical expressions.

What is a solution of an equation? _____

_____

Example	**Try it**

Example

Is –1 a solution of $4x - 2 = x^2 - 7$?

    Solution

      Replace x by –1.

      Evaluate the numerical expressions.

      If the results are equal, –1 is a solution of the equation.

      If the results are not equal, –1 is not a solution of the equation.

$$\begin{array}{c|c} \multicolumn{2}{c}{4x - 2 = x^2 - 7} \\ \hline 4(-1) - 2 & (-1)^2 - 7 \\ -4 - 2 & 1 - 7 \\ \multicolumn{2}{c}{-6 = -6} \end{array}$$

      Yes, –1 is a solution of the equation.

Try it

1. Is –4 a solution of
$3x + 8 = x^2 - 20$?

Example

Is 3 a solution of $5 - 2x = x - 2$?

    Solution

$$\begin{array}{c|c} \multicolumn{2}{c}{5 - 2x = x - 2} \\ \hline 5 - 2(3) & 3 - 2 \\ 5 - 6 & 1 \\ \multicolumn{2}{c}{-1 \neq 1} \end{array}$$

No, 3 is not a solution of the equation.

Try it

2. Is 2 a solution of $7 - 3x = x - 4$?

What does it mean to **solve an equation**? _____

_____

The simplest equation to solve is an equation of the form *variable* = *constant*, because the constant is the solution.

Equations that have the same solution are called **equivalent equations**.

Addition Property of Equations

The same number can be added to each side of an equation without changing its solution. In symbols, the equation $a = b$ has the same solution as the equation $a + c = b + c$.

In solving an equation, the goal is to rewrite the given equation in the form *variable = constant*.

The Addition Property of Equations is used to remove a *term* from one side of the equation
    by adding the opposite of that term to each side of the equation.

Example

Solve: $\dfrac{2}{3} = z - \dfrac{1}{4}$

    Solution

        Goal: rewrite the equation in the form *variable = constant.*

$$\frac{2}{3} = z - \frac{1}{4} \qquad \text{Copy the equation}$$

$$\frac{2}{3} + \frac{1}{4} = z - \frac{1}{4} + \frac{1}{4} \qquad \text{Add } \frac{1}{4} \text{ to each side}$$

$$\frac{8}{12} + \frac{3}{12} = z + 0 \qquad \text{Simplify}$$

$$\frac{11}{12} = z$$

    The solution is $\dfrac{11}{12}$.

Try it

3. Solve: $16 = x - 9$

Because subtraction is defined in terms of addition, the Addition Property of Equations also makes it possible to
    subtract the same number from each side of an equation without changing the solution of the equation.

Example

Solve: $x + 4 = 2$

    Solution

        Goal: rewrite the equation in the form *variable = constant.*

$$x + 4 = 2 \qquad \text{Copy the equation}$$

$$x + 4 - 4 = 2 - 4 \qquad \text{Subtract 4 from each side}$$

$$x + 0 = -2 \qquad \text{Simplify}$$

$$x = -2$$

    Check:

$$\begin{array}{r} x + 4 = 2 \\ \hline -2 + 4 \mid 2 \\ 2 = 2 \end{array}$$

    The solution is –2.

Try it

4. Solve: $y + \dfrac{5}{8} = \dfrac{1}{4}$

Multiplication Property of Equations

Each side of an equation can be multiplied by the same nonzero number without changing the solution of the equation. In symbols, if $c \neq 0$, then the equation $a = b$ has the same solutions as the equation $ac = bc$.

The Multiplication Property of Equations is used to remove a coefficient
by multiplying each side of the equation by the reciprocal of the coefficient.

Example	**Try it**
Solve: $-\dfrac{3}{5}x = 6$	5. Solve: $-\dfrac{2}{3}x = 8$

Solution

Goal: rewrite the equation in the form *variable = constant*.

$$-\frac{3}{5}x = 6 \qquad \text{Copy the equation}$$

$$\left(-\frac{5}{3}\right)\left(-\frac{3}{5}x\right) = \left(-\frac{5}{3}\right)6 \qquad \begin{array}{l}\text{Multiply each side of} \\ \text{the equation by } -\dfrac{5}{3}\end{array}$$

$$1 \cdot x = -10 \qquad \text{Simplify}$$

$$x = -10$$

The solution is –10.

Because division is defined in terms of multiplication, *each side* of an equation
can be divided by the same nonzero number without changing the solution of the equation.

Example	**Try it**
Solve: $6 = 4x$	6. Solve: $10x = 16$

Solution

Goal: rewrite the equation in the form *variable = constant*.

$$6 = 4x \qquad \text{Copy the equation}$$

$$\frac{6}{4} = \frac{4x}{4} \qquad \text{Divide each side by 4}$$

$$\frac{3}{2} = x$$

The solution is $\dfrac{3}{2}$.

Note   When using the Multiplication Property of Equations, multiply each side of the equation by the *reciprocal of the coefficient* when the coefficient is a *fraction*.
    Divide each side of the equation by the *coefficient* when the coefficient is an *integer* or *decimal*.

Reflect on it

- An equation expresses the equality of two mathematical expressions.
- A solution of an equation is a number that, when substituted for the variable, results in a true equation.
- To solve an equation means to find a solution of the equation.
- Addition Property of Equations: $a = b$ has the same solution as $a + c = b + c$.
- Multiplication Property of Equations: as long as $c \neq 0$, $a = b$ has the same solution as $ac = bc$.

Quiz Yourself 3.2A

1. Is -3 a solution of $3 - z = z^2 - 3$?

2. Solve: $x - 6 = -9$

3. Solve: $-7 = x + \dfrac{1}{2}$

4. Solve and check: $-2.1 = \dfrac{z}{9.2}$

5. Solve and check: $-\dfrac{2}{3}z = -16$

6. Solve and check: $4x - 11x = -14$

Practice Sheet 3.2A

Solve.

1. Is 4 a solution of $3x + 4 = 4x$?

2. Is 0 a solution of $5 - m = 3m + 3$?

3. Is 2 a solution of $x^2 = x + 2$?

4. Is -3 a solution of $y^2 - 1 = y + 11$?

Solve and check.

5. $x + 3 = 8$

6. $x + 4 = 7$

7. $a - 4 = 11$

8. $y - 5 = 9$

9. $1 + a = 10$

10. $m + 7 = 2$

11. $t - 5 = -3$

12. $9 + a = 15$

13. $-6 = n + 2$

14. $x - \dfrac{1}{5} = \dfrac{2}{5}$

15. $b + \dfrac{1}{2} = -\dfrac{1}{3}$

16. $x + \dfrac{1}{3} = -\dfrac{5}{6}$

17. $3a = -15$

18. $-6m = 18$

19. $-2x = -24$

20. $\dfrac{2}{3}y = 8$

21. $\dfrac{3}{5}x = 12$

22. $-\dfrac{3}{4}d = 9$

23. $\dfrac{2x}{3} = 1$

24. $6d - 3d = 6$

25. $9x - 5x = 20$

1. _____

2. _____

3. _____

4. _____

5. _____

6. _____

7. _____

8. _____

9. _____

10. _____

11. _____

12. _____

13. _____

14. _____

15. _____

16. _____

17. _____

18. _____

19. _____

20. _____

21. _____

22. _____

23. _____

24. _____

25. _____

Answers

Try it 3.2A
1. Yes
2. No
3. $x = 25$
4. $y = -\dfrac{3}{8}$
5. $x = -12$
6. $x = \dfrac{8}{5}$

Quiz 3.2A
1. Yes
2. $x = -3$
3. $x = -\dfrac{15}{2}$
4. $z = -19.32$
5. $z = 24$
6. $x = 2$

Solutions to Practice Sheet 3.2A

1.
$$3x + 4 = 4x$$

$3(4) + 4$	$4(4)$
$12 + 4$	16

$$16 = 16$$
Yes, 4 is a solution.

2.
$$5 - m = 3m + 3$$

$5 - 0$	$3(0) + 3$
5	$0 + 3$

$$5 \neq 3$$
No, 0 is not a solution.

3.
$$x^2 = x + 2$$

2^2	$2 + 2$

$$4 = 4$$
Yes, 2 is a solution.

4.
$$y^2 - 1 = y + 11$$

$(-3)^2 - 1$	$-3 + 11$
$9 - 1$	8

$$8 = 8$$
Yes, −3 is a solution.

5.
$$x + 3 = 8$$
$$x + 3 - 3 = 8 - 3$$
$$x = 5$$
The solution is 5.

6.
$$x + 4 = 7$$
$$x + 4 - 4 = 7 - 4$$
$$x = 3$$
The solution is 3.

7.
$$a - 4 = 11$$
$$a - 4 + 4 = 11 + 4$$
$$a = 15$$
The solution is 15.

8.
$$y - 5 = 9$$
$$y - 5 + 5 = 9 + 5$$
$$y = 14$$
The solution is 14.

9.
$$1 + a = 10$$
$$1 - 1 + a = 10 - 1$$
$$a = 9$$
The solution is 9.

10.
$$m + 7 = 2$$
$$m + 7 - 7 = 2 - 7$$
$$m = -5$$
The solution is −5.

11.
$$t - 5 = -3$$
$$t - 5 + 5 = -3 + 5$$
$$t = 2$$
The solution is 2.

12.
$$9 + a = 15$$
$$9 - 9 + a = 15 - 9$$
$$a = 6$$
The solution is 6.

13.
$$-6 = n + 2$$
$$-6 - 2 = n + 2 - 2$$
$$-8 = n$$
The solution is −8.

14.
$$x - \frac{1}{5} = \frac{2}{5}$$
$$x - \frac{1}{5} + \frac{1}{5} = \frac{2}{5} + \frac{1}{5}$$
$$x = \frac{3}{5}$$
The solution is $\dfrac{3}{5}$.

15.
$$b + \frac{1}{2} = -\frac{1}{3}$$
$$b + \frac{1}{2} - \frac{1}{2} = -\frac{1}{3} - \frac{1}{2}$$
$$b = -\frac{2}{6} - \frac{3}{6} = -\frac{5}{6}$$
The solution is $-\dfrac{5}{6}$.

16. $x + \dfrac{1}{3} = -\dfrac{5}{6}$

$x + \dfrac{1}{3} - \dfrac{1}{3} = -\dfrac{5}{6} - \dfrac{1}{3}$

$x = -\dfrac{5}{6} - \dfrac{2}{6} = -\dfrac{7}{6}$

The solution is $-\dfrac{7}{6}$.

17. $3a = -15$

$\dfrac{3a}{3} = \dfrac{-15}{3}$

$a = -5$

The solution is -5.

18. $-6m = 18$

$\dfrac{-6m}{-6} = \dfrac{18}{-6}$

$m = -3$

The solution is -3.

19. $-2x = -24$

$\dfrac{-2x}{-2} = \dfrac{-24}{-2}$

$x = 12$

The solution is 12.

20. $\dfrac{2}{3}y = 8$

$\dfrac{3}{2}\left(\dfrac{2}{3}y\right) = \dfrac{3}{2}(8)$

$y = 12$

The solution is 12.

21. $\dfrac{3}{5}x = 12$

$\dfrac{5}{3}\left(\dfrac{3}{5}x\right) = \dfrac{5}{3}(12)$

$x = 20$

The solution is 20.

22. $-\dfrac{3}{4}d = 9$

$-\dfrac{4}{3}\left(-\dfrac{3}{4}d\right) = -\dfrac{4}{3}(9)$

$d = -12$

The solution is -12.

23. $\dfrac{2x}{3} = 1$

$\dfrac{3}{2}\left(\dfrac{2x}{3}\right) = \dfrac{3}{2}(1)$

$x = \dfrac{3}{2}$

The solution is $\dfrac{3}{2}$.

24. $6d - 3d = 6$

$3d = 6$

$\dfrac{3d}{3} = \dfrac{6}{3}$

$d = 2$

The solution is 2.

25. $9x - 5x = 20$

$4x = 20$

$\dfrac{4x}{4} = \dfrac{20}{4}$

$x = 5$

The solution is 5.

Objective 3.2B – Solve proportions

Key Words and Concepts

What is a ratio? _____

_____

A ratio can be written in three ways:
1. As a fraction
2. As two numbers separated by a colon
3. As two number s separated by the word *to*

A ratio is in **simplest form** when the two numbers do not have a common factor.

> *Note*   The units are not written in a ratio.

Example	Try it
Write the comparison 9 miles to 12 miles as a ratio in simplest form using a fraction, a colon, and the word *to*.	1. Write the comparison 20 pints to 5 pints as a ratio in simplest form using a fraction, a colon, and the word *to*.

Solution

Fraction

$$\frac{9 \text{ miles}}{12 \text{ miles}} = \frac{9}{12} = \frac{3}{4}$$

Colon

$$9 \text{ miles} : 12 \text{ miles} = 9 : 12 = 3 : 4$$

Word *to*

$$9 \text{ miles to } 12 \text{ miles} = 9 \text{ to } 12 = 3 \text{ to } 4$$

A **rate** is the comparison of two quantities with *different* units.

A rate is in _____ **form** when the numbers have no common factors.

> *Note*   The units are written as part of the rate.

Many rates are written as **unit rates**.
A unit rate is a rate in which the number in the denominator is 1.

Example	Try it
Write "190 km in 2 h" as a unit rate.	2. Write "726 words in 12 min" as a unit rate.

Solution
Write the rate as a fraction.
Divide the numerator by the denominator.

$$\frac{190 \text{ km}}{2 \text{ h}}$$

$$190 \div 2 = 85$$

The unit rate is 85 km/h.

A **proportion** states the equality of two ratios or rates.

Definition of Proportion

If $\frac{a}{b}$ and $\frac{c}{d}$ are equal ratios or rates, then $\frac{a}{b} = \frac{c}{d}$, is a proportion.

Each of the four numbers in a proportion is called a _____.

In the proportion $\frac{a}{b} = \frac{c}{d}$,

      the terms a and d are called the **extremes**;
      the terms b and c are called the **means**.

In any true proportion,
      the product of the means equals the product of the extremes.

      Note   This is sometimes phrased as "the cross products are equal."

Example

Determine whether the proportion $\frac{18 \text{ mi}}{10 \text{ gal}} = \frac{40.5 \text{ mi}}{22.5 \text{ gal}}$

is a true proportion.

    Solution
      The product of the means: The product of the extremes:
        $(10)(40.5) = 405$ $(18)(22.5) = 405$

      The proportion is true because $405 = 405$.

Try it

3. Determine whether the proportion

$\frac{606 \text{ words}}{10 \text{ minutes}} = \frac{302 \text{ words}}{5 \text{ minutes}}$ is a true proportion.

When three terms of a proportion are given, the fourth term can be found.

To solve a proportion for an unknown term, use the fact that
    the product of the means equals the product of the extremes.

Example

Solve: $\frac{r+3}{12} = \frac{3}{2}$

    Solution
      The product of the means equals the product of the extremes.
      Then solve for r.

        $\frac{r+3}{12} = \frac{3}{2}$ Copy the equation

        $12(3) = 2(r+3)$ The cross products are equal

          $36 = 2r + 6$ Use the Distributive Property

           $30 = 2r$ Subtract 6 from each side

           $15 = r$ Divide each side by 2

Try it

4. Solve: $\frac{42}{15} = \frac{n}{12}$

In setting up a proportion, keep the same units in the *numerators* and the same units in the *denominators*.

Example

Suppose that 2 gallons of paint can cover 580 square feet of wall space. How many square feet of wall space can 7 gallons cover?

> Solution
>
> Let n represent the number of square feet of wall space that 7 gallons can cover.
>
> Write a proportion.
>
> $$\frac{580 \text{ sq ft}}{2 \text{ gal}} = \frac{n \text{ sq ft}}{7 \text{ gal}}$$
>
> $7 \cdot 580 = 2n$ The cross products are equal
>
> $$\frac{4060}{2} = \frac{2n}{2}$$ Divide both sides by 2
>
> $2030 = n$
>
> 7 gallons of paint can cover 2030 sq ft of wall space.

Try it

5. A transistor company expects that 3 out of 245 transistors will be defective. How many defective transistors will be found in a batch of 184,485 transistors?

Reflect on it

- A ratio is the quotient or comparison of two quantities with the *same* unit.
- A rate is the comparison of two quantities with *different* units.
- A rate is in simplest form when the numbers have *no common factors*.
- **Definition of Proportion** If $\frac{a}{b}$ and $\frac{c}{d}$ are equal ratios or rates, then $\frac{a}{b} = \frac{c}{d}$, is a proportion.
- Each of the four numbers in a proportion is called a term.
- The product of the means equals the product of the extremes, or the cross products are equal.
- In setting up a proportion, keep the same units in the *numerators* and the same units in the *denominators*.

Quiz Yourself 3.2B

1. Write 30 mi in 8 h as a unit rate.

2. Write $78 for 15 oz. as a unit rate.

3. A vehicle uses 15 gallons of gasoline to travel 537 miles. How many miles per gallon did the car get?

4. Determine whether the proportion $\dfrac{40 \text{ mi}}{2 \text{ hr}} = \dfrac{50 \text{ mi}}{3 \text{ hr}}$ is a true proportion.

5. Solve: $\dfrac{15}{45} = \dfrac{72}{n}$

6. Solve: $\dfrac{10}{y-1} = \dfrac{15}{24}$

7. An automobile was driven 84 miles and used 3 gal of gasoline. At the same rate of consumption, how far would the car travel on 14.5 gallons of gasoline?

8. A building contractor estimates that 5 overhead lights are needed for every 400 sq ft of office space. Using this estimate, how many light fixtures are necessary for an office building of 36,000 sq ft?

Practice Sheet 3.2B

Write the comparison as a ratio in simplest form using a fraction, a colon (:), and the word *to*.

1.   43 years to 86 years   2.   27 feet to 12 feet   1. _____

                                                                             2. _____

Write as a rate in simplest form.

3.   106 miles in 4 hours   4.   72 feet in 9 seconds   3. _____

                                                                             4. _____

5. $117 for 18 boards   6. $10,816 earned in 26 weeks   5. _____

                                                                              6. _____

Solve. Round to the nearest hundredth, if necessary.

7. $\dfrac{3}{5} = \dfrac{n}{10}$ 8. $\dfrac{n}{8} = \dfrac{20}{32}$ 9. $\dfrac{14}{24} = \dfrac{7}{n}$ 7. _____

                                                                              8. _____

                                                                               9. _____

10. $\dfrac{8}{n} = \dfrac{9}{27}$ 11. $\dfrac{13}{36} = \dfrac{39}{n}$ 12. $\dfrac{n}{18} = \dfrac{5}{9}$ 10. _____

                                                                              11. _____

13. $\dfrac{n}{15} = \dfrac{0.8}{5.6}$ 14. $\dfrac{1.8}{18} = \dfrac{n}{12}$ 15. $\dfrac{3.4}{20} = \dfrac{5.1}{n}$ 12. _____

                                                                              13. _____

                                                                              14. _____

                                                                              15. _____

Solve.

16. A $19.75 sales tax is charged for a $395 purchase. At this rate, what is the sales tax for a $621 purchase?

17. The scale on the plans for a new office building is 1 inch equals 4 feet. How long is a room that measures $8\frac{1}{2}$ inches on the drawing?

16. _____

17. _____

18. For every 10 people who work in a city, 7 of them commute by public transportation. If 34,600 people work in the city, how many of them do not take public transportation?

19. For every 15 gallons of water pumped into the holding tank, 8 gallons were pumped out. After 930 gallons had been pumped in, how much water remained in the tank?

18. _____

19. _____

Answers

Try it 3.2B

1. $\frac{4}{1}$; 4:1; 4 to 1

2. 60.5 words/min

3. Not true

4. 33.6

5. 2259 defective transistors

Quiz 3.2B

1. 3.75 mi/h

2. $5.20/oz

3. 35.8 mi/gal

4. No

5. $n = 216$

6. $y = 17$

7. 406 miles

8. 450 lights

Solutions to Practice Sheet 3.2B

1. $\dfrac{43 \text{ years}}{86 \text{ years}} = \dfrac{43}{86} = \dfrac{1}{2}$

   $43 \text{ years}: 86 \text{ years} = 43:86 = 1:2$

   $43 \text{ years to } 86 \text{ years} = 43 \text{ to } 86 = 1 \text{ to } 2$

2. $\dfrac{27 \text{ feet}}{12 \text{ feet}} = \dfrac{27}{12} = \dfrac{9}{4}$

   $27 \text{ feet}: 12 \text{ feet} = 27:12 = 9:4$

   $27 \text{ feet to } 12 \text{ feet} = 27 \text{ to } 12 = 9 \text{ to } 4$

3. $\dfrac{106 \text{ miles}}{4 \text{ hours}} = 26.5 \text{ mph}$

4. $\dfrac{72 \text{ ft}}{9 \text{ s}} = 8 \text{ ft/s}$

5. $\dfrac{\$117}{18 \text{ boards}} = \$6.50 \,/\, \text{board}$

6. $\dfrac{\$10,816}{26 \text{ weeks}} = \$416 \,/\, \text{week}$

7. $\dfrac{3}{5} = \dfrac{n}{10}$

   $3 \cdot 10 = 5 \cdot n$

   $30 = 5n$

   $\dfrac{30}{5} = \dfrac{5n}{5}$

   $6 = n$

8. $\dfrac{n}{8} = \dfrac{20}{32}$

   $n \cdot 32 = 8 \cdot 20$

   $32n = 160$

   $\dfrac{32n}{32} = \dfrac{160}{32}$

   $n = 5$

9. $\dfrac{14}{24} = \dfrac{7}{n}$

   $14 \cdot n = 24 \cdot 7$

   $14n = 168$

   $\dfrac{14n}{14} = \dfrac{168}{14}$

   $n = 12$

10. $\dfrac{8}{n} = \dfrac{9}{27}$

    $8 \cdot 27 = n \cdot 9$

    $216 = 9n$

    $\dfrac{216}{9} = \dfrac{9n}{9}$

    $24 = n$

11. $\dfrac{13}{36} = \dfrac{39}{n}$

    $13 \cdot n = 36 \cdot 39$

    $13n = 1404$

    $\dfrac{13n}{13} = \dfrac{1404}{13}$

    $n = 108$

12. $\dfrac{n}{18} = \dfrac{5}{9}$

    $n \cdot 9 = 18 \cdot 5$

    $9n = 90$

    $\dfrac{9n}{9} = \dfrac{90}{9}$

    $n = 10$

13. $\dfrac{n}{15} = \dfrac{0.8}{5.6}$

    $n \cdot 5.6 = 15 \cdot 0.8$

    $5.6n = 12$

    $\dfrac{5.6n}{5.6} = \dfrac{12}{5.6}$

    $n \approx 2.14$

14. $\dfrac{1.8}{18} = \dfrac{n}{12}$

    $1.8 \cdot 12 = 18 \cdot n$

    $21.6 = 18n$

    $\dfrac{21.6}{18} = \dfrac{18n}{18}$

    $1.2 = n$

15.
$$\frac{3.4}{20} = \frac{5.1}{n}$$
$$3.4 \cdot n = 20 \cdot 5.1$$
$$3.4n = 102$$
$$\frac{3.4n}{3.4} = \frac{102}{3.4}$$
$$n = 30$$

16. **STRATEGY**
To find the sales tax, write and solve a proportion using x to represent the amount of tax.

SOLUTION
$$\frac{19.75}{395} = \frac{x}{621}$$
$$19.75 \cdot 621 = 395 \cdot x$$
$$12,264.75 = 395x$$
$$\frac{12,264.75}{395} = x$$
$$31.05 = x$$
The sales tax is $31.05.

17. **STRATEGY**
To find the length of the room, write and solve a proportion using n to represent the length of the room.

SOLUTION
$$\frac{1 \text{ in.}}{4 \text{ ft}} = \frac{8.5 \text{ in}}{x}$$
$$1 \cdot x = 4 \cdot 8.5$$
$$x = 34$$
The room is 34 ft long.

18. **STRATEGY**
If 7 out of 10 people take public transportation, then $10 - 7$, or 3 out of 10 people do not take public transportation. To find the number of people who do not take public transportation, write and solve a proportion using x to represent the people not taking public transportation.

SOLUTION
$$\frac{10}{3} = \frac{34,600}{x}$$
$$10 \cdot x = 3 \cdot 34,600$$
$$10x = 103,800$$
$$x = 10,380$$
10,380 people do not take public transportation.

19. **STRATEGY**
If 8 out of 15 gallons are pumped out, then $15 - 8$, or 7 out of 15 gallons remain in the tank. To find the number of gallons of water that remains in the tank, write and solve proportion using x to represent the gallons remaining in the tank.

SOLUTION
$$\frac{15}{7} = \frac{930}{x}$$
$$15 \cdot x = 7 \cdot 930$$
$$15x = 6510$$
$$x = \frac{6510}{15}$$
$$x = 434$$
Then 434 gallons of water remain in the tank.

Objective 3.2C – Solve the basic percent equation

Key Words and Concepts

Percent means "part of _____."

The symbol % is the **percent sign**.

To write a percent as a decimal,
      remove the percent sign and multiply by 0.01.

To write a percent as a fraction,
    remove the percent sign and multiply by $\frac{1}{100}$.

Example
Write 135% as a decimal and as a fraction.

    Solution
        To write a percent as a decimal,
        remove the percent sign and multiply by 0.01.
$$135\% = 135 \times 0.01 = 1.35$$

        To write a percent as a fraction,

        remove the percent sign and multiply by $\frac{1}{100}$.

$$135\% = 135 \times \frac{1}{100} = \frac{135}{100} = \frac{27}{20} = 1\frac{7}{20}$$

Try it
1. Write 160% as a decimal and as a fraction.

Example
Write $7\frac{8}{9}\%$ as a fraction.

    Solution
        To write a percent as a fraction,

        remove the percent sign and multiply by $\frac{1}{100}$.

$$7\frac{8}{9}\% = 7\frac{8}{9} \times \frac{1}{100} = \frac{71}{9} \times \frac{1}{100} = \frac{71}{900}$$

Try it
2. Write $3\frac{4}{7}\%$ as a fraction.

A decimal or a fraction can be written as a percent by multiplying by 100%.

Example

Write 0.875 as a percent.

Solution

A decimal can be written as a percent by multiplying by 100%

$$0.875 = 0.875 \times 100\% = 87.5\%$$

Try it

3. Write 0.467 as a percent.

When changing a fraction to a percent,

if the fraction can be written as a _____ decimal, the percent is written in decimal form.
                                                        (terminating/repeating)

If the decimal representation of the fraction is a _____ decimal, the answer is written with a fraction.
                                                                        (terminating/repeating)

Example

Write $\dfrac{3}{16}$ as a percent.

Solution

$\dfrac{3}{16} = 0.1875$ is a terminating decimal.

$$\dfrac{3}{16} = \dfrac{3}{16} \times \dfrac{100\%}{1}$$
$$= \dfrac{300\%}{16}$$
$$= 18.75\%$$

The answer, 18.75% is written in decimal form.

Try it

4. Write $\dfrac{6}{25}$ as a percent.

Example

Write $\dfrac{1}{15}$ as a percent.

Solution

$\dfrac{1}{15} = 0.0\overline{6}$ is a repeating decimal.

$$\dfrac{1}{15} = \dfrac{1}{15} \times \dfrac{100\%}{1}$$
$$= \dfrac{100\%}{15}$$
$$= 6\dfrac{2}{3}\%$$

The answer, $6\dfrac{2}{3}\%$ is written with a fraction.

Try it

5. Write $\dfrac{5}{12}$ as a percent.

The Basic Percent Equation

    Percent · base = amount

Example	**Try it**
What is 16% of 30?	6. What is 27% of 52?

Solution
    Use the basic percent equation.
        Percent = 16% = 0.16
        base = 30
        amount = n

$$\text{Percent} \cdot \text{base} = \text{amount}$$
$$0.16 \cdot 30 = n$$
$$4.8 = n$$
16% of 30 is 4.8.

Note   Look for the number or phrase that follows the word *of*
          when determining the base in the basic percent equation.

In most cases, the percent is written as a decimal before the basic percent equation is solved. However, some percents are more easily written as a fraction than as a decimal.

The three elements of the basic percent equation are _____, _____, and

_____ .

    If any two elements of the basic percent equation are given, the third element can be found.

Example	**Try it**
45% of what number is 18?	7. 63% of what number is 44.1?

Solution
    Use the basic percent equation.
        Percent = 45% = 0.45
        base = n
        amount = 18

$$\text{Percent} \cdot \text{base} = \text{amount}$$
$$0.45 \cdot n = 18$$
$$\frac{0.45n}{0.45} = \frac{18}{0.45}$$
$$n = 40$$
45% of 40 is 18.

Example

What percent of 5 is 4?

Solution

Use the basic percent equation.

Percent = n
base = 5
amount = 4

Percent · base = amount

$$n \cdot 5 = 4$$

$$\frac{5n}{5} = \frac{4}{5}$$

$$n = 0.80$$

$$n = 80\% \qquad \text{Write the decimal as a percent}$$

80% of 5 is 4.

Try it

8. What percent of 64 is 48?

Reflect on it

- Percent means parts of 100.
- To write a percent as a decimal, move the decimal point two places to the left and remove the percent sign.
- When changing a fraction to a percent, if the fraction can be written as a terminating decimal, the percent is written in decimal form.
- If the decimal representation of the fraction is a repeating decimal, the answer is written with a fraction.
- The three elements of the basic percent equation are the percent, the base and the amount.

Quiz Yourself 3.2C

1. Write 8.5% as a decimal and as a fraction.

2. Write 0.5% as a decimal and as a fraction.

3. Write 28% as a decimal and as a fraction.

4. Write $3\frac{1}{8}\%$ as a fraction.

5.   Write 0.094 as a percent.

6.   Write $\frac{1}{8}$ as a percent.

7.   Write $\frac{21}{40}$ as a percent.

Use the percent equation to solve 8–10.

8.   8 is what percent of 20?

9.   Find 34% of 60.

10. 0.5% of what number is 9?

Practice Sheet 3.2C

Write as a decimal and as a fraction.

1. 39%	**2.** 64%	**3.** 125%	

4. 26% **5.** 85% **6.** 20%

Write as a fraction.

7. $64\frac{1}{2}\%$ **8.** $43\frac{1}{3}\%$ **9.** $99\frac{3}{5}\%$

Write as a percent.

10. 0.32 **11.** 0.04 **12.** 1.97

Write as a percent.

13. $\frac{5}{32}$ **14.** $\frac{3}{5}$ **15.** $\frac{3}{8}$

Solve using the basic percent equation.

16. What is 30% of 90? **17.** 15% of what is 60? **18.** 25% of what is 8?

19. What is 12.5% of 48? **20.** 20 is what percent of 50? **21.** 20% of what is 8?

22. 25% of what is 16? **23.** 60 is what percent of 12? **24.** 24 is what percent of 6?

Answer lines:

1. _____
2. _____
3. _____
4. _____
5. _____
6. _____
7. _____
8. _____
9. _____
10. _____
11. _____
12. _____
13. _____
14. _____
15. _____
16. _____
17. _____
18. _____
19. _____
20. _____
21. _____
22. _____
23. _____
24. _____

Answers

Try it 3.2C

1. $1.6, 1\frac{3}{5}$

2. $\frac{1}{28}$

3. 46.7%

4. 24%

5. $41\frac{2}{3}\%$

6. 14.04

7. 70

8. 75%

Quiz 3.2C

1. $0.085, \frac{17}{200}$

2. $0.005, \frac{1}{200}$

3. $0.28, \frac{7}{25}$

4. $\frac{1}{32}$

5. 9.4%

6. 12.5%

7. 52.5%

8. 40%

9. 20.4

10. 1800

Solutions to Practice Sheet 3.2C

1. $39\% = 39 \times 0.01 = 0.39$
   $39\% = 39 \times \frac{1}{100} = \frac{39}{100}$

2. $64\% = 64 \times 0.01 = 0.64$
   $64\% = 64 \times \frac{1}{100} = \frac{64}{100} = \frac{16}{25}$

3. $125\% = 125 \times 0.01 = 1.25$
   $125\% = 125 \times \frac{1}{100} = \frac{125}{100} = \frac{5}{4}$

4. $26\% = 26 \times 0.01 = 0.26$
   $26\% = 26 \times \frac{1}{100} = \frac{26}{100} = \frac{13}{50}$

5. $85\% = 85 \times 0.01 = 0.85$
   $85\% = 85 \times \frac{1}{100} = \frac{85}{100} = \frac{17}{20}$

6. $20\% = 20 \times 0.01 = 0.20$
   $20\% = 20 \times \frac{1}{100} = \frac{20}{100} = \frac{1}{5}$

7. $64\frac{1}{2}\% = 64\frac{1}{2} \times \frac{1}{100} = \frac{129}{2} \times \frac{1}{100}$
   $= \frac{129}{200}$

8. $43\frac{1}{3}\% = 43\frac{1}{3} \times \frac{1}{100} = \frac{130}{3} \times \frac{1}{100}$
   $= \frac{130}{300} = \frac{13}{30}$

9. $99\frac{3}{5}\% = 99\frac{3}{5} \times \frac{1}{100} = \frac{498}{5} \times \frac{1}{100}$
   $= \frac{498}{500} = \frac{249}{250}$

10. $0.32 = 0.32 \times 100\% = 32\%$

11. $0.04 = 0.04 \times 100\% = 4\%$

12. $1.97 = 1.97 \times 100\% = 197\%$

13. $\frac{5}{32} = \frac{5}{32} \times 100\% = \frac{500}{32}\% = 15.625\%$

14. $\frac{3}{5} = \frac{3}{5} \times 100\% = \frac{300}{5}\% = 60\%$

15. $\frac{3}{8} = \frac{3}{8} \times 100\% = \frac{300}{8}\% = 37.5\%$

16. Use the basic percent equation.
    Percent = 30% = 0.30, base = 90, amount = n
    Percent $\cdot$ base = amount
    $0.30 \cdot 90 = n$
    $27 = n$
    30% of 90 is 27.

17. Use the basic percent equation.
    Percent = 15% = 0.15, base = n, amount = 60
    Percent $\cdot$ base = amount
    $0.15 \cdot n = 60$
    $n = \frac{60}{0.15}$
    $n = 400$
    15% of 400 is 60.

18. Use the basic percent equation.
    Percent = 25% = 0.25, base = n, amount = 8
    Percent $\cdot$ base = amount
    $0.25 \cdot n = 8$
    $n = \frac{8}{0.25}$
    $n = 32$
    25% of 32 is 8.

19. Use the basic percent equation.

Percent $= 12\frac{1}{2}\% = 0.125$, base $= 48$,

amount $= n$

Percent $\cdot$ base $=$ amount

$0.125 \cdot 48 = n$

$6 = n$

$12\frac{1}{2}\%$ of 48 is 6.

20. Use the basic percent equation.
Percent $= n$, base $= 50$, amount $= 20$

Percent $\cdot$ base $=$ amount

$n \cdot 50 = 20$

$\dfrac{50n}{50} = \dfrac{20}{50}$

$n = 0.4 = 40\%$

20 is 40% of 50.

21. Use the basic percent equation.
Percent $= 20\% = 0.20$, base $= n$, amount $= 8$

Percent $\cdot$ base $=$ amount

$0.20 \cdot n = 8$

$n = \dfrac{8}{0.20}$

$n = 40$

20% of 40 is 8.

22. Use the basic percent equation.
Percent $= 25\% = 0.25$, base $= n$, amount $= 16$

Percent $\cdot$ base $=$ amount

$0.25 \cdot n = 16$

$n = \dfrac{16}{0.25}$

$n = 64$

25% of 64 is 16.

23. Use the basic percent equation.
Percent $= n$, base $= 12$, amount $= 60$

Percent $\cdot$ base $=$ amount

$n \cdot 12 = 60$

$\dfrac{12n}{12} = \dfrac{60}{12}$

$n = 5 = 500\%$

60 is 500% of 12.

24. Use the basic percent equation.
Percent $= n$, base $= 6$, amount $= 24$

Percent $\cdot$ base $=$ amount

$n \cdot 6 = 24$

$\dfrac{6n}{6} = \dfrac{24}{6}$

$n = 4 = 400\%$

24 is 400% of 6.

Objective 3.2D – Solve applications of the basic percent equation

Key Words and Concepts

Example

An 18-carat yellow-gold necklace contains 75% gold, 16% silver, and 9% copper. If the necklace weighs 25 grams, how many grams of copper are in the necklace?

> Solution
>> To find the number of grams of copper in the necklace, write and solve the basic percent equation using n to represent the amount of copper.
>> The percent is 9% and the base is 25 grams.

$$\text{Percent} \cdot \text{base} = \text{amount}$$
$$9\% \cdot 25 = n$$
$$0.09 \cdot 25 = n$$
$$2.25 = n$$

2.25 grams of copper are in the necklace.

Try it

1. A used car dealer estimates that 65% of the company's sales will occur on Thursday, Friday, or Saturday. Using this estimate, how many cars out of the 80 sold last month were sold on Thursday, Friday, or Saturday?

Example

In a survey, 1236 adults nationwide were asked, "What irks you most about the actions of other motorists?" 293 people gave the response "tailgaters." What percent of those surveyed were most irked by tailgaters? Round to the nearest tenth of a percent.

> Solution
>> To find what percent were most irked by tailgaters, write and solve the basic percent equation using n to represent the unknown percent.
>> The base is 1236 and the amount is 293.

$$\text{Percent} \cdot \text{base} = \text{amount}$$
$$1236n = 293$$
$$\frac{1236n}{1236} = \frac{293}{1236}$$
$$n \approx 0.237$$

Approximately 23.7% of those surveyed were most irked by tailgaters.

Try it

2. There are 40 part-time employees at a business who attend college. There are a total of 50 part-time employees. What percent of part-time employees attend college?

The simple interest that an investment earns is given by the **simple interest equation**, $I = Prt$, where
   I is the simple interest;
   P is the principal, or amount invested;
   r is the simple interest rate; and
   t is the time.

Example

A rancher borrowed $120,000 for 180 days at an annual simple interest rate of 8.75%. What is the simple interest due on the loan?

   Solution
      Because an annual interest rate is given, the time must be
      converted from days to years. 180 days $= \dfrac{180}{365}$ year

      P is 120,000
      $r = 8.75\% = 0.0875$
      Find I.
         $I = Prt$
         $I = 120,000 \cdot 0.0875 \cdot \dfrac{180}{365}$
         $I \approx 5178.08$
      The simple interest due on the loan is $5178.08.

Try it

3. A $12,000 investment earned $462 in 6 months. Find the annual simple interest rate on the loan.

Percent increase is used to show how much a quantity has increased over its original value.

To find the percent increase:
   Find the amount of increase.
   Use the basic percent equation, where the amount is the amount of increase.

Example

A car's gas mileage increases from 17.5 miles per gallon to 18.2 miles per gallon. Find the percent increase in gas mileage.

   Solution
      The amount of increase is $18.2 - 17.5 = 0.7$.
      Use the basic percent equation.
      Percent $= n$
      base $= 17.5$
         Percent $\cdot$ base $=$ amount
            $n \cdot 17.5 = 0.7$
            $\dfrac{17.5n}{17.5} = \dfrac{0.7}{17.5}$
               $n = 0.04$
               $n = 4\%$
      The percent increase in gas mileage is 4%.

Try it

4. The value of a $7000 investment increased to $8750. What percent increase does this represent?

Percent decrease is used to show how much a quantity has decreased from its original value.

To find the percent decrease:
    Find the amount of decrease.
    Use the basic percent equation, where the amount is the amount of decrease.

Example

A new production method reduced the time needed to clean a piece of metal from 8 min to 5 min. What percent decrease does this represent?

    Solution
        The amount of decrease is $8 - 5 = 3$.
        Use the basic percent equation.
        Percent $= n$
        base $= 8$

$$\text{Percent} \cdot \text{base} = \text{amount}$$
$$n \cdot 8 = 3$$
$$\frac{8n}{8} = \frac{3}{8}$$
$$n = 0.375 = 37.5\%$$

The percent decrease is 37.5%.

Try it

5. A new bypass around a small town reduced the normal 40-minute driving time between two cities by 12 minutes. What percent decrease does this represent?

Reflect on it

- The simple interest equation is $I = Prt$,
- Percent increase is used to show how much a quantity has increased over its original value.
- Percent decrease is used to show how much a quantity has decreased from its original value.

Quiz Yourself 3.2D

1. In a test of the breaking strength of concrete slabs for freeway construction, 3 of the 200 slabs tested did not meet safety requirements. What percent of the slabs *did* meet safety requirements?

2. There are approximately 11,289,900 dog owners in the United States. This number is about 3.55% of the U.S. population. Based on this data, what is the population of the U.S.? Round your answer to the nearest million.

3. The concentration of platinum in a necklace is 15%. If the necklace weighs 12 g, find the amount of platinum in the necklace.

4. A mechanic borrowed $15,000 for 90 days at an annual simple interest rate of 7.2%. What is the simple interest due on the loan?

5. The rent on an apartment increased from $925/month to $950/month. What the percent increase in the monthly rent? Round to the nearest tenth of a percent.

6. A child grew from 50 in. to 60 in. in one year. What was the percent increase in the child's height?

7. The annual precipitation in a city dropped from 20 in. to 17 in. in one year. What is the percent decrease in precipitation?

8. The voter turnout in a midterm city election was 204,000 people. During the previous election, the voter turnout was 220,000 people. What was the percent decrease in voter turnout? Round to the nearest tenth of a percent.

Practice Sheet 3.2D

Solve.

1. A company built and equipped a new suite of offices for $200,000. Of this amount, $40,000 is for equipment and furnishings. What percent of the total expenditure was for equipment and furnishings?

2. Approximately 70% of the employees in a company purchase supplementary dental insurance. Using this estimate, how many of the company's 200 employees purchased supplementary dental insurance?

1. _____

2. _____

3. A computer dealer estimates that 12.5% of the computers sold will require service during the one-year warranty period. Using this estimate, how many computers were sold in a year in which 75 new computers were serviced?

4. The number of building permits issued in a city this year was 72,000. This represents 125% of last year's number of permits. How many building permits were issued last year?

3. _____

4. _____

5. A mobile home dealer borrowed $160,000 at a 15.5% annual interest rate for four years. What is the simple interest due on the loan?

6. An executive was offered a $34,000 loan at a 14.5% annual interest rate for three years. Find the simple interest due on the loan.

5. _____

6. _____

7. A college increased its number of parking spaces from 1000 to 1050.
   a. How many new spaces were added?
   b. What percent increase does this represent?

8. A town plans to increase its 4000 water meters by 7.5%.
   a. How many more water meters is this?
   b. What will the total number of water meters be after this increase?

7. a. _____

   b. _____

8. a. _____

   b. _____

9. Last year a company earned a profit of $285,000. This year, the company's profits were 6% less than last year's.
   a. What was the amount of decrease?
   b. What was the profit this year?

10. Because of a decrease in orders for telephones, a telephone center reduced the orders for phones from 140 per month to 91 per month.
    a. What is the amount of decrease?
    b. What percent decrease does this represent?

9. a. _____

   b. _____

10. a. _____

    b. _____

Answers

Try it 3.2D

1. 52 cars
2. 80%
3. 7.7%
4. 25%
5. 30%

Quiz 3.2D

1. 98.5%
2. 318,000,000
3. 1.8 g
4. $266.30
5. 2.7%
6. 20%
7. 15%
8. 7.8%

Solutions to Practice Sheet 3.2D

1. **STRATEGY**
   To find the percent, solve the basic percent
   equation $P \cdot B = A$ using $B = 200$ and $A = 40$.

   SOLUTION
   $$P \cdot B = A$$
   $$P \cdot 200 = 40$$
   $$P = \frac{40}{200} = 0.20$$

   The percent of expenditures for equipment
   and furnishings was 20%.

2. **STRATEGY**
   To find the number of cars sold, solve the
   basic percent equation $P \cdot B = A$ using
   $P = 70\% = 0.70$ and $B = 200$.

   SOLUTION
   $$P \cdot B = A$$
   $$0.70 \cdot 200 = A$$
   $$A = 140$$

   There were 140 employees purchased
   supplementary dental insurance.

3. **STRATEGY**
   To find the number of computers sold, solve
   the basic percent
   equation $P \cdot B = A$ using $P = 12.5\% = 0.125$,
   and $A = 75$.

   SOLUTION
   $$P \cdot B = A$$
   $$0.125 \cdot B = 75$$
   $$B = \frac{75}{0.125} = 600$$

   There were 600 computers sold.

4. **STRATEGY**
   To find the number of computers sold, solve
   the basic percent
   equation $P \cdot B = A$ using $P = 125\% = 1.25$,
   and $A = 72,000$.

   SOLUTION
   $$P \cdot B = A$$
   $$1.25 \cdot B = 72,000$$
   $$B = \frac{72,000}{1.25} = 57,600$$

   Last year 57,600 permits were issued.

5. **STRATEGY**
   To find the simple interest due, solve the
   simple interest formula $I = Prt$ for I.
   $P = 160,000$, $r = 0.155$, $t = 4$

   SOLUTION
   $$I = Prt$$
   $$I = (160,000)(0.155)(4)$$
   $$I = 99,200$$

   The interest due on the loan is $99,200.

6. **STRATEGY**
   To find the simple interest due, solve the
   simple interest formula $I = Prt$ for I.
   $P = 34,000$, $r = 0.145$, $t = 3$

   SOLUTION
   $$I = Prt$$
   $$I = (34,000)(0.145)(3)$$
   $$I = 14,790$$

   The interest due on the loan is $14,790.

7. STRATEGY
   a. Find the increase in parking spaces.
   b. Use the basic percent equation. Percent = n, base = 1000, amount = amount of increase

   SOLUTION
   a. $1050 - 1000 = 50$
   There are 50 new parking spaces added.
   b. Percent $\cdot$ base = amount
   $$P \cdot B = A$$
   $$P \cdot 1000 = 50$$
   $$P = \frac{50}{1000} = 0.05$$
   The percent increase is 5%.

8. STRATEGY
   a. To find the amount of increase:
   Use the basic percent equation.
   Percent = 7.5%, base = 4000, amount = n
   b. To find the total number of water meters, add the amount of increase to the base of 4000.

   SOLUTION
   a. Percent $\cdot$ base = amount
   $$P \cdot B = A$$
   $$0.075 \cdot 4000 = A$$
   $$A = 300$$
   The amount of increase is 300 water meters.
   b. $4000 + 300 = 4300$
   The total number of water meters is 4300.

9. STRATEGY
   a. To find the amount of decrease:
   Use the basic percent equation.
   Percent = 6%, base = 285,000, amount = n
   b. To find the profit, subtract the amount of decrease from the base of $285,000.

   SOLUTION
   a. Percent $\cdot$ base = amount
   $$P \cdot B = A$$
   $$0.06 \cdot 285,000 = A$$
   $$A = 17,100$$
   The amount of decrease is $17,100.
   b. $285,000 - $17,100 = $267,900$
   The profit this year was $267,900.

10. STRATEGY
   a. Find the amount of decrease.
   b. Use the basic percent equation. Percent = n, base = 140, amount = amount of decrease.

   SOLUTION
   a. $140 - 91 = 49$
   The amount of decrease is 49 orders.
   b. Percent $\cdot$ base = amount
   $$P \cdot B = A$$
   $$P \cdot 140 = 49$$
   $$P = \frac{49}{140} = 0.35$$
   The percent decrease is 35%.

Objective 3.3A – Solve an equation of the form $ax + b = c$

Key Words and Concepts

In solving an equation of the form $ax + b = c$, the goal is to rewrite the equation in the form *variable = constant*.

Example	**Try it**
Solve: $\frac{2}{3}x - 5 = 7$	1. Solve: $\frac{3}{5}x - 2 = 10$

Solution
> The goal is to write the equation in the form
> *variable = constant*

$\frac{2}{3}x - 5 = 7$ Copy the equation.

$\frac{2}{3}x - 5 + 5 = 7 + 5$ Add 5 to each side

$\frac{2}{3}x = 12$ Simplify

$\frac{3}{2} \cdot \frac{2}{3}x = \frac{3}{2} \cdot 12$ Multiply each side by $\frac{3}{2}$

$x = 18$ The equation is in the form
                    variable = constant

The solution is 18.

It may be easier to solve an equation containing two or more fractions by multiplying each side of the equation by the least common multiple (LCM) of the denominators. This method is called **clearing denominators**.

Example	**Try it**
Solve: $\frac{3}{4}x - \frac{1}{5} = \frac{1}{2}$	2. Solve: $\frac{2}{3}x - \frac{1}{4} = \frac{1}{3}$

Solution
> Multiply each side by 20, the LCM of 4, 5, and 2.
> Then use the properties of equality.

$\frac{3}{4}x - \frac{1}{5} = \frac{1}{2}$ Copy the equation

$20\left(\frac{3}{4}x - \frac{1}{5}\right) = 20\left(\frac{1}{2}\right)$ Multiply by 20, the LCM

$20\left(\frac{3}{4}x\right) - 20\left(\frac{1}{5}\right) = 20\left(\frac{1}{2}\right)$ Distributive property

$15x - 4 = 10$ Multiply

$15x - 4 + 4 = 10 + 4$ Add 4 to each side

$15x = 14$ Simplify

$\frac{15x}{15} = \frac{14}{15}$ Divide each side by 15

$x = \frac{14}{15}$

The solution is $\frac{14}{15}$.

Reflect on it
- In solving an equation of the form $ax + b = c$, the goal is to rewrite the equation in the form *variable = constant*.

Quiz Yourself 3.3A

1. Solve: $4x - 1 = -25$

2. Solve: $3 - 9x = 12$

3. Solve: $-\dfrac{2}{5}x + 4 = 0$

4. Solve: $\dfrac{3}{4}x - \dfrac{2}{3} = -\dfrac{1}{2}$

5. Solve: $3x - 7 - 5x = -4$

Practice Sheet 3.3A

Solve.

1. $2x + 3 = 15$

2. $3y + 5 = 26$

3. $4a - 7 = 9$

4. $5m - 16 = 19$

5. $3 = 4a + 15$

6. $3x - 7 = -22$

7. $6n - 9 = -51$

8. $9 = 5 + 2d$

9. $17 = 9 + 4z$

10. $8 - c = 7$

11. $6 - 2w = -4$

12. $7 - 5x = -23$

13. $9 - 4t = 1$

14. $16 - 7x = 2$

15. $5y - 35 = 0$

16. $12 + 3b = 0$

17. $14 + 2m = 0$

18. $-3x + 7 = -5$

19. $-4d + 5 = -31$

20. $-9x - 4 = -22$

21. $-11x + 20 = -2$

22. $-15 = -12y + 21$

23. $3 = 7 - 4a$

24. $2 = 12 - 5n$

25. $-33 = -6b + 3$

26. $-7x + 2 = -26$

27. $-4x - 36 = 0$

1. _____
2. _____
3. _____
4. _____
5. _____
6. _____
7. _____
8. _____
9. _____
10. _____
11. _____
12. _____
13. _____
14. _____
15. _____
16. _____
17. _____
18. _____
19. _____
20. _____
21. _____
22. _____
23. _____
24. _____
25. _____
26. _____
27. _____

Answers

Try it 3.3A
1. 20
2. $\dfrac{7}{8}$

Quiz 3.3A
1. -6
2. -1
3. 10
4. $\dfrac{2}{9}$
5. $-\dfrac{3}{2}$

Solutions to Practice Sheet 3.3A

1. $2x + 3 = 15$
$2x + 3 - 3 = 15 - 3$
$2x = 12$
$\dfrac{2x}{2} = \dfrac{12}{2}$
$x = 6$
The solution is 6.

2. $3y + 5 = 26$
$3y + 5 - 5 = 26 - 5$
$3y = 21$
$\dfrac{3y}{3} = \dfrac{21}{3}$
$y = 7$
The solution is 7.

3. $4a - 7 = 9$
$4a - 7 + 7 = 9 + 7$
$4a = 16$
$\dfrac{4a}{4} = \dfrac{16}{4}$
$a = 4$
The solution is 4.

4. $5m - 16 = 19$
$5m - 16 + 16 = 19 + 16$
$5m = 35$
$\dfrac{5m}{5} = \dfrac{35}{5}$
$m = 7$
The solution is 7.

5. $3 = 4a + 15$
$3 - 15 = 4a + 15 - 15$
$-12 = 4a$
$\dfrac{-12}{4} = \dfrac{4a}{4}$
$-3 = a$ or $a = -3$
The solution is -3.

6. $3x - 7 = -22$
$3x - 7 + 7 = -22 + 7$
$3x = -15$
$\dfrac{3x}{3} = \dfrac{-15}{3}$
$x = -5$
The solution is -5.

7. $6n - 9 = -51$
$6n - 9 + 9 = -51 + 9$
$6n = -42$
$\dfrac{6n}{6} = \dfrac{-42}{6}$
$n = -7$
The solution is -7.

8. $9 = 5 + 2d$
$9 - 5 = 5 - 5 + 2d$
$4 = 2d$
$\dfrac{4}{2} = \dfrac{2d}{2}$
$2 = d$ or $d = 2$
The solution is 2.

9. $17 = 9 + 4z$
$17 - 9 = 9 - 9 + 4z$
$8 = 4z$
$\dfrac{8}{4} = \dfrac{4z}{4}$
$2 = z$
The solution is 2.

10. $8 - c = 7$
$8 - 8 - c = 7 - 8$
$-c = -1$
$\dfrac{-c}{-1} = \dfrac{-1}{-1}$
$c = 1$
The solution is 1.

11. $6 - 2w = -4$
$6 - 6 - 2w = -4 - 6$
$-2w = -10$
$\dfrac{-2w}{-2} = \dfrac{-10}{-2}$
$w = 5$
The solution is 5.

12. $7 - 5x = -23$
$7 - 7 - 5x = -23 - 7$
$-5x = -30$
$\dfrac{-5x}{-5} = \dfrac{-30}{-5}$
$x = 6$
The solution is 6.

13.
$$9 - 4t = 1$$
$$9 - 9 - 4t = 1 - 9$$
$$-4t = -8$$
$$\frac{-4t}{-4} = \frac{-8}{-4}$$
$$t = 2$$
The solution is 2.

20.
$$-9x - 4 = -22$$
$$-9x - 4 + 4 = -22 + 4$$
$$-9x = -18$$
$$\frac{-9x}{-9} = \frac{-18}{-9}$$
$$x = 2$$
The solution is 2.

14.
$$16 - 7x = 2$$
$$16 - 16 - 7x = 2 - 16$$
$$-7x = -14$$
$$\frac{-7x}{-7} = \frac{-14}{-7}$$
$$x = 2$$
The solution is 2.

21.
$$-11x + 20 = -2$$
$$-11x + 20 - 20 = -2 - 20$$
$$-11x = -22$$
$$\frac{-11x}{-11} = \frac{-22}{-11}$$
$$x = 2$$
The solution is 2.

15.
$$5y - 35 = 0$$
$$5y - 35 + 35 = 0 + 35$$
$$5y = 35$$
$$\frac{5y}{5} = \frac{35}{5}$$
$$y = 7$$
The solution is 7.

22.
$$-15 = -12y + 21$$
$$-15 - 21 = -12y + 21 - 21$$
$$-36 = -12y$$
$$\frac{-36}{-12} = \frac{-12y}{-12}$$
$$3 = y \quad \text{or} \quad y = 3$$
The solution is 3.

16.
$$12 + 3b = 0$$
$$12 - 12 + 3b = 0 - 12$$
$$3b = -12$$
$$\frac{3b}{3} = \frac{-12}{3}$$
$$b = -4$$
The solution is –4.

23.
$$3 = 7 - 4a$$
$$3 - 7 = 7 - 7 - 4a$$
$$-4 = -4a$$
$$\frac{-4}{-4} = \frac{-4a}{-4}$$
$$1 = a \quad \text{or} \quad a = 1$$
The solution is 1.

17.
$$14 + 2m = 0$$
$$14 - 14 + 2m = 0 - 14$$
$$2m = -14$$
$$\frac{2m}{2} = \frac{-14}{2}$$
$$m = -7$$
The solution is –7.

24.
$$2 = 12 - 5n$$
$$2 - 12 = 12 - 12 - 5n$$
$$-10 = -5n$$
$$\frac{-10}{-5} = \frac{-5n}{-5}$$
$$2 = n$$
The solution is 2.

18.
$$-3x + 7 = -5$$
$$-3x + 7 - 7 = -5 - 7$$
$$-3x = -12$$
$$\frac{-3x}{-3} = \frac{-12}{-3}$$
$$x = 4$$
The solution is 4.

25.
$$-33 = -6b + 3$$
$$-33 - 3 = -6b + 3 - 3$$
$$-36 = -6b$$
$$\frac{-36}{-6} = \frac{-6b}{-6}$$
$$6 = b$$
The solution is 6.

19.
$$-4d + 5 = -31$$
$$-4d + 5 - 5 = -31 - 5$$
$$-4d = -36$$
$$\frac{-4d}{-4} = \frac{-36}{-4}$$
$$d = 9$$
The solution is 9.

26.
$$-7x + 2 = -26$$
$$-7x + 2 - 2 = -26 - 2$$
$$-7x = -28$$
$$\frac{-7x}{-7} = \frac{-28}{-7}$$
$$x = 4$$
The solution is 4.

27. $\quad -4x - 36 = 0$

$-4x - 36 + 36 = 0 + 36$

$-4x = 36$

$\dfrac{-4x}{-4} = \dfrac{36}{-4}$

$x = -9$

The solution is -9.

Objective 3.3B – Solve an equation of the form $ax + b = cx + d$

Key Words and Concepts

In solving an equation of the form $ax + b = cx + d$,

    the goal is to rewrite the equation in the form *variable = constant*.

Begin by rewriting the equation so that there is only one variable term in the equation.

    Then rewrite the equation so that there is only one constant term.

Example		**Try it**
Solve: $4x - 8 - 7x = 2 - 5x$		1. Solve: $6x - 8 - 3x = 4 - 9x$

    Solution

$4x - 8 - 7x = 2 - 5x$	Copy the equation
$-3x - 8 = 2 - 5x$	Combine like terms
$-3x + 5x - 8 = 2 - 5x + 5x$	Add $5x$ to each side to get all x-terms on one side
$2x - 8 = 2$	Simplify
$2x - 8 + 8 = 2 + 8$	Add 8 to each side to get all constants on other side
$2x = 10$	Simplify
$\dfrac{2x}{2} = \dfrac{10}{2}$	Divide each side by 2
$x = 5$	

    The solution is 5.

Reflect on it

- To solve equations of the form $ax + b = cx + d$,

    Add the same x-term to both sides (so that the x-term is eliminated from one side)
       Simplify
    Add the same constant term to both sides (so that the constant term is eliminated from the x-term side)
       Simplify
    Divide both sides by the coefficient of the x-term
       Simplify
    State the solution

- To check your solution, substitute the solution in the original equation and simplify each side separately.

Quiz Yourself 3.3B

    1.   Solve and check: $4x - 7 = 6x - 1$

2.   Solve and check: $0.5x - 2 = 0.1x + 14$

3.   Solve and check: $2x - 12 = 9x - 4 - 11x$

4.   Solve and check: $4x - 1 = x - 11$

5.   If $2 - x = -2x - 2$, evaluate $1 - 2x$.

Practice Sheet 3.3B

Solve.

1. $7x + 4 = 3x + 32$ 2. $5y + 1 = y + 17$ 3. $10m + 2 = 9m + 10$ 1. _____

2. _____

3. _____

4. $4x - 3 = 2x + 7$ 5. $8a - 9 = 2a + 15$ 6. $11y - 3 = 8y - 9$ 4. _____

5. _____

6. _____

7. $12b - 3 = 4b - 27$ 8. $14x - 1 = 3x - 23$ 9. $6a - 4 = a - 19$ 7. _____

8. _____

9. _____

10. $4x + 3 = 13 - x$ 11. $3x - 4 = -22 - 3x$ 12. $5y - 1 = -17 - 3y$ 10. _____

11. _____

12. _____

13. $2b + 9 = 5b + 15$ 14. $m + 3 = 3m + 11$ 15. $5d - 3 = 7d + 9$ 13. _____

14. _____

15. _____

16. $5y - 7 = 2y - 7$ 17. $4a + 11 = a + 11$ 18. $5 - 4x = 7 - 2x$ 16. _____

17. _____

18. _____

19. $9 - 5n = 15 - 2n$ 20. $4 + 7x = 12 + 3x$ 21. $3x - 5 = 8x$ 19. _____

20. _____

21. _____

22. $2a - 15 = 7a$ 23. $7m = 4m + 24$ 24. $8y = 4y + 40$ 22. _____

23. _____

24. _____

25. $-2x - 3 = 3x + 7$ 26. $-6a - 1 = 2a + 23$ 27. $-7n + 4 = -4n - 14$ 25. _____

26. _____

27. _____

Answers

Try it 3.3B
1. 1

Quiz 3.3B
1. −3
2. 40
3. 2
4. $-\dfrac{10}{3}$
5. 9

Solutions to Practice Sheet 3.3B

1.
$$7x + 4 = 3x + 32$$
$$7x - 3x + 4 = 3x - 3x + 32$$
$$4x + 4 = 32$$
$$4x + 4 - 4 = 32 - 4$$
$$4x = 28$$
$$\frac{4x}{4} = \frac{28}{4}$$
$$x = 7$$
The solution is 7.

2.
$$5y + 1 = y + 17$$
$$5y - y + 1 = y - y + 17$$
$$4y + 1 = 17$$
$$4y + 1 - 1 = 17 - 1$$
$$4y = 16$$
$$\frac{4y}{4} = \frac{16}{4}$$
$$y = 4$$
The solution is 4.

3.
$$10m + 2 = 9m + 10$$
$$10m - 9m + 2 = 9m - 9m + 10$$
$$m + 2 = 10$$
$$m + 2 - 2 = 10 - 2$$
$$m = 8$$
The solution is 8.

4.
$$4x - 3 = 2x + 7$$
$$4x - 2x - 3 = 2x - 2x + 7$$
$$2x - 3 = 7$$
$$2x - 3 + 3 = 7 + 3$$
$$2x = 10$$
$$\frac{2x}{2} = \frac{10}{2}$$
$$x = 5$$
The solution is 5.

5.
$$8a - 9 = 2a + 15$$
$$8a - 2a - 9 = 2a - 2a + 15$$
$$6a - 9 = 15$$
$$6a - 9 + 9 = 15 + 9$$
$$6a = 24$$
$$\frac{6a}{6} = \frac{24}{6}$$
$$a = 4$$
The solution is 4.

6.
$$11y - 3 = 8y - 9$$
$$11y - 8y - 3 = 8y - 8y - 9$$
$$3y - 3 = -9$$
$$3y - 3 + 3 = -9 + 3$$
$$3y = -6$$
$$\frac{3y}{3} = \frac{-6}{3}$$
$$y = -2$$
The solution is −2.

7.
$$12b - 3 = 4b - 27$$
$$12b - 4b - 3 = 4b - 4b - 27$$
$$8b - 3 = -27$$
$$8b - 3 + 3 = -27 + 3$$
$$8b = -24$$
$$\frac{8b}{8} = \frac{-24}{8}$$
$$b = -3$$
The solution is −3.

8.
$$14x - 1 = 3x - 23$$
$$14x - 3x - 1 = 3x - 3x - 23$$
$$11x - 1 = -23$$
$$11x - 1 + 1 = -23 + 1$$
$$11x = -22$$
$$\frac{11x}{11} = \frac{-22}{11}$$
$$x = -2$$
The solution is −2.

9.
$$6a - 4 = a - 19$$
$$6a - a - 4 = a - a - 19$$
$$5a - 4 = -19$$
$$5a - 4 + 4 = -19 + 4$$
$$5a = -15$$
$$\frac{5a}{5} = \frac{-15}{5}$$
$$a = -3$$
The solution is −3.

10.
$$4x+3=13-x$$
$$4x+x+3=13-x+x$$
$$5x+3=13$$
$$5x+3-3=13-3$$
$$5x=10$$
$$\frac{5x}{5}=\frac{10}{5}$$
$$x=2$$
The solution is 2.

11.
$$3x-4=-22-3x$$
$$3x+3x-4=-22-3x+3x$$
$$6x-4=-22$$
$$6x-4+4=-22+4$$
$$6x=-18$$
$$\frac{6x}{6}=\frac{-18}{6}$$
$$x=-3$$
The solution is -3.

12.
$$5y-1=-17-3y$$
$$5y+3y-1=-17-3y+3y$$
$$8y-1=-17$$
$$8y-1+1=-17+1$$
$$8y=-16$$
$$\frac{8y}{8}=\frac{-16}{8}$$
$$y=-2$$
The solution is -2.

13.
$$2b+9=5b+15$$
$$2b-5b+9=5b-5b+15$$
$$-3b+9=15$$
$$-3b+9-9=15-9$$
$$-3b=6$$
$$\frac{-3b}{-3}=\frac{6}{-3}$$
$$b=-2$$
The solution is -2.

14.
$$m+3=3m+11$$
$$m-3m+3=3m-3m+11$$
$$-2m+3=11$$
$$-2m+3-3=11-3$$
$$-2m=8$$
$$\frac{-2m}{-2}=\frac{8}{-2}$$
$$m=-4$$
The solution is -4.

15.
$$5d-3=7d+9$$
$$5d-7d-3=7d-7d+9$$
$$-2d-3=9$$
$$-2d-3+3=9+3$$
$$-2d=12$$
$$\frac{-2d}{-2}=\frac{12}{-2}$$
$$d=-6$$
The solution is -6.

16.
$$5y-7=2y-7$$
$$5y-2y-7=2y-2y-7$$
$$3y-7=-7$$
$$3y-7+7=-7+7$$
$$3y=0$$
$$\frac{3y}{3}=\frac{0}{3}$$
$$y=0$$
The solution is 0.

17.
$$4a+11=a+11$$
$$4a-a+11=a-a+11$$
$$3a+11=11$$
$$3a+11-11=11-11$$
$$3a=0$$
$$\frac{3a}{3}=\frac{0}{3}$$
$$a=0$$
The solution is 0.

18.
$$5-4x=7-2x$$
$$5-4x+2x=7-2x+2x$$
$$5-2x=7$$
$$5-5-2x=7-5$$
$$-2x=2$$
$$\frac{-2x}{-2}=\frac{2}{-2}$$
$$x=-1$$
The solution is -1.

19.
$$9-5n=15-2n$$
$$9-5n+2n=15-2n+2n$$
$$9-3n=15$$
$$9-9-3n=15-9$$
$$-3n=6$$
$$\frac{-3n}{-3}=\frac{6}{-3}$$
$$n=-2$$
The solution is -2.

20.
$$4 + 7x = 12 + 3x$$
$$4 + 7x - 3x = 12 + 3x - 3x$$
$$4 + 4x = 12$$
$$4 - 4 + 4x = 12 - 4$$
$$4x = 8$$
$$\frac{4x}{4} = \frac{8}{4}$$
$$x = 2$$

The solution is 2.

21.
$$3x - 5 = 8x$$
$$3x - 8x - 5 = 8x - 8x$$
$$-5x - 5 = 0$$
$$-5x - 5 + 5 = 0 + 5$$
$$-5x = 5$$
$$\frac{-5x}{-5} = \frac{5}{-5}$$
$$x = -1$$

The solution is −1.

22.
$$2a - 15 = 7a$$
$$2a - 7a - 15 = 7a - 7a$$
$$-5a - 15 = 0$$
$$-5a - 15 + 15 = 0 + 15$$
$$-5a = 15$$
$$\frac{-5a}{-5} = \frac{15}{-5}$$
$$a = -3$$

The solution is −3.

23.
$$7m = 4m + 24$$
$$7m - 4m = 4m - 4m + 24$$
$$3m = 24$$
$$\frac{3m}{3} = \frac{24}{3}$$
$$m = 8$$

The solution is 8.

24.
$$8y = 4y + 40$$
$$8y - 4y = 4y - 4y + 40$$
$$4y = 40$$
$$\frac{4y}{4} = \frac{40}{4}$$
$$y = 10$$

The solution is 10.

25.
$$-2x - 3 = 3x + 7$$
$$-2x - 3x - 3 = 3x - 3x + 7$$
$$-5x - 3 = 7$$
$$-5x - 3 + 3 = 7 + 3$$
$$-5x = 10$$
$$\frac{-5x}{-5} = \frac{10}{-5}$$
$$x = -2$$

The solution is −2.

26.
$$-6a - 1 = 2a + 23$$
$$-6a - 2a - 1 = 2a - 2a + 23$$
$$-8a - 1 = 23$$
$$-8a - 1 + 1 = 23 + 1$$
$$-8a = 24$$
$$\frac{-8a}{-8} = \frac{24}{-8}$$
$$a = -3$$

The solution is −3.

27.
$$-7n + 4 = -4n - 14$$
$$-7n + 4n + 4 = -4n + 4n - 14$$
$$-3n + 4 = -14$$
$$-3n + 4 - 4 = -14 - 4$$
$$-3n = -18$$
$$\frac{-3n}{-3} = \frac{-18}{-3}$$
$$n = 6$$

The solution is 6.

Objective 3.3C – Solve an equation containing parentheses

Key Words and Concepts

When an equation contains parentheses, one of the steps in solving the equation is to use the Distributive Property. The Distributive Property is used to remove parentheses from a variable expression.

Example

Solve: $4-7(2x-1)=3(3-5x)$

Solution

$4-7(2x-1)=3(3-5x)$	Copy the equation
$4-14x+7=9-15x$	Use the Distributive Property
$11-14x=9-15x$	Combine like terms
$11-14x+15x=9-15x+15x$	Add $15x$ to each side to get all x-terms on one side
$11+x=9$	Simplify
$11-11+x=9-11$	Subtract 11 from each side to get all constants on other side
$x=-2$	Simplify

The solution is -2.

Try it

1. Solve: $10-2(3x-4)=5(6-2x)$

Example

Solve: $2\left[5-3(3x-7)\right]=-3x-8$

Solution

$2\left[5-3(3x-7)\right]=-3x-8$	Copy the equation
$2[5-9x+21]=-3x-8$	Use the Distributive Property
$2[26-9x]=-3x-8$	Combine like terms
$52-18x=-3x-8$	Use the Distributive Property
$52-18x+18x=-3x+18x-8$	Add $18x$ to each side (x-terms to one side)
$52=15x-8$	Simplify
$52+8=15x-8+8$	Add 8 to each side (constants to other side)
$60=15x$	Simplify
$\dfrac{60}{15}=\dfrac{15x}{15}$	Divide each side by 15
$4=x$	

The solution is 4.

Try it

2. Solve: $3\left[4-2(2x-6)\right]=-9x-9$

Reflect on it
- You will need to use the Distributive Property once for each set of parentheses in the equation.

Quiz Yourself 3.3C

1. Solve and check: $4x - 2(x - 4) = 10$

2. Solve and check: $2(4 - 2y) + 2y = -5y - 1$

3. Solve and check : $0.02(x - 5) + 0.1x = 2.9$

4. Solve and check: $2x + 1 = 3 - (5 - 9x)$

5. Solve and check: $2 - 4[1 + 3(2x - 1)] = 8x - 2$

Practice Sheet 3.3C

Solve.

1. $3x + 2(x-1) = 8$

2. $5y + 3(y-3) = 15$

3. $8n - 5(2n+1) = 7$

4. $10x - 3(2x+5) = 13$

5. $8m - 5(m-2) = 7$

6. $5a - (2a-7) = 10$

7. $7n - 3(4n-1) = 13$

8. $3(2x-1) - 2 = 13$

9. $2(3b+2) - 7 = 9$

10. $4(2-3y) + 3y = 26$

11. $5(1-2x) + 5 = 80$

12. $9x + 1 = 3(2x+7) - 2$

13. $4y - 5 = 6 + 3(y-1)$

14. $11x + 4 = 2(3x-2) - 7$

15. $8 - 7x = 14 - (5x+6)$

16. $2x - 3 = 5(3x+2)$

17. $4n - 9 = 3(2n+5)$

18. $4b + 2(3b-5) = 2b+6$

19. $x + 2(3x-1) = 6x - 5$

20. $3 - 2a = 10 - 3(2a-3)$

21. $8 - 4x = 11 - (5x+6)$

22. $4b + 3(2b-1) = 2b+13$

23. $x + 2(3x-2) = 6x-2$

24. $2y - 5 = 4(y-7) + 5$

25. $3a - 4 = 3(2a+1) + 5$

26. $4 - (8-5x) = 3x-8$

27. $6 - (4-7x) = 3x+2$

1. _____

2. _____

3. _____

4. _____

5. _____

6. _____

7. _____

8. _____

9. _____

10. _____

11. _____

12. _____

13. _____

14. _____

15. _____

16. _____

17. _____

18. _____

19. _____

20. _____

21. _____

22. _____

23. _____

24. _____

25. _____

26. _____

27. _____

Answers

Try it 3.3C
1. 3
2. 19

Quiz 3.3C:
1. 1
2. −3
3. 25
4. $\dfrac{3}{7}$
5. $\dfrac{3}{8}$

Solutions to Practice Sheet 3.3C

1. $3x + 2(x - 1) = 8$
$3x + 2x - 2 = 8$
$5x - 2 = 8$
$5x - 2 + 2 = 8 + 2$
$5x = 10$
$\dfrac{5x}{5} = \dfrac{10}{5}$
$x = 2$
The solution is 2.

2. $5y + 3(y - 3) = 15$
$5y + 3y - 9 = 15$
$8y - 9 = 15$
$8y - 9 + 9 = 15 + 9$
$8y = 24$
$\dfrac{8y}{8} = \dfrac{24}{8}$
$y = 3$
The solution is 3.

3. $8n - 5(2n + 1) = 7$
$8n - 10n - 5 = 7$
$-2n - 5 = 7$
$-2n - 5 + 5 = 7 + 5$
$-2n = 12$
$\dfrac{-2n}{-2} = \dfrac{12}{-2}$
$n = -6$
The solution is −6.

4. $10x - 3(2x + 5) = 13$
$10x - 6x - 15 = 13$
$4x - 15 = 13$
$4x - 15 + 15 = 13 + 15$
$4x = 28$
$\dfrac{4x}{4} = \dfrac{28}{4}$
$x = 7$
The solution is 7.

5. $8m - 5(m - 2) = 7$
$8m - 5m + 10 = 7$
$3m + 10 = 7$
$3m + 10 - 10 = 7 - 10$
$3m = -3$
$\dfrac{3m}{3} = \dfrac{-3}{3}$
$m = -1$
The solution is −1.

6. $5a - (2a - 7) = 10$
$5a - 2a + 7 = 10$
$3a + 7 = 10$
$3a + 7 - 7 = 10 - 7$
$3a = 3$
$\dfrac{3a}{3} = \dfrac{3}{3}$
$a = 1$
The solution is 1.

7. $7n - 3(4n - 1) = 13$
$7n - 12n + 3 = 13$
$-5n + 3 = 13$
$-5n + 3 - 3 = 13 - 3$
$-5n = 10$
$\dfrac{-5n}{-5} = \dfrac{10}{-5}$
$n = -2$
The solution is −2.

8. $3(2x - 1) - 2 = 13$
$6x - 3 - 2 = 13$
$6x - 5 = 13$
$6x - 5 + 5 = 13 + 5$
$6x = 18$
$\dfrac{6x}{6} = \dfrac{18}{6}$
$x = 3$
The solution is 3.

9. $2(3b + 2) - 7 = 9$
$6b + 4 - 7 = 9$
$6b - 3 = 9$
$6b - 3 + 3 = 9 + 3$
$6b = 12$
$\dfrac{6b}{6} = \dfrac{12}{6}$
$b = 2$
The solution is 2.

10.
$$4(2-3y)+3y=26$$
$$8-12y+3y=26$$
$$8-9y=26$$
$$8-8-9y=26-8$$
$$-9y=18$$
$$\frac{-9y}{-9}=\frac{18}{-9}$$
$$y=-2$$

The solution is -2.

11.
$$5(1-2x)+5=80$$
$$5-10x+5=80$$
$$10-10x=80$$
$$10-10-10x=80-10$$
$$-10x=70$$
$$\frac{-10x}{-10}=\frac{70}{-10}$$
$$x=-7$$

The solution is -7.

12.
$$9x+1=3(2x+7)-2$$
$$9x+1=6x+21-2$$
$$9x+1=6x+19$$
$$9x-6x+1=6x-6x+19$$
$$3x+1=19$$
$$3x+1-1=19-1$$
$$3x=18$$
$$\frac{3x}{3}=\frac{18}{3}$$
$$x=6$$

The solution is 6.

13.
$$4y-5=6+3(y-1)$$
$$4y-5=6+3y-3$$
$$4y-5=3y+3$$
$$4y-3y-5=3y-3y+3$$
$$y-5=3$$
$$y-5+5=3+5$$
$$y=8$$

The solution is 8.

14.
$$11x+4=2(3x-2)-7$$
$$11x+4=6x-4-7$$
$$11x+4=6x-11$$
$$11x-6x+4=6x-6x-11$$
$$5x+4=-11$$
$$5x+4-4=-11-4$$
$$5x=-15$$
$$\frac{5x}{5}=\frac{-15}{5}$$
$$x=-3$$

The solution is -3.

15.
$$8-7x=14-(5x+6)$$
$$8-7x=14-5x-6$$
$$8-7x=8-5x$$
$$8-7x+5x=8-5x+5x$$
$$8-2x=8$$
$$8-8-2x=8-8$$
$$-2x=0$$
$$\frac{-2x}{-2}=\frac{0}{-2}$$
$$x=0$$

The solution is 0.

16.
$$2x-3=5(3x+2)$$
$$2x-3=15x+10$$
$$2x-3-15x=15x-15x+10$$
$$-13x-3=10$$
$$-13x-3+3=10+3$$
$$-13x=13$$
$$\frac{-13x}{-13}=\frac{13}{-13}$$
$$x=-1$$

The solution is -1.

17.
$$4n-9=3(2n+5)$$
$$4n-9=6n+15$$
$$4n-6n-9=6n-6n+15$$
$$-2n-9=15$$
$$-2n-9+9=15+9$$
$$-2n=24$$
$$\frac{-2n}{-2}=\frac{24}{-2}$$
$$n=-12$$

The solution is -12.

18.
$$4b+2(3b-5)=2b+6$$
$$4b+6b-10=2b+6$$
$$10b-10=2b+6$$
$$10b-2b-10=2b-2b+6$$
$$8b-10=6$$
$$8b-10+10=6+10$$
$$8b=16$$
$$\frac{8b}{8}=\frac{16}{8}$$
$$b=2$$

The solution is 2.

19.
$$x+2(3x-1)=6x-5$$
$$x+6x-2=6x-5$$
$$7x-2=6x-5$$
$$7x-6x-2=6x-6x-5$$
$$x-2=-5$$
$$x-2+2=-5+2$$
$$x=-3$$

The solution is -3.

20.
$$3 - 2a = 10 - 3(2a - 3)$$
$$3 - 2a = 10 - 6a + 9$$
$$3 - 2a = -6a + 19$$
$$3 - 2a + 6a = -6a + 6a + 19$$
$$3 + 4a = 19$$
$$3 - 3 + 4a = 19 - 3$$
$$4a = 16$$
$$\frac{4a}{4} = \frac{16}{4}$$
$$a = 4$$

The solution is 4.

21.
$$8 - 4x = 11 - (5x + 6)$$
$$8 - 4x = 11 - 5x - 6$$
$$8 - 4x = 5 - 5x$$
$$8 - 4x + 5x = 5 - 5x + 5x$$
$$8 + x = 5$$
$$8 - 8 + x = 5 - 8$$
$$x = -3$$

The solution is –3.

22.
$$4b + 3(2b - 1) = 2b + 13$$
$$4b + 6b - 3 = 2b + 13$$
$$10b - 3 = 2b + 13$$
$$10b - 2b - 3 = 2b - 2b + 13$$
$$8b - 3 = 13$$
$$8b - 3 + 3 = 13 + 3$$
$$8b = 16$$
$$\frac{8b}{8} = \frac{16}{8}$$
$$b = 2$$

The solution is 2.

23.
$$x + 2(3x - 2) = 6x - 2$$
$$x + 6x - 4 = 6x - 2$$
$$7x - 4 = 6x - 2$$
$$7x - 6x - 4 = 6x - 6x - 2$$
$$x - 4 = -2$$
$$x - 4 + 4 = -2 + 4$$
$$x = 2$$

The solution is 2.

24.
$$2y - 5 = 4(y - 7) + 5$$
$$2y - 5 = 4y - 28 + 5$$
$$2y - 5 = 4y - 23$$
$$2y - 4y - 5 = 4y - 4y - 23$$
$$-2y - 5 = -23$$
$$-2y - 5 + 5 = -23 + 5$$
$$-2y = -18$$
$$\frac{-2y}{-2} = \frac{-18}{-2}$$
$$y = 9$$

The solution is 9.

25.
$$3a - 4 = 3(2a + 1) + 5$$
$$3a - 4 = 6a + 3 + 5$$
$$3a - 4 = 6a + 8$$
$$3a - 6a - 4 = 6a - 6a + 8$$
$$-3a - 4 = 8$$
$$-3a - 4 + 4 = 8 + 4$$
$$-3a = 12$$
$$\frac{-3a}{-3} = \frac{12}{-3}$$
$$a = -4$$

The solution is –4.

26.
$$4 - (8 - 5x) = 3x - 8$$
$$4 - 8 + 5x = 3x - 8$$
$$-4 + 5x = 3x - 8$$
$$-4 + 5x - 3x = 3x - 3x - 8$$
$$-4 + 2x = -8$$
$$-4 + 4 + 2x = -8 + 4$$
$$2x = -4$$
$$\frac{2x}{2} = \frac{-4}{2}$$
$$x = -2$$

The solution is –2.

27.
$$6 - (4 - 7x) = 3x + 2$$
$$6 - 4 + 7x = 3x + 2$$
$$2 + 7x = 3x + 2$$
$$2 + 7x - 3x = 3x - 3x + 2$$
$$2 + 4x = 2$$
$$2 - 2 + 4x = 2 - 2$$
$$4x = 0$$
$$\frac{4x}{4} = \frac{0}{4}$$
$$x = 0$$

The solution is 0.

Objective 3.4A – Solve a literal equation

Key Words and Concepts

What is a literal equation? _____

_____

What is a formula? _____

_____

The Addition and Multiplication Properties can be used to solve a literal equation for one of the variables. The goal is to rewrite the equation so that the variable being solved for is alone on one side of the equation, and all the other numbers and variables are on the other side.

Example

Solve $7x - 8y = 16$ for y.

    Solution

        The goal is to solve for y.

$7x - 8y = 16$	Copy the equation
$7x - 7x - 8y = 16 - 7x$	Subtract $7x$ from each side
$-8y = 16 - 7x$	Simplify
$\dfrac{-8y}{-8} = \dfrac{16-7x}{-8}$	Divide each side by -8
$y = \dfrac{16}{-8} - \dfrac{7x}{-8}$	Simplify each fraction
$y = -2 + \dfrac{7}{8}x$	

    Note that it may later prove convenient
    to write this last line as

$$y = \frac{7}{8}x - 2.$$

Try it

1. Solve $2x - 5y = 10$ for y.

Example

Solve $y = 5(6 - 2x)$ for x.

Try it

2. Solve $L = \dfrac{P - 2W}{2}$ for P.

Solution

The goal is to solve for x.

$y = 5(6 + 2x)$	Copy the equation
$y = 30 + 10x$	Distribute
$y - 30 = 30 - 30 + 10x$	Subtract 30 from each side
$y - 30 = 10x$	Simplify
$\dfrac{y - 30}{10} = \dfrac{10x}{10}$	Divide each side by 10
$\dfrac{y}{10} - \dfrac{30}{10} = x$	Simplify each fraction
$\dfrac{y}{10} - 3 = x$	

So far, we have used the Distributive Property to remove parentheses.

$$a(b + c) = ab + ac$$

We can use the Distributive Property to rewrite two or more terms containing a common factor.

$$ab + ac = a(b + c)$$

In this case we can say that we have *factored* the common factor from the expression.

Example

Solve $A = P + Prt$ for t.

Try it

3. Solve $P = sL + sW$ for W.

Solution

P is a common factor. Use the Distributive Property to factor P from the two terms, recalling that $P = P \cdot 1$:

$$A = P + Prt$$
$$A = P(1 + rt)$$

$\dfrac{A}{1 + rt} = \dfrac{P(1 + rt)}{1 + rt}$	Divide each side by $(1 + rt)$
$P = \dfrac{A}{1 + rt}$	

Reflect on it

- A literal equation is an equation that contains more than one variable.
- A formula is a literal equation that states a rule about measurements.
- What formulas have you used in the last month?
- Using the Distributive Property $a(b + c) = ab + ac$ "in reverse" $ab + ac = a(b + c)$ is also known as factoring.

Quiz Yourself 3.4A

1. Solve $3x - 5y = 15$ for y.

2. Solve $F = ma$ for a.

3. Solve $y = \dfrac{3}{4}x - 5$ for x.

4. Solve $2A = b_1 h + b_2 h$ for h. (Note that b_1 and b_2 are different variables.)

5. Solve $a = b(c - d)$ for d.

Practice Sheet 3.4A

Solve for _y_.

1. $2x + y = 8$

2. $5x - y = 2$

3. $2x + 5y = 2$

4. $5x - 2y = 10$

5. $3x + 6y = 18$

6. $x - 2y = 6$

7. $x + 4y = 8$

8. $2x - 3y - 18 = 0$

9. $3x - y + 9 = 0$

Solve for _x_.

10. $x + 4y = 6$

11. $x - 5y = 7$

12. $2x - y = 4$

13. $3x + y = 12$

14. $2x + 3y = 8$

15. $2x + 5y = 10$

16. $3x - 5y - 30 = 0$

17. $2x - 7y - 14 = 0$

18. $5x - 4y + 5 = 0$

Solve the formula for the given variable.

19. $P = a + b + c; \ a$

20. $R = \dfrac{C - S}{t}; \ C$

21. $R = \dfrac{MB - T}{K}; \ M$

22. $K = mr - mv; \ m$

23. $S = \pi r(r - I); \ I$

24. $A = \dfrac{P}{1 - c}; \ c$

1. _____

2. _____

3. _____

4. _____

5. _____

6. _____

7. _____

8. _____

9. _____

10. _____

11. _____

12. _____

13. _____

14. _____

15. _____

16. _____

17. _____

18. _____

19. _____

20. _____

21. _____

22. _____

23. _____

24. _____

Answers

Try it 3.4A

1. $y = \dfrac{2}{5}x - 2$

2. $P = 2L + 2W$

3. $W = \dfrac{P - sL}{s}$ or $W = \dfrac{P}{s} - L$

Quiz 3.4A

1. $y = \dfrac{3}{5}x - 3$

2. $a = \dfrac{F}{m}$

3. $x = \dfrac{4}{3}(y + 5)$ or $x = \dfrac{4y + 20}{3}$

4. $h = \dfrac{2A}{b_1 + b_2}$

5. $d = -\dfrac{a - bc}{b}$ or $d = \dfrac{bc - a}{b}$

Solutions to Practice Sheet 3.4A

1.
$$2x + y = 8$$
$$2x - 2x + y = 8 - 2x$$
$$y = -2x + 8$$

2.
$$5x - y = 2$$
$$5x - 5x - y = 2 - 5x$$
$$-y = -5x + 2$$
$$y = 5x - 2$$

3.
$$2x + 5y = 2$$
$$2x - 2x + 5y = 2 - 2x$$
$$5y = -2x + 2$$
$$\dfrac{5y}{5} = \dfrac{-2x + 2}{5}$$
$$y = -\dfrac{2}{5}x + \dfrac{2}{5}$$

4.
$$5x - 2y = 10$$
$$5x - 5x - 2y = 10 - 5x$$
$$-2y = -5x + 10$$
$$\dfrac{-2y}{-2} = \dfrac{-5x + 10}{-2}$$
$$y = \dfrac{5}{2}x - 5$$

5.
$$3x + 6y = 18$$
$$3x - 3x + 6y = 18 - 3x$$
$$6y = -3x + 18$$
$$\dfrac{6y}{6} = \dfrac{-3x + 18}{6}$$
$$y = -\dfrac{1}{2}x + 3$$

6.
$$x - 2y = 6$$
$$x - x - 2y = 6 - x$$
$$-2y = -x + 6$$
$$\dfrac{-2y}{-2} = \dfrac{-x + 6}{-2}$$
$$y = \dfrac{1}{2}x - 3$$

7.
$$x + 4y = 8$$
$$x - x + 4y = 8 - x$$
$$4y = -x + 8$$
$$\dfrac{4y}{4} = \dfrac{-x + 8}{4}$$
$$y = -\dfrac{1}{4}x + 2$$

8.
$$2x - 3y - 18 = 0$$
$$2x - 2x - 3y - 18 + 18 = -2x + 18$$
$$-3y = -2x + 18$$
$$\dfrac{-3y}{-3} = \dfrac{-2x + 18}{-3}$$
$$y = \dfrac{2}{3}x - 6$$

9.
$$3x - y + 9 = 0$$
$$3x - 3x - y + 9 - 9 = -3x - 9$$
$$-y = -3x - 9$$
$$\dfrac{-y}{-1} = \dfrac{-3x - 9}{-1}$$
$$y = 3x + 9$$

10.
$$x + 4y = 6$$
$$x + 4y - 4y = 6 - 4y$$
$$x = -4y + 6$$

11.
$$x - 5y = 7$$
$$x - 5y + 5y = 7 + 5y$$
$$x = 5y + 7$$

12.
$$2x - y = 4$$
$$2x - y + y = 4 + y$$
$$2x = y + 4$$
$$\dfrac{2x}{2} = \dfrac{y + 4}{2}$$
$$x = \dfrac{1}{2}y + 2$$

13.
$$3x + y = 12$$
$$3x + y - y = 12 - y$$
$$3x = -y + 12$$
$$\frac{3x}{3} = \frac{-y + 12}{3}$$
$$x = -\frac{1}{3}y + 4$$

14.
$$2x + 3y = 8$$
$$2x + 3y - 3y = 8 - 3y$$
$$2x = -3y + 8$$
$$\frac{2x}{2} = \frac{-3y + 8}{2}$$
$$x = -\frac{3}{2}y + 4$$

15.
$$2x + 5y = 10$$
$$2x + 5y - 5y = 10 - 5y$$
$$2x = -5y + 10$$
$$\frac{2x}{2} = \frac{-5y + 10}{2}$$
$$x = -\frac{5}{2}y + 5$$

16.
$$3x - 5y - 30 = 0$$
$$3x - 5y + 5y - 30 + 30 = 5y + 30$$
$$3x = 5y + 30$$
$$\frac{3x}{3} = \frac{5y + 30}{3}$$
$$x = \frac{5}{3}y + 10$$

17.
$$2x - 7y - 14 = 0$$
$$2x - 7y + 7y - 14 + 14 = 7y + 14$$
$$2x = 7y + 14$$
$$\frac{2x}{2} = \frac{7y + 14}{2}$$
$$x = \frac{7}{2}y + 7$$

18.
$$5x - 4y + 5 = 0$$
$$5x - 4y + 4y + 5 - 5 = 4y - 5$$
$$5x = 4y - 5$$
$$\frac{5x}{5} = \frac{4y - 5}{5}$$
$$x = \frac{4}{5}y - 1$$

19.
$$P = a + b + c$$
$$P - b - c = a + b - b + c - c$$
$$P - b - c = a$$

20.
$$R = \frac{C - S}{t}$$
$$t(R) = t\left(\frac{C - S}{t}\right)$$
$$Rt = C - S$$
$$Rt + S = C - S + S$$
$$Rt + S = C$$

21.
$$R = \frac{MB - T}{K}$$
$$K(R) = K\left(\frac{MB - T}{K}\right)$$
$$KR = MB - T$$
$$KR + T = MB - T + T$$
$$KR + T = MB$$
$$\frac{KR + T}{B} = \frac{MB}{B}$$
$$\frac{KR + T}{B} = M$$

22.
$$K = mr - mv$$
$$K = m(r - v)$$
$$\frac{K}{r - v} = \frac{m(r - v)}{r - v}$$
$$\frac{K}{r - v} = m$$

23.
$$S = \pi r(r - I)$$
$$\frac{S}{\pi r} = \frac{\pi r(r - I)}{\pi r}$$
$$\frac{S}{\pi r} = r - I$$
$$\frac{S}{\pi r} + I = r - I + I - \frac{S}{\pi r}$$
$$I = r - \frac{S}{\pi r}$$

24.
$$A = \frac{P}{1 - c}$$
$$(1 - c)A = (1 - c)\frac{P}{1 - c}$$
$$A - Ac = P$$
$$A - A - Ac = P - A$$
$$-Ac = P - A$$
$$\frac{-Ac}{-A} = \frac{P - A}{-A}$$
$$c = \frac{P - A}{-A}, \text{ or } c = -\frac{P}{A} + 1$$

Objective 3.4B – Solve an absolute value equation

Key Words and Concepts

Recall that the *absolute value* of a number is its distance from zero on the number line.

    Distance is always a positive number or zero.

    Therefore, the absolute value of a number is always a positive number or zero.

Absolute value can be used to represent the distance between any two points on the number line.

The **distance between two points** on the number line is the absolute value of the difference between the coordinates of the two points.

    The distance between point a and point b is given by $|b-a|$.

    Note   For any two numbers a and b, $|a-b|=|b-a|$.

An equation containing a variable within an absolute value symbol is called an **absolute value equation**.

Solutions of an Absolute Value Equation

    If $a>0$ and $|x|=a$, then $x=-a$ or $x=a$.

    If $|x|=0$, then $x=0$.

    If $a<0$, then $|x|=a$ has no solution.

Example

Solve: $|2x-1|=5$

    Solution

        Remove the absolute value sign and rewrite as two equations.
        Then solve each equation.

$$|2x-1|=5$$
$$2x-1=5 \qquad 2x-1=-5$$
$$2x=6 \qquad 2x=-4$$
$$x=3 \qquad x=-2$$

    The solutions are 3 and −2.

Try it

1. Solve: $|7-x|=17$

Example

Solve: $4 - |5 - 3x| = 1$

Solution

$$4 - |5 - 3x| = 1$$

$$-|5 - 3x| = -3 \qquad \text{Subtract 3 from each side}$$

$$|5 - 3x| = 3 \qquad \text{Multiply each side by } -1$$

$$5 - 3x = 3 \qquad 5 - 3x = -3$$

$$-3x = -2 \qquad -3x = -8$$

$$x = \frac{2}{3} \qquad\;\; x = \frac{8}{3}$$

The solutions are $\frac{2}{3}$ and $\frac{8}{3}$.

Try it

2. Solve: $|3 - 2x| + 6 = 9$

Reflect on it

- The *absolute value* of a number is its distance from zero on the number line.
- If an absolute value is equal to 0, there is one solution.
- If an absolute value is equal to a negative number, there is no solution.
- If an absolute value is equal to a positive number, there are two solutions.

Quiz Yourself 3.4B

1. Solve: $|x + 5| = 0$

2. Solve: $|3x - 2| = 6$

3. Solve: $2 - |3x - 4| = -7$

4. Solve: $2 - |x - 1| = 5$

Name:

Class:

Module 3: Applications of Equations and Inequalities

Section 3.4: Literal, Absolute Value, and Radical Equations

Practice Sheet 3.4B

Solve.

1. $|x| = 6$

2. $|a| = 3$

3. $|b| = 5$

4. $|c| = 11$

5. $|-y| = 7$

6. $|-t| = 2$

7. $|-a| = 9$

8. $|-x| = 1$

9. $|x| = -3$

10. $|-x| = -5$

11. $|5x - 10| = 0$

12. $|2x - 6| = 8$

13. $|4 - 2x| = 8$

14. $|a - 6| = 0$

15. $|3x - 2| = 0$

16. $|3 - 2x| = 5$

17. $|2 - 3x| = 7$

18. $|2a + 7| = 1$

19. $|5x - 1| - 1 = 3$

20. $|4x - 3| + 2 = 5$

21. $|3x + 2| - 1 = 3$

22. $|4x + 3| - 1 = 8$

23. $|5 - 3x| + 4 = 1$

24. $3 - |3x + 1| = 3$

25. $4 + |2x - 1| = 7$

26. $7 - |3x - 2| = 4$

27. $3 - |2x - 5| = 3$

1.	_____
2.	_____
3.	_____
4.	_____
5.	_____
6.	_____
7.	_____
8.	_____
9.	_____
10.	_____
11.	_____
12.	_____
13.	_____
14.	_____
15.	_____
16.	_____
17.	_____
18.	_____
19.	_____
20.	_____
21.	_____
22.	_____
23.	_____
24.	_____
25.	_____
26.	_____
27.	_____

Answers

Try it 3.4B

1. -10 and 24
2. 0 and 3

Quiz 3.4B

1. -5
2. $-\dfrac{4}{3}$ and $\dfrac{8}{3}$
3. $-\dfrac{5}{3}$ and $\dfrac{13}{3}$
4. No solution

Solutions to Practice Sheet 3.4B

1. $|x| = 6$
   $x = -6$ or $x = 6$

2. $|a| = 3$
   $a = -3$ or $a = 3$

3. $|b| = 5$
   $b = -5$ or $b = 5$

4. $|c| = 11$
   $c = -11$ or $c = 11$

5. $|-y| = 7$
   $y = -7$ or $y = 7$

6. $|-t| = 2$
   $t = -2$ or $t = 2$

7. $|-a| = 9$
   $a = -9$ or $a = 9$

8. $|-x| = 1$
   $x = -1$ or $x = 1$

9. $|x| = -3$
   The is no solution to this equation because the absolute value of a number must be nonnegative.

10. $|-x| = -5$
    The is no solution to this equation because the absolute value of a number must be nonnegative.

11. $|5x - 10| = 0$
    $5x - 10 = 0$
    $5x = 10$
    $x = 2$
    The solution is 2.

12. $|2x - 6| = 8$
    $2x - 6 = 8$ or $2x - 6 = -8$
    $2x = 14$ $\quad$ $2x = -2$
    $x = 7$ $\quad$ $x = -1$
    The solutions are -1 and 7.

13. $|4 - 2x| = 8$
    $4 - 2x = 8$ or $4 - 2x = -8$
    $-2x = 4$ $\quad$ $-2x = -12$
    $x = -2$ $\quad$ $x = 6$
    The solutions are -2 and 6.

14. $|a - 6| = 0$
    $a - 6 = 0$
    $a = 6$
    The solution is 6.

15. $|3x - 2| = 0$
    $3x - 2 = 0$
    $3x = 2$
    $x = \dfrac{2}{3}$
    The solution is $\dfrac{2}{3}$.

16. $|3 - 2x| = 5$
    $3 - 2x = 5$ or $3 - 2x = -5$
    $-2x = 2$ $\quad$ $-2x = -8$
    $x = -1$ $\quad$ $x = 4$
    The solutions are -1 and 4.

17. $|2 - 3x| = 7$
    $2 - 3x = 7$ or $2 - 3x = -7$
    $-3x = 5$ $\quad$ $-3x = -9$
    $x = -\dfrac{5}{3}$ $\quad$ $x = 3$
    The solutions are $-\dfrac{5}{3}$ and 3.

18. $|2a + 7| = 1$
    $2a + 7 = 1$ or $2a + 7 = -1$
    $2a = -6$ $\quad$ $2a = -8$
    $a = -3$ $\quad$ $a = -4$
    The solutions are -4 and -3.

19. $|5x - 1| - 1 = 3$
    $|5x - 1| = 4$
    $5x - 1 = 4$ or $5x - 1 = -4$
    $5x = 5$ $\quad$ $5x = -3$
    $x = 1$ $\quad$ $x = -\dfrac{3}{5}$
    The solutions are $-\dfrac{3}{5}$ and 1.

20. $|4x-3|+2=5$
     $|4x-3|=3$
     $4x-3=3$ or $4x-3=-3$
     $\quad 4x=6 \qquad\qquad 4x=0$
     $\quad\quad x=\dfrac{3}{2} \qquad\qquad x=0$

The solutions are 0 and $\dfrac{3}{2}$.

21. $|3x+2|-1=3$
     $|3x+2|=4$
     $3x+2=4$ or $3x+2=-4$
     $\quad 3x=2 \qquad\qquad 3x=-6$
     $\quad\quad x=\dfrac{2}{3} \qquad\qquad x=-2$

The solutions are -2 and $\dfrac{2}{3}$.

22. $|4x+3|-1=8$
     $|4x+3|=9$
     $4x+3=9$ or $4x+3=-9$
     $\quad 4x=6 \qquad\qquad 4x=-12$
     $\quad\quad x=\dfrac{3}{2} \qquad\qquad x=-3$

The solutions are -3 and $\dfrac{3}{2}$.

23. $|5-3x|+4=1$
     $|5-3x|=-3$

The is no solution to this equation because the absolute value of a number must be nonnegative.

24. $3-|3x+1|=3$
     $-|3x+1|=0$
     $|3x+1|=0$
     $3x+1=0$
     $3x=-1$
     $x=-\dfrac{1}{3}$

The solution is $-\dfrac{1}{3}$.

25. $4+|2x-1|=7$
     $|2x-1|=3$
     $2x-1=3$ or $2x-1=-3$
     $\quad 2x=4 \qquad\qquad 2x=-2$
     $\quad\quad x=2 \qquad\qquad x=-1$

The solutions are -1 and 2.

26. $7-|3x-2|=4$
     $-|3x-2|=-3$
     $|3x-2|=3$
     $3x-2=3$ or $3x-2=-3$
     $\quad 3x=5 \qquad\qquad 3x=-1$
     $\quad\quad x=\dfrac{5}{3} \qquad\qquad x=-\dfrac{1}{3}$

The solutions are $-\dfrac{1}{3}$ and $\dfrac{5}{3}$.

27. $3-|2x-5|=3$
     $-|2x-5|=0$
     $|2x-5|=0$
     $2x-5=0$
     $2x=5$
     $x=\dfrac{5}{2}$

The solution is $\dfrac{5}{2}$.

Objective 3.4C – Solve a radical equation

Key Words and Concepts

An equation that contains a variable expression in a radicand is a **radical equation**.

Property of Squaring Both Sides of an Equation

If a and b are real numbers and $a = b$, then $a^2 = b^2$.

Any time each side of an equation is squared, you must check the proposed solution of the equation.

Example

Solve: $\sqrt{3x} - 6 = -4$

Solution

Rewrite the equation so that the radical is alone on one side of the equation.

Then square both sides and solve for x.

$$\sqrt{3x} - 6 = -4$$
$$\sqrt{3x} = -4 + 6 = 2$$
$$\left(\sqrt{3x}\right)^2 = 2^2$$
$$3x = 4$$
$$x = \frac{4}{3}$$

Check:

$$\sqrt{3x} - 6 = -4$$

$$\begin{array}{c|c} \sqrt{3\left(\frac{4}{3}\right)} - 6 & -4 \\ \sqrt{4} - 6 & -4 \\ 2 - 6 & -4 \end{array}$$

$-4 = -4$ This is a true equation. The solution checks.

The solution is $\frac{4}{3}$

Try it

1. Solve $\sqrt{2x + 5} - 3 = 0$

Reflect on it

- To solve radical equations
  Rewrite the equation so that the radical is alone on one side of the equation.
  Then square both sides and solve for x.
  Check the solution.

Quiz Yourself 3.4C

1. Solve: $\sqrt{4x+1} = 5$

2. Solve: $\sqrt{3x} - 2 = 4$

Practice Sheet 3.4C

Solve and check.

1. $\sqrt{y} = 8$ 2. $\sqrt{a} = 16$ 3. $\sqrt{a} = 10$

4. $\sqrt{3x} = 3$ 5. $\sqrt{2x} = 4$ 6. $\sqrt{3x} = 6$

7. $5 - \sqrt{5x} = 0$ 8. $\sqrt{6x} - 12 = 0$ 9. $3 - \sqrt{2x} = 0$

10. $\sqrt{2x-3} = 5$ 11. $\sqrt{4x+5} = 1$ 12. $\sqrt{2x+7} = 3$

13. $\sqrt{2x-1} = 3$ 14. $\sqrt{2x-3} = 1$ 15. $\sqrt{6x+1} = 5$

16. $\sqrt{5x-4} = 6$ 17. $0 = 3 - \sqrt{2+x}$ 18. $0 = 5 - \sqrt{5+x}$

19. $\sqrt{2x} - 1 = 3$ 20. $5 = 7 - \sqrt{2c}$ 21. $10 = 7 + \sqrt{2x+1}$

22. $\sqrt{2x+5} = 0$ 23. $\sqrt{3x-2} = 0$ 24. $0 = \sqrt{3x-6} - 9$

1. _____
2. _____
3. _____
4. _____
5. _____
6. _____
7. _____
8. _____
9. _____
10. _____
11. _____
12. _____
13. _____
14. _____
15. _____
16. _____
17. _____
18. _____
19. _____
20. _____
21. _____
22. _____
23. _____
24. _____

Answers

Try it 3.4C
1. 2

Quiz 3.4C
1. 6
2. 12

Solutions to Practice Sheet 3.4C

1. $\sqrt{y} = 8$

   $\left(\sqrt{y}\right)^2 = 8^2$

   $y = 64$

   Check: $\sqrt{y} = 8$

   $\sqrt{64} \overset{?}{=} 8$

   $8 = 8$

   The solution is 64.

2. $\sqrt{a} = 16$

   $\left(\sqrt{a}\right)^2 = 16^2$

   $a = 256$

   Check: $\sqrt{a} = 16$

   $\sqrt{256} \overset{?}{=} 16$

   $16 = 16$

   The solution is 256.

3. $\sqrt{a} = 10$

   $\left(\sqrt{a}\right)^2 = 10^2$

   $a = 100$

   Check: $\sqrt{a} = 10$

   $\sqrt{100} \overset{?}{=} 10$

   $10 = 10$

   The solution is 100.

4. $\sqrt{3x} = 3$

   $\left(\sqrt{3x}\right)^2 = 3^2$

   $3x = 9$

   $x = 3$

   Check: $\sqrt{3x} = 3$

   $\sqrt{3(3)} \overset{?}{=} 3$

   $\sqrt{9} \overset{?}{=} 3$

   $3 = 3$

   The solution is 3.

5. $\sqrt{2x} = 4$

   $\left(\sqrt{2x}\right)^2 = 4^2$

   $2x = 16$

   $x = 8$

   Check: $\sqrt{2x} = 4$

   $\sqrt{2(8)} \overset{?}{=} 4$

   $\sqrt{16} \overset{?}{=} 4$

   $4 = 4$

   The solution is 8.

6. $\sqrt{3x} = 6$

   $\left(\sqrt{3x}\right)^2 = 6^2$

   $3x = 36$

   $x = 12$

   Check: $\sqrt{3x} = 6$

   $\sqrt{3(12)} \overset{?}{=} 6$

   $\sqrt{36} \overset{?}{=} 6$

   $6 = 6$

   The solution is 12.

7. $5 - \sqrt{5x} = 0$

   $\sqrt{5x} = 5$

   $\left(\sqrt{5x}\right)^2 = 5^2$

   $5x = 25$

   $x = 5$

   Check: $5 - \sqrt{5x} = 0$

   $5 - \sqrt{5(5)} \overset{?}{=} 0$

   $5 - \sqrt{25} \overset{?}{=} 0$

   $5 - 5 \overset{?}{=} 0$

   $0 = 0$

   The solution is 5.

8. $\sqrt{6x} - 12 = 0$

   $\sqrt{6x} = 12$

   $\left(\sqrt{6x}\right)^2 = 12^2$

   $6x = 144$

   $x = 24$

   Check: $\sqrt{6x} - 12 = 0$

   $\sqrt{6(24)} - 12 \overset{?}{=} 0$

   $\sqrt{144} - 12 \overset{?}{=} 0$

   $12 - 12 \overset{?}{=} 0$

   $0 = 0$

   The solution is 24.

9. $3 - \sqrt{2x} = 0$

$\sqrt{2x} = 3$

$\left(\sqrt{2x}\right)^2 = 3^2$

$2x = 9$

$x = \dfrac{9}{2}$ or 4.5

Check: $3 - \sqrt{2x} = 0$

$3 - \sqrt{2(4.5)} \overset{?}{=} 0$

$3 - \sqrt{9} \overset{?}{=} 0$

$3 - 3 \overset{?}{=} 0$

$0 = 0$

The solution is 4.5.

10. $\sqrt{2x - 3} = 5$

$\left(\sqrt{2x - 3}\right)^2 = 5^2$

$2x - 3 = 25$

$2x = 28$

$x = 14$

Check: $\sqrt{2x - 3} = 5$

$\sqrt{2(14) - 3} \overset{?}{=} 5$

$\sqrt{28 - 3} \overset{?}{=} 5$

$\sqrt{25} \overset{?}{=} 5$

$5 = 5$

The solution is 14.

11. $\sqrt{4x + 5} = 1$

$\left(\sqrt{4x + 5}\right)^2 = 1^2$

$4x + 5 = 1$

$4x = -4$

$x = -1$

Check: $\sqrt{4x + 5} = 1$

$\sqrt{4(-1) + 5} \overset{?}{=} 1$

$\sqrt{-4 + 5} \overset{?}{=} 1$

$\sqrt{1} \overset{?}{=} 1$

$1 = 1$

The solution is -1.

12. $\sqrt{2x + 7} = 3$

$\left(\sqrt{2x + 7}\right)^2 = 3^2$

$2x + 7 = 9$

$2x = 2$

$x = 1$

Check: $\sqrt{2x + 7} = 3$

$\sqrt{2(1) + 7} \overset{?}{=} 3$

$\sqrt{2 + 7} \overset{?}{=} 3$

$\sqrt{9} \overset{?}{=} 3$

$3 = 3$

The solution is 1.

13. $\sqrt{2x - 1} = 3$

$\left(\sqrt{2x - 1}\right)^2 = 3^2$

$2x - 1 = 9$

$2x = 10$

$x = 5$

Check: $\sqrt{2x - 1} = 3$

$\sqrt{2(5) - 1} \overset{?}{=} 3$

$\sqrt{10 - 1} \overset{?}{=} 3$

$\sqrt{9} \overset{?}{=} 3$

$3 = 3$

The solution is 5.

14. $\sqrt{2x - 3} = 1$

$\left(\sqrt{2x - 3}\right)^2 = 1^2$

$2x - 3 = 1$

$2x = 4$

$x = 2$

Check: $\sqrt{2x - 3} = 1$

$\sqrt{2(2) - 3} \overset{?}{=} 1$

$\sqrt{4 - 3} \overset{?}{=} 1$

$\sqrt{1} \overset{?}{=} 1$

$1 = 1$

The solution is 2.

15.
$$\sqrt{6x+1} = 5$$
$$\left(\sqrt{6x+1}\right)^2 = 5^2$$
$$6x+1 = 25$$
$$6x = 24$$
$$x = 4$$

Check:
$$\sqrt{6x+1} = 5$$
$$\sqrt{6(4)+1} \overset{?}{=} 5$$
$$\sqrt{24+1} \overset{?}{=} 5$$
$$\sqrt{25} \overset{?}{=} 5$$
$$5 = 5$$

The solution is 4.

16.
$$\sqrt{5x-4} = 6$$
$$\left(\sqrt{5x-4}\right)^2 = 6^2$$
$$5x-4 = 36$$
$$5x = 40$$
$$x = 8$$

Check:
$$\sqrt{5x-4} = 6$$
$$\sqrt{5(8)-4} \overset{?}{=} 6$$
$$\sqrt{40-4} \overset{?}{=} 6$$
$$\sqrt{36} \overset{?}{=} 6$$
$$6 = 6$$

The solution is 8.

17.
$$0 = 3-\sqrt{2+x}$$
$$\sqrt{2+x} = 3$$
$$\left(\sqrt{2+x}\right)^2 = 3^2$$
$$2+x = 9$$
$$x = 7$$

Check:
$$0 = 3-\sqrt{2+x}$$
$$0 \overset{?}{=} 3-\sqrt{2+7}$$
$$0 \overset{?}{=} 3-\sqrt{9}$$
$$0 \overset{?}{=} 3-3$$
$$0 = 0$$

The solution is 7.

18.
$$0 = 5-\sqrt{5+x}$$
$$\sqrt{5+x} = 5$$
$$\left(\sqrt{5+x}\right)^2 = 5^2$$
$$5+x = 25$$
$$x = 20$$

Check:
$$0 = 5-\sqrt{5+x}$$
$$0 \overset{?}{=} 5-\sqrt{5+20}$$
$$0 \overset{?}{=} 5-\sqrt{25}$$
$$0 \overset{?}{=} 5-5$$
$$0 = 0$$

The solution is 20.

19.
$$\sqrt{2x}-1 = 3$$
$$\sqrt{2x} = 4$$
$$\left(\sqrt{2x}\right)^2 = 4^2$$
$$2x = 16$$
$$x = 8$$

Check:
$$\sqrt{2x}-1 = 3$$
$$\sqrt{2(8)}-1 \overset{?}{=} 3$$
$$\sqrt{16}-1 \overset{?}{=} 3$$
$$4-1 \overset{?}{=} 3$$
$$3 = 3$$

The solution is 8.

20.
$$5 = 7-\sqrt{2c}$$
$$\sqrt{2c} = 2$$
$$\left(\sqrt{2c}\right)^2 = 2^2$$
$$2c = 4$$
$$c = 2$$

Check:
$$5 = 7-\sqrt{2c}$$
$$5 \overset{?}{=} 7-\sqrt{2(2)}$$
$$5 \overset{?}{=} 7-\sqrt{4}$$
$$5 \overset{?}{=} 7-2$$
$$5 = 5$$

The solution is 2.

21. $10 = 7 + \sqrt{2x+1}$
$3 = \sqrt{2x+1}$
$3^2 = \left(\sqrt{2x+1}\right)^2$
$9 = 2x+1$
$8 = 2x$
$4 = x$

Check: $10 = 7 + \sqrt{2x+1}$
$10 \overset{?}{=} 7 + \sqrt{2(4)+1}$
$10 \overset{?}{=} 7 + \sqrt{8+1}$
$10 \overset{?}{=} 7 + \sqrt{9}$
$10 \overset{?}{=} 7 + 3$
$10 = 10$

The solution is 4.

22. $\sqrt{2x+5} = 0$
$\left(\sqrt{2x+5}\right)^2 = 0^2$
$2x+5 = 0$
$2x = -5$
$x = -\dfrac{5}{2}, \text{ or } -2.5$

Check: $\sqrt{2x+5} = 0$
$\sqrt{2(-2.5)+5} \overset{?}{=} 0$
$\sqrt{-5+5} \overset{?}{=} 0$
$\sqrt{0} \overset{?}{=} 0$
$0 = 0$

The solution is −2.5.

23. $\sqrt{3x-2} = 0$
$\left(\sqrt{3x-2}\right)^2 = 0^2$
$3x-2 = 0$
$3x = 2$
$x = \dfrac{2}{3}$

Check: $\sqrt{3x-2} = 0$
$\sqrt{3\left(\frac{2}{3}\right)-2} \overset{?}{=} 0$
$\sqrt{2-2} \overset{?}{=} 0$
$\sqrt{0} \overset{?}{=} 0$
$0 = 0$

The solution is $\dfrac{2}{3}$.

24. $0 = \sqrt{3x-6} - 9$
$9 = \sqrt{3x-6}$
$9^2 = \left(\sqrt{3x-6}\right)^2$
$81 = 3x-6$
$87 = 3x$
$29 = x$

Check: $0 = \sqrt{3x-6} - 9$
$0 \overset{?}{=} \sqrt{3(29)-6} - 9$
$0 \overset{?}{=} \sqrt{87-6} - 9$
$0 \overset{?}{=} \sqrt{81} - 9$
$0 \overset{?}{=} 9 - 9$
$0 = 0$

The solution is 29.

Objective 3.5A – Graph an inequality on a number line

Key Words and Concepts

A *set* is a collection of objects, which are called the *elements* of the set.

A set can be represented using **set-builder notation**.

For example, $\{x \mid x > 4,\ x \in \text{real numbers}\}$ is read "the set of all real numbers x that are greater than 4."

For the remainder of this section, all variables will represent real numbers.

Given this convention, $\{x \mid x > 4,\ x \in \text{real numbers}\}$ can be written $\{x \mid x > 4\}$.

Some sets of real numbers that are written in set-builder notation can also be written in **interval notation**.

For example, the interval notation $[-3, 2)$ represents the set of real numbers between -3 and 2.

Note   The *bracket* means that -3 is included in the set.

             The *parenthesis* means that 2 is *not* included in the set.

Interval notation	Set-builder notation	Read
$[-3, 2)$	$\{x \mid -3 \leq x < 2\}$	"The set of al real numbers x between -3 and 2, including -3 but excluding 2."

To indicate an interval that extends forever in the positive direction, we use the **infinity symbol** ∞.

To indicate an interval that extends forever in the negative direction, we use the **negative infinity symbol** $-\infty$.

Note   When writing a set in interval notation, we always
             use a parenthesis to the right of ∞ and to the left of $-\infty$.

Example

Write $[6, \infty)$ in set-builder notation.

Solution

    The interval $[6, \infty)$ is the set of real numbers greater than 6.

    Because of the bracket, 6 is included in the set.

    This set extends forever in the positive direction.

    In set-builder notation, this set is written $\{x \mid x \geq 6\}$.

Try it

1. Write $(-\infty, -2]$ in set-builder notation.

Example

Write $\{x \mid -1 < x \leq 4\}$ in interval notation.

Solution

    This is the set of real numbers between -1 and 4, excluding 4.

    -1 is not included in the set, so use a parenthesis.

    4 is included in the set, so use a bracket.

    In interval notation, this set is written $(-1, 4]$.

Try it

2. Write $\{x \mid -5 \leq x < 2\}$ in interval notation.

We can graph sets of real numbers given in set-builder notation or in interval notation.

Example

Graph (–4, 4).

    Solution

        The graph is the set of all real numbers between –4 and 4.

        Use parentheses at –4 and 4 and shade between.

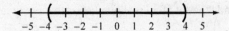

Try it

3. Graph [–2,–1].

Example

Graph $\{x \mid x \geq -2\}$.

    Solution

        The graph is the set of all real numbers greater than
        and including –2.
        Use a bracket at –2 and shade to the right.

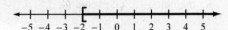

Try it

4. Graph $\{x \mid x < 3\}$.

Reflect on it

- Sets can be represented in set-builder notation and in interval notation.
- Use a parenthesis when an endpoint is included.
- Use a bracket when an endpoint is not included.
- When writing a set in interval notation, always use a parenthesis to the right of ∞ and to the left of $-\infty$.

Quiz Yourself 3.5A

1. Use set-builder notation to write the set of integers greater than –5.

2. Use set-builder notation to write the set of real numbers less than or equal to 5.

3. Write $\{x \mid x < 2\}$ in interval notation.

4. Write $[-3, 5)$ in set-builder notation.

5. Graph: $[3, \infty)$

6. Graph: $\{x \mid x < -1\}$

Practice Sheet 3.5A

Use set builder notation to write the set.

1. Negative integers greater than –6

2. Positive integers less than 8

3. Even integers greater than 1

4. The real numbers greater than –2

5. The real numbers less than –21

6. Odd integers greater than –4

1. _____

2. _____

3. _____

4. _____

5. _____

6. _____

Write the set in interval notation.

7. $\{x|x < -2\}$

8. $\{x|x \geq 0\}$

9. $\{x|-6 \leq x < -4\}$

10. $\{x|1 < x \leq 2\}$

11. $\{x|-5 < x \leq 7\}$

12. $\{x|x > -4\}$

7. _____

8. _____

9. _____

10. _____

11. _____

12. _____

Write the interval in set-builder notation.

13. $(-3,-1]$

14. $[6, \infty)$

15. $(-2, 0)$

16. $[4, 6]$

13. _____

14. _____

15. _____

16. _____

Answers

Try it 3.5A

1. $\{x \mid x \le -2\}$
2. $[-5, 2)$
3. $[-2, -1]$

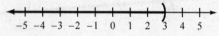

4. $\{x \mid x < 3\}$

Quiz 3.5A

1. $\{x \mid x > -5,\ x \in \text{integers}\}$
2. $\{x \mid x \le 5\}$
3. $(-\infty, 2)$
4. $\{x \mid -3 \le x < 5\}$
5. $[3, \infty)$

6. $\{x \mid x < -1\}$

Solutions to Practice Sheet 3.5A

1. $\{x \mid -6 < x < 0,\ x \in \text{integers}\}$

2. $\{x \mid 0 < x < 8,\ x \in \text{integers}\}$

3. $\{x \mid x > 1,\ x \in \text{even integers}\}$

4. $\{x \mid x > -2,\ x \in \text{real numbers}\}$

5. $\{x \mid x < -21,\ x \in \text{real numbers}\}$

6. $\{x \mid x > -4,\ x \in \text{odd integers}\}$

7. $\{x \mid x < -2\} = (-\infty, -2)$

8. $\{x \mid x \ge 0\} = [0, \infty)$

9. $\{x \mid -6 \le x < -4\} = [-6, -4)$

10. $\{x \mid 1 < x \le 2\} = (1, 2]$

11. $\{x \mid -5 < x \le 7\} = (-5, 7]$

12. $\{x \mid x > -4\} = (-4, \infty)$

13. $(-3, -1] = \{x \mid -3 < x \le -1\}$

14. $[6, \infty) = \{x \mid x \ge 6\}$

15. $(-2, 0) = \{x \mid -2 < x < 0\}$

16. $[4, 6] = \{x \mid 4 \le x \le 6\}$

Objective 3.5B – Solve a first-degree inequality

Key Words and Concepts

The **solution set of an inequality** is a set of numbers each element of which, when substituted for the variable, results in a true inequality.

When solving an inequality, we use the Addition and Multiplication Properties of Inequalities to rewrite the inequality in the form *variable < constant* or in the form *variable > constant*.

Addition Property of Inequalities
> If $a < b$, then $a + c < b + c$.
> If $a > b$, then $a + c > b + c$.

The Addition Property of Inequalities is used to remove a term from one side of an inequality by adding the additive inverse of that term to each side of the inequality.

Multiplication Property of Inequalities
> Rule 1
>> If $a < b$ and $c > 0$, then $ac < bc$.
>> If $a > b$ and $c > 0$, then $ac > bc$.
> Rule 2
>> If $a < b$ and $c < 0$, then $ac > bc$.
>> If $a > b$ and $c < 0$, then $ac < bc$.

Write Rule 1 and Rule 2 in your own words. _____

_____

_____

Example

Solve and graph the solution set of $2x - 4 < 2$.
Write the solution set in set-builder notation.

    Solution

$2x - 4 < 2$	Copy the inequality
$2x - 4 + 4 < 2 + 4$	Add 4 to each side
$2x < 6$	Simplify
$\dfrac{2x}{2} < \dfrac{6}{2}$	Divide each side by 2, because 2 is positive, do not change inequality
$x < 3$	

The solution set is $\{x \mid x < 3\}$.

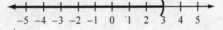

Try it

1. Solve and graph the solution set of $6x + 11 < 5$. Write the solution set in set-builder notation.

Example

Solve and graph the solution set of $2 - \frac{1}{4}x \le 3$.

Write the solution set in interval notation.

Solution

$$2 - \frac{1}{4}x \le 3 \qquad \text{Copy the inequality}$$

$$2 - 2 - \frac{1}{4}x \le 3 - 2 \qquad \text{Subtract 2 from each side}$$

$$-\frac{1}{4}x \le 1 \qquad \text{Simplify}$$

$$-4\left(-\frac{1}{4}x\right) \ge -4(1) \qquad \begin{array}{l}\text{Multiply each side by } -4 \\ \text{and reverse the inequality}\end{array}$$

$$x \ge -4$$

The solution set is $[-4, \infty)$.

Try it

2. Solve and graph the solution set of
   $5 - \frac{1}{2}x \le 7$. Write the solution set
   in interval notation.

Reflect on it

• The Addition Property of Inequalities is used to remove a term from one side of an inequality by adding the additive inverse of that term to each side of the inequality.
• For the Multiplication Property of Inequalities, when each side of an inequality is multiplied by
    the same *positive* constant, the inequality symbol is *unchanged*.
    the same *negative* constant, the inequality symbol is *reversed*.
• The Multiplication Property of Inequalities is used to remove a coefficient from one side of an inequality by multiplying each side of the inequality by the reciprocal of the coefficient.

Quiz Yourself 3.5B

1. Solve and graph the solution set of $-3x \le 6$. Write the solution in set-builder notation.

2. Solve $2x + 3 < 9$. Write the solution in set-builder notation.

3. Solve $4 - 2x \ge x + 7$. Write the solution set in interval notation.

4. Solve $\frac{2}{3}x - \frac{1}{4} > x + \frac{1}{5}$. Write the solution set in interval notation.

Practice Sheet 3.5B

Solve.

1. $x - 4 < 2$

2. $x + 5 \geq 2$

    1. _____

    2. _____

3. $2x \leq 10$

4. $3x > 15$

    3. _____

    4. _____

5. $-4x > 12$

6. $-5x \leq -20$

    5. _____

    6. _____

7. $7x - 1 > 2x + 9$

8. $6x + 1 \geq 4x - 3$

    7. _____

    8. _____

9. $3x - 2 > 7$

10. $2x + 3 < 15$

    9. _____

    10. _____

11. $4x - 3 \leq 9$

12. $5x + 1 \leq -9$

    11. _____

    12. _____

13. $7x + 2 > 4x - 7$

14. $6x + 5 < 4x - 5$

    13. _____

    14. _____

15. $5x + 7 > 4x - 2$

16. $9x + 6 < 2x - 8$

    15. _____

    16. _____

17. $7x + 3 \geq 3x + 19$

18. $6x - 1 < 2x + 7$

    17. _____

    18. _____

19. $7 - 6x \geq 31$

20. $2 - 7x \leq 23$

    19. _____

    20. _____

21. $-4 - 3x > -7$

22. $-3 - x < 5$

    21. _____

    22. _____

23. $7x - 3 < 5x - 11$

24. $5x + 4 \geq x - 16$

    23. _____

    24. _____

Answers

Try it 3.5B

1. $\{x \mid x < -1\}$

2. $[-4, \infty)$

Quiz 3.5B

1. $\{x \mid x \geq -2\}$

2. $\{x \mid x < 3\}$

3. $(-\infty, -1]$

4. $\left(-\infty, -\dfrac{27}{20}\right)$

Solutions to Practice Sheet 3.5B

1. $x - 4 < 2$
   $x < 6$
   $\{x \mid x < 6\}$

2. $x + 5 \geq 2$
   $x \geq -3$
   $\{x \mid x \geq -3\}$

3. $2x \leq 10$
   $\dfrac{2x}{2} \leq \dfrac{10}{2}$
   $x \leq 5$
   $\{x \mid x \leq 5\}$

4. $3x > 15$
   $\dfrac{3x}{3} > \dfrac{15}{3}$
   $x > 5$
   $\{x \mid x > 5\}$

5. $-4x > 12$
   $\dfrac{-4x}{-4} < \dfrac{12}{-4}$
   $x < -3$
   $\{x \mid x < -3\}$

6. $-5x \leq -20$
   $\dfrac{-5x}{-5} \geq \dfrac{-20}{-5}$
   $x \geq 4$
   $\{x \mid x \geq 4\}$

7. $7x - 1 > 2x + 9$
   $5x - 1 > 9$
   $5x > 10$
   $\dfrac{5x}{5} > \dfrac{10}{5}$
   $x > 2$
   $\{x \mid x > 2\}$

8. $6x + 1 \geq 4x - 3$
   $2x + 1 \geq -3$
   $2x \geq -4$
   $\dfrac{2x}{2} \geq \dfrac{-4}{2}$
   $x \geq -2$
   $\{x \mid x \geq -2\}$

9. $3x - 2 > 7$
   $3x > 9$
   $\dfrac{3x}{3} > \dfrac{9}{3}$
   $x > 3$
   $\{x \mid x > 3\}$

10. $2x + 3 < 15$
    $2x < 12$
    $\dfrac{2x}{2} < \dfrac{12}{2}$
    $x < 6$
    $\{x \mid x < 6\}$

11. $4x - 3 \leq 9$
    $4x \leq 12$
    $\dfrac{4x}{4} \leq \dfrac{12}{4}$
    $x \leq 3$
    $\{x \mid x \leq 3\}$

12. $5x + 1 \leq -9$
    $5x \leq -10$
    $\dfrac{5x}{5} \leq \dfrac{-10}{5}$
    $x \leq -2$
    $\{x \mid x \leq -2\}$

13. $7x + 2 > 4x - 7$
    $3x + 2 > -7$
    $3x > -9$
    $\dfrac{3x}{3} > \dfrac{-9}{3}$
    $x > -3$
    $\{x \mid x > -3\}$

14. $6x+5 < 4x-5$
    $2x+5 < -5$
    $\quad 2x < -10$
    $\quad \dfrac{2x}{2} < \dfrac{-10}{2}$
    $\quad x < -5$

    $\{x \mid x < -5\}$

15. $5x+7 > 4x-2$
    $\quad x+7 > -2$
    $\quad\quad x > -9$

    $\{x \mid x > -9\}$

16. $9x+6 < 2x-8$
    $7x+6 < -8$
    $\quad 7x < -14$
    $\quad \dfrac{7x}{7} < \dfrac{-14}{7}$
    $\quad x < -2$

    $\{x \mid x < -2\}$

17. $7x+3 \ge 3x+19$
    $4x+3 \ge 19$
    $\quad 4x \ge 16$
    $\quad \dfrac{4x}{4} \ge \dfrac{16}{4}$
    $\quad x \ge 4$

    $\{x \mid x \ge 4\}$

18. $6x-1 < 2x+7$
    $4x-1 < 7$
    $\quad 4x < 8$
    $\quad \dfrac{4x}{4} < \dfrac{8}{4}$
    $\quad x < 2$

    $\{x \mid x < 2\}$

19. $7-6x \ge 31$
    $\quad -6x \ge 24$
    $\quad \dfrac{-6x}{-6} \le \dfrac{24}{-6}$
    $\quad x \le -4$

    $\{x \mid x \le -4\}$

20. $2-7x \le 23$
    $\quad -7x \le 21$
    $\quad \dfrac{-7x}{-7} \ge \dfrac{21}{-7}$
    $\quad x \ge -3$

    $\{x \mid x \ge -3\}$

21. $-4-3x > -7$
    $\quad -3x > -3$
    $\quad \dfrac{-3x}{-3} < \dfrac{-3}{-3}$
    $\quad x < 1$

    $\{x \mid x < 1\}$

22. $-3-x < 5$
    $\quad -x < 8$
    $\quad \dfrac{-1x}{-1} > \dfrac{8}{-1}$
    $\quad x > -8$

    $\{x \mid x > -8\}$

23. $7x-3 < 5x-11$
    $2x-3 < -11$
    $\quad 2x < -8$
    $\quad \dfrac{2x}{2} < \dfrac{-8}{2}$
    $\quad x < -4$

    $\{x \mid x < -4\}$

24. $5x+4 \ge x-16$
    $4x+4 \ge -16$
    $\quad 4x \ge -20$
    $\quad \dfrac{4x}{4} \ge \dfrac{-20}{4}$
    $\quad x \ge -5$

    $\{x \mid x \ge -5\}$

Objective 3.5C – Solve an absolute value inequality

Key Words and Concepts

Define an absolute value inequality. _____

_____

Solutions of Absolute Value Inequalities

1. To solve an absolute value inequality of the form $|ax+b| < c,\ c > 0,$

    solve the equivalent compound inequality $-c < ax + b < c.$

2. To solve an absolute value inequality of the form $|ax+b| > c,\ c > 0,$

    solve the equivalent compound inequality $ax + b < -c$ or $ax + b > c.$

Note   All solutions in this objective are written in set-builder notation.

Example

Solve: $|5 - 4x| > 9$

Solution

   To solve the inequality $|5 - 4x| > 9,$

   first rewrite the inequality as
          $5 - 4x < -9$ or $5 - 4x > 9.$

   Solve each inequality.

   $$5 - 4x < -9 \qquad \text{or} \qquad 5 - 4x > 9$$
   $$-4x < -14 \qquad\qquad -4x > 4$$
   $$\frac{-4x}{-4} > \frac{-14}{-4} \qquad\qquad \frac{-4x}{-4} < \frac{4}{-4}$$
   $$x > \frac{7}{2} \qquad\qquad x < -1$$
   $$\left\{ x \mid x > \frac{7}{2} \right\} \qquad\qquad \{ x \mid x < -1 \}$$

   The solution set of a compound inequality with the word
   OR is the union of the solution sets of the two inequalities.

   The solution set is $\left\{ x \mid x < -1 \right\} \cup \left\{ x \mid x > \frac{7}{2} \right\}.$

Try it

1. Solve: $|2 - 3x| > 4$

Example

Solve: $|7+2x| \le 3$

Solution

To solve the inequality $|7+2x| \le 3$,

first rewrite the inequality as $-3 \le 7+2x \le 3$.

$$-3 \le 7+2x \le 3$$
$$-3-7 \le 7-7+2x \le 3-7 \quad \text{Subtract 7 from each part}$$
$$-10 \le 2x \le -4 \quad \text{Simplify}$$
$$\frac{-10}{2} \le \frac{2x}{2} \le \frac{-4}{2} \quad \text{Divide each part by 3}$$
$$-5 \le x \le -2 \quad \text{Simplify}$$

The solution set is $\{x \mid -5 \le x \le -2\}$.

Try it

2. Solve: $|8+3x| \le 5$

Reflect on it

- An absolute value inequality is an inequality that contains a variable within an absolute value symbol.

Quiz Yourself 3.5C

1. Solve: $|4x-1| \ge 12$

2. Solve: $|3-8x| > 5$

3. Solve: $|4-2x| \le 8$

4. Solve: $|6x-5| < 13$

5. Solve: $|4x-9| < -2$

Practice Sheet 3.5C

Solve.

1. $|x| > 2$

2. $|x| < 4$

3. $|3 - x| \geq 1$

4. $|2 - x| \geq 5$

5. $|2x - 1| < 7$

6. $|x + 7| \geq 3$

7. $|x - 3| < 2$

8. $|x - 4| \leq 7$

9. $|3x + 5| > 11$

10. $|5x - 1| > 14$

11. $|2x - 3| \geq -3$

12. $|4 - 3x| \leq 1$

13. $|3 - 4x| > 11$

14. $|3x - 2| < 10$

15. $|18 - 2x| \leq 0$

16. $|5x - 3| < 12$

17. $|3 - 2x| > 7$

18. $|4x - 1| < 3$

1. _____

2. _____

3. _____

4. _____

5. _____

6. _____

7. _____

8. _____

9. _____

10. _____

11. _____

12. _____

13. _____

14. _____

15. _____

16. _____

17. _____

18. _____

Answers

Try it 3.5C

1. $\left\{x\mid x<\dfrac{2}{3}\right\}\cup\{x\mid x>2\}$

2. $\left\{x\mid -\dfrac{13}{3}\le x\le -1\right\}$

Quiz 3.5C

1. $\left\{x\mid x\le -\dfrac{11}{4}\right\}\cup\left\{x\mid x\ge \dfrac{13}{4}\right\}$

2. $\left\{x\mid x<-\dfrac{1}{4}\right\}\cup\{x\mid x>1\}$

3. $\{x\mid -2\le x\le 6\}$

4. $\left\{x\mid -\dfrac{4}{3}<x<3\right\}$

5. No solution

Solutions to Practice Sheet 3.5C

1. $|x|>2$

$x>2 \quad \text{or} \quad x<-2$
$\{x\mid x>2\} \quad \{x\mid x<-2\}$
$\{x\mid x>2\}\cup\{x\mid x<-2\}$

2. $|x|<4$
$-4<x<4$
$\{x\mid -4<x<4\}$

3. $|3-x|\ge 1$

$3-x\ge 1 \quad \text{or} \quad 3-x\le -1$
$-x\ge -2 \qquad\qquad -x\le -4$
$x\le 2 \qquad\qquad\quad x\ge 4$
$\{x\mid x\le 2\} \qquad \{x\mid x\ge 4\}$
$\{x\mid x\le 2\}\cup\{x\mid x\ge 4\}$

4. $|2-x|\ge 5$

$2-x\ge 5 \quad \text{or} \quad 2-x\le -5$
$-x\ge 3 \qquad\qquad -x\le -7$
$x\le -3 \qquad\qquad\quad x\ge 7$
$\{x\mid x\le -3\} \qquad \{x\mid x\ge 7\}$
$\{x\mid x\le -3\}\cup\{x\mid x\ge 7\}$

5. $|2x-1|<7$
$-7<2x-1<7$
$-6<2x<8$
$-3<x<4$
$\{x\mid -3<x<4\}$

6. $|x+7|\ge 3$

$x+7\ge 3 \quad \text{or} \quad x+7\le -3$
$x\ge -4 \qquad\qquad x\le -10$
$\{x\mid x\ge -4\} \qquad \{x\mid x\le -10\}$
$\{x\mid x\ge -4\}\cup\{x\mid x\le -10\}$

7. $|x-3|<2$
$-2<x-3<2$
$1<x<5$
$\{x\mid 1<x<5\}$

8. $|x-4|\le 7$
$-7\le x-4\le 7$
$-3\le x\le 11$
$\{x\mid -3\le x\le 11\}$

9. $|3x+5|>11$

$3x+5>11 \quad \text{or} \quad 3x+5<-11$
$3x>6 \qquad\qquad\quad 3x<-16$
$x>2 \qquad\qquad\quad x<-\dfrac{16}{3}$
$\{x\mid x>2\} \qquad \left\{x\mid x<-\dfrac{16}{3}\right\}$

$\{x\mid x>2\}\cup\left\{x\mid x<-\dfrac{16}{3}\right\}$

10. $|5x-1|>14$

$5x-1>14 \quad \text{or} \quad 5x-1<-14$
$5x>15 \qquad\qquad\quad 5x<-13$
$x>3 \qquad\qquad\quad x<-\dfrac{13}{5}$
$\{x\mid x>3\} \qquad \left\{x\mid x<-\dfrac{13}{5}\right\}$

$\{x\mid x>3\}\cup\left\{x\mid x<-\dfrac{13}{5}\right\}$

11. $|2x-3|\ge -3$

$2x-3\ge -3 \quad \text{or} \quad 2x-3\le 3$
$2x\ge 0 \qquad\qquad\quad 2x\le 6$
$x\ge 0 \qquad\qquad\quad x\le 3$
$\{x\mid x\ge 0\} \qquad \{x\mid x\le 3\}$
$\{x\mid x\ge 0\}\cup\{x\mid x\ge 3\}$

The solution set is the set of all real numbers.

12. $|4-3x| \leq 1$

$-1 \leq 4-3x \leq 1$

$-5 \leq -3x \leq -3$

$\dfrac{5}{3} \geq x \geq 1$ or $1 \leq x \leq \dfrac{5}{3}$

$\left\{ x \middle| 1 \leq x \leq \dfrac{5}{3} \right\}$

13. $|3-4x| > 11$

$3-4x > 11$ or $3-4x < -11$

$-4x > 8$ $-4x < -14$

$x < -2$ $x > \dfrac{7}{2}$

$\{x | x < -2\}$ $\left\{ x \middle| x > \dfrac{7}{2} \right\}$

$\{x | x < -2\} \cup \left\{ x \middle| x > \dfrac{7}{2} \right\}$

14. $|3x-2| < 10$

$-10 < 3x-2 < 10$

$-8 < 3x < 12$

$-\dfrac{8}{3} < x < 4$

$\left\{ x \middle| -\dfrac{8}{3} < x < 4 \right\}$

15. $|18-2x| \leq 0$

$18-2x = 0$

$-2x = -18$

$x = 9$

$\{x | x = 9\}$

16. $|5x-3| < 12$

$-12 < 5x-3 < 12$

$-9 < 5x < 15$

$-\dfrac{9}{5} < x < 3$

$\left\{ x \middle| -\dfrac{9}{5} < x < 3 \right\}$

17. $|3-2x| > 7$

$3-2x > 7$ or $3-2x < -7$

$-2x > 4$ $-2x < -10$

$x < -2$ $x > 5$

$\{x | x < -2\}$ $\{x | x > 5\}$

$\{x | x < -2\} \cup \{x | x > 5\}$

18. $|4x-1| < 3$

$-3 < 4x-1 < 3$

$-2 < 4x < 4$

$-\dfrac{1}{2} < x < 1$

$\left\{ x \middle| -\dfrac{1}{2} < x < 1 \right\}$

Application and Activities

Discussion and reflection questions

1. If you slam on your brakes in the car, do you know how far you will skid? Do you know what affects the length of the skid?

2. Do you know what the legal blood alcohol content limit is?

3. Do you think it is important for car parts to have a set limit on the weight of certain parts?

Group Activity (2-3 people)

Objectives for the lesson:

In this lesson you will use and enhance your understanding of the following:

- Evaluating a variable expression
- Solving a radical equation
- Solving a literal equation
- Solving applications of the basic percent equation
- Solving a linear equation
- Solving a percent equation
- Solving a proportion
- Solving an absolute value inequality

1. Ralph was involved in an automobile accident. After the accident the specialists measured the skid marks to have an average length of 145 feet. The road was an asphalt road, which has a drag factor of about 0.75. The formula, $S = \sqrt{30df}$ will enable the specialists to find the approximate speed Ralph's vehicle was moving before the collision. In the formula, S is the minimum speed, in miles per hour, d is the skid distance in feet, and f is the drag factor (which depends on the surface of the road).

   Ralph told the police officer that he was traveling under the speed limit of 50 miles per hour. Use the formula to find his estimated speed, based on the skid marks.   Was Ralph speeding?

2. If a car was traveling at 50mph, what would the length of the skid mark be?

3. Solve the formula $S = \sqrt{30df}$ for d, and use the new formula to find the length of a skid mark of a car traveling 45 miles per hour on the same road.

4. The police at the scene of the accident found Ralph's blood alcohol content (BAC) level to be 0.092. If the legal limit is 0.08, what percent of the legal limit is this? (Hint: If his level is above the legal limit then the percent should be above 100%.)

5. Using data from the NHSTA (National Highway Traffic Safety Administration), the following model was created to represent the number of alcohol related fatalities in the nation: $y = -0.34x + 14.2$, where x is the number of years after 2000 and y is the number of alcohol related fatalities in the nation in thousands.

    According to this model, in which year will alcohol related fatalities reach 7,400?

6. According to the NHSTA (National Highway Traffic Safety Administration), in 2013 the total number of drivers involved in fatal crashes was 44,574. The number of those drivers with a blood-alcohol content level of 0.8 or higher was 9,461. If there were 65,000 drivers involved in fatal crashes, how many would you expect to have a blood alcohol level of 0.8 or higher?

7. Ralph believes the car company is at fault for the accident. He is blaming the weight of a particular part in the car. A regulation part must weigh 21.5 ounces, with a tolerance of less than 0.05 ounces.
    If w = the weight of the part, then weights that fall within $|w - 21.5| > 0.05$ should be rejected. Find the range of weights that should be rejected by the car company for the part.

8. When the part was taken from the vehicle, it weighted 21.455 ounces. Should it have been rejected?

Objective 4.1A – Define and describe lines and angles

Key Words and Concepts

A **plane** is a flat surface. Figures that lie totally in a plane are called **plane figures**.

Space extends in all directions. Objects in space are called **solids**.

A **line** extends indefinitely in two directions in a plane.

What is the width of a line? _____

A **line segment** is part of a line and has _____ endpoints.
                                            (one/two/three)

The line segment between two points A and B would be denoted $\overline{AB}$. The length of a line segment is the distance between the endpoints of the line segment. The length of line segment $\overline{AB}$ is denoted AB.

Example	**Try it**
Given $AC = 52$ and $AB = 14$. Find BC.	1. Given $AC = 63$ and $AB = 18$. Find BC.

A — B ——— C

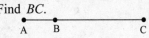

Solution
Since $AB + BC = AC$, substitute the known values in the equation and solve for BC.
$$AB + BC = AC$$
$$14 + BC = 52$$
$$BC = 38$$

Lines in a plane can be parallel or intersecting.

_____ **lines** never meet.   _____ **lines** cross at a point in the plane.
(Parallel/Intersecting)   (Parallel/Intersecting)

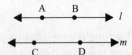

In the figure above, $l \square m$, and $\overline{AB} \square \overline{BC}$.

> *Note*   The symbol $\square$ means "is parallel to."

A **ray** starts at one point and extends indefinitely in one direction. An **angle** is formed when two rays starts from the same point. The common endpoint is called the **vertex** of the angle.

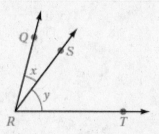

In the diagram above, $\angle x = \angle QRS = \angle SRQ$. When an angle is named using the names of three points, the name of the point representing the vertex must be in the middle.

A unit in which angles are measured is the **degree**. One complete revolution is 360 degrees ($360°$).
One quarter of a revolution is $90°$. A $90°$ angle is called a **right angle**. A right angle, with right angle is denoted by the ⌐ symbol.

Perpendicular lines are intersecting lines that form right angles.

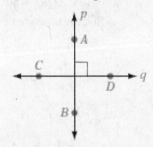

In the diagram above, p is perpendicular to q ($p \perp q$). Perpendicular lines contain perpendicular segments. Thus in the above diagram $\overline{AB} \perp \overline{CD}$.

What is the difference between a complementary angle, a straight angle, and a supplementary angle?

_____

_____

_____

Example
Find the complement of a 37° angle.

    Solution
        Let x represent the complement of 37°.
$$x + 37° = 90°$$
$$x + 37° - 37° = 90° - 37°$$
$$x = 53°$$
      53° is the complement of 37°.

Try it
2. Find the complement of a 67° angle.

Example

Find the supplement of a 37° angle.

Solution

Let x represent the supplement of 37°.
$$x + 37° = 180°$$
$$x + 37° - 37° = 180° - 37°$$
$$x = 143°$$
143° is the supplement of 37°.

Try it

3. Find the supplement of a 67° angle.

An _____ **angle** is an angle whose measure is between 0° and 90°.
   (acute/obtuse)

An _____ **angle** is an angle whose measure is between 90° and 180°.
   (acute/obtuse)

Example

Find the measure of $\angle a$ in the diagram.

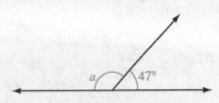

Solution

Because the 47° angle and $\angle a$ together form a
straight angle, $\angle a + 47° = 180°$.
$$\angle a + 47° = 180°$$
$$\angle a + 47° - 47° = 180° - 47°$$
$$\angle a = 133°$$

Try it

4. Given $\angle QRT = 80°$ and $\angle x = 15°$,
find the measure of $\angle y$ in the
diagram.

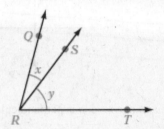

Reflect on it

- A line segment is part of a line and has two endpoints
- An angle is formed when two rays starts from the same point, called the vertex of the angle.
- Complementary angles are two angles whose sum is 90°.
- A 180° angle is called a straight angle.
- Supplementary angles are two angles whose sum is 180°.
- An acute angle is an angle whose measure is between 0° and 90°.
- An obtuse angle is an angle whose measure is between 90° and 180°.

Quiz Yourself 4.1A

1. In the figure, $EF = 13$ and $EG = 17$. Find FG.

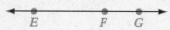

2. In the figure, $AB = 5$, $BC = 7$, and $AD = 16$. Find AC and CD.

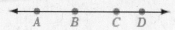

3. Find the measure of $\angle AOB$.

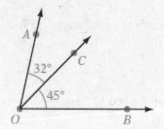

4. Find the measure of $\angle a$.

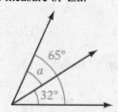

5. Find the complement of a 19° angle.

6. Find the supplement of a 24° angle.

Practice Sheet 4.1A

Solve.

1. How many degrees are in one-half of a revolution?

2. Find the supplement of a 116° angle.

1. _____

2. _____

3. Find the complement of a 75° angle.

4. Find the complement of a 22° angle.

3. _____

4. _____

5. Find the supplement of a 128° angle.

6. Find the complement of a 49° angle.

5. _____

6. _____

7. Find the complement of a 66° angle.

8. Find the supplement of a 138° angle.

7. _____

8. _____

9. Find the supplement of a 42° angle.

10. Find the complement of a 12 ° angle.

9. _____

10. _____

11. Find the complement of a 81° angle.

12. Find the supplement of a 171° angle.

11. _____

12. _____

13. Find the supplement of a 157° angle.

14. Find the complement of an 8° angle.

13. _____

14. _____

15. Find the supplement of a 22° angle.

16. Find the supplement of a 163° angle.

15. _____

16. _____

17. Find the supplement of a 174° angle.

18. Find the complement of a 78° angle.

17. _____

18. _____

19. Find the complement of a 39° angle.

20. Find the supplement of a 93° angle.

19. _____

20. _____

Answers

Try it 4.1A
1. $BC = 45$
2. $23°$
3. $113°$
4. $65°$

Quiz 4.1A
1. 4
2. $AC = 12$; $CD = 4$
3. $77°$
4. $33°$
5. $71°$
6. $156°$

Solutions to Practice Sheet 4.1A

1. There are $180°$ in one-half of a revolution.
2. Let x represent the supplement of $116°$.
$$x + 116° = 180°$$
$$x + 116° - 116° = 180° - 116°$$
$$x = 64°$$
The supplement of $116°$ is $64°$.
3. Let x represent the complement of $75°$.
$$x + 75° = 90°$$
$$x + 75° - 75° = 90° - 75°$$
$$x = 15°$$
The complement of $75°$ is $15°$.
4. Let x represent the complement of $22°$.
$$x + 22° = 90°$$
$$x + 22° - 22° = 90° - 22°$$
$$x = 68°$$
The complement of $22°$ is $68°$.
5. Let x represent the supplement of $128°$.
$$x + 128° = 180°$$
$$x + 128° - 128° = 180° - 128°$$
$$x = 52°$$
The supplement of $128°$ is $52°$.
6. Let x represent the complement of $49°$.
$$x + 49° = 90°$$
$$x + 49° - 49° = 90° - 49°$$
$$x = 41°$$
The complement of $49°$ is $41°$.
7. Let x represent the complement of $66°$.
$$x + 66° = 90°$$
$$x + 66° - 66° = 90° - 66°$$
$$x = 24°$$
The complement of $66°$ is $24°$.

8. Let x represent the supplement of $138°$.
$$x + 138° = 180°$$
$$x + 138° - 138° = 180° - 138°$$
$$x = 42°$$
The supplement of $138°$ is $42°$.
9. Let x represent the supplement of $42°$.
$$x + 42° = 180°$$
$$x + 42° - 42° = 180° - 42°$$
$$x = 138°$$
The supplement of $42°$ is $138°$.
10. Let x represent the complement of $12°$.
$$x + 12° = 90°$$
$$x + 12° - 12° = 90° - 12°$$
$$x = 78°$$
The complement of $12°$ is $78°$.
11. Let x represent the complement of $81°$.
$$x + 81° = 90°$$
$$x + 81° - 81° = 90° - 81°$$
$$x = 9°$$
The complement of $81°$ is $9°$.
12. Let x represent the supplement of $171°$.
$$x + 171° = 180°$$
$$x + 171° - 171° = 180° - 171°$$
$$x = 9°$$
The supplement of $171°$ is $9°$.
13. Let x represent the supplement of $157°$.
$$x + 157° = 180°$$
$$x + 157° - 157° = 180° - 157°$$
$$x = 23°$$
The supplement of $157°$ is $23°$.
14. Let x represent the complement of $8°$.
$$x + 8° = 90°$$
$$x + 8° - 8° = 90° - 8°$$
$$x = 82°$$
The complement of $8°$ is $82°$.
15. Let x represent the supplement of $22°$.
$$x + 22° = 180°$$
$$x + 22° - 22° = 180° - 22°$$
$$x = 158°$$
The supplement of $22°$ is $158°$.
16. Let x represent the supplement of $163°$.
$$x + 163° = 180°$$
$$x + 163° - 163° = 180° - 163°$$
$$x = 17°$$
The supplement of $163°$ is $17°$.
17. Let x represent the supplement of $174°$.
$$x + 174° = 180°$$
$$x + 174° - 174° = 180° - 174°$$
$$x = 6°$$
The supplement of $174°$ is $6°$.

18. Let x represent the complement of 78°.
$$x + 78° = 90°$$
$$x + 78° - 78° = 90° - 78°$$
$$x = 12°$$
The complement of 78° is 12°.

19. Let x represent the complement of 39°.
$$x + 39° = 90°$$
$$x + 39° - 39° = 90° - 39°$$
$$x = 51°$$
The complement of 39° is 51°.

20. Let x represent the supplement of 93°.
$$x + 93° = 180°$$
$$x + 93° - 93° = 180° - 93°$$
$$x = 87°$$
The supplement of 93° is 87°.

Objective 4.1B – Define and describe geometric figures

Key Words and Concepts

A triangle is a closed, three-sided plane figure. Figure ABC is a triangle. $\overline{AB}$ is called the **base**. $\overline{CD}$, perpendicular to the base, is called the **height**.

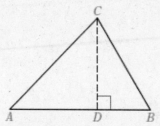

What is the sum of the three angles in a triangle? _____

Example

Two angles of a triangle measure 17° and 97°. Find the measure of the third angle.

Solution

If the measure of the third angle is $x°$,

then $17 + 97 + x = 180$.

$$17 + 97 + x = 180$$
$$114 + x = 180$$
$$114 - 114 + x = 180 - 114$$
$$x = 66$$

The missing angle measures 66°.

Try it

1. Two angles of a triangle measure 23° and 114°. Find the measure of the third angle.

A **right triangle** contains one right angle. The side opposite the right angle is called the **hypotenuse**. The **legs of a right triangle** are its other two sides. In a right triangle, the two acute angles are complementary. In the right triangle pictured below, $\angle A + \angle B = 90°$.

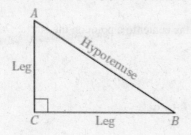

Example

One of the acute angles of a right triangle measures 30º. What is the measure of the other acute angle?

> Solution
>
> If the measure of the other acute angle is $x°$,
>
> then $x+30+90=180$.
>
> $$x+30+90=180$$
> $$x+120=180$$
> $$x+120-120=180-120$$
> $$x=60$$
>
> The other acute angle measures 60°.

Try it

2. One of the acute angles of a right triangle measures 45º. What is the measure of the other acute angle?

Give the definition of a quadrilateral. _____

A **parallelogram** is a quadrilateral with opposite sides parallel and equal. In the parallelogram pictured below, the perpendicular distance AE between the parallel sides is called the **height**.

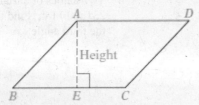

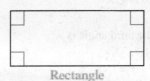

Rectangle   Square

A **rectangle** is a parallelogram that has four right angles.

What is the difference between a square and a rectangle? _____

_____

A **circle** is a plane figure in which all points are the same distance from point O, which is called the **center** of the circle.

The **diameter of a circle** (d) is the length of a line segment through the center of the circle with endpoints on the circle. AB is a diameter of the circle shown.

The **radius of a circle** (r) is the length of a line segment from the center to a point on the circle. OC is a radius of the circle shown.

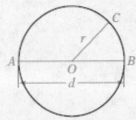

> *Note* $r=\frac{1}{2}d$ and $d=2r$.

Example
The radius of a circle is 3 in. Find the diameter of the circle.

Solution
The diameter is twice the radius.
$$d = 2r$$
$$d = 2(3 \text{ in.}) = 6 \text{ in.}$$
The diameter of the circle is 6 in.

Try it
3. The radius of a circle is 8 cm.
    Find the diameter of the circle.

Example
The diameter of a circle is 4 ft. Find the radius of the circle.

Solution
The radius is one-half the diameter.
$$r = \frac{1}{2}d$$
$$r = \frac{1}{2}(4 \text{ ft}) = 2 \text{ ft}$$
The radius of the circle is 2 ft.

Try it
4. The diameter of a circle is 14 in.
    Find the radius of the circle.

A **geometric solid** is a figure in space, or space figure. Four common space figures are the rectangular solid, cube, sphere, and cylinder.

A **rectangular solid** is a solid in which all six faces are rectangles.

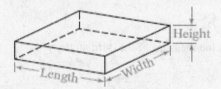

A **cube** is a rectangular solid in which all six faces are squares.

A **sphere** is a solid in which all points on the surface are the same distance from point O, which is called the **center** of the sphere.

What is the difference between a circle and a sphere? _____

_____

The **diameter of a sphere** is the length of a line segment going through the center with endpoints on the sphere. AB is a diameter of the sphere shown.

The **radius of a sphere** is the length of a line segment from the center to a point on the sphere. OC is a radius of the sphere shown below.

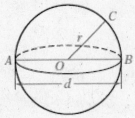

Note as with a circle, $r = \frac{1}{2}d$ and $d = 2r$.

The most common **cylinder** is one in which the bases are circles and are perpendicular to the side.

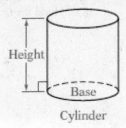

Cylinder

Reflect on it
- The sum of the three angles in a triangle is 180°.
- A quadrilateral is a four-sided plane figure.
- A rectangle is a parallelogram that has four right angles. A square is a rectangle that has four equal sides.
- The diameter of a circle is the length of a line segment through the center of the circle with endpoints on the circle.
- The radius of a circle is the length of a line segment from the center to a point on the circle.

Quiz Yourself 4.1B
1. A triangle has a 42° angle and a 71° angle. Find the measure of the third angle.

2. A right triangle has an angle measuring 29°. Find the measures of the other two angles.

3. The radius of a circle is 8 cm. Find the diameter of the circle.

4. The diameter of a circle is 16 ft. Find the radius of the circle.

5. The diameter of a sphere is 4 cm. Find the radius of the sphere.

Practice Sheet 4.1B

Solve.

1. A triangle has a 34° angle and a 83° angle. Find the measure of the other angle.

2. A triangle has a 103° angle and a 24° angle. Find the measure of the other angle.

3. A right triangle has a 30° angle. Find the measure of the other two angles.

4. A triangle has a 49° angle and a 121° angle. Find the measure of the other angle.

5. A triangle has a 64° angle and a 76° angle. Find the measure of the other angle.

6. A right triangle has a 33° angle. Find the measure of the other two angles.

7. A triangle has an 18° angle and a 99° angle. Find the measure of the other angle.

8. A triangle has a 15° angle and a 43° angle. Find the measure of the other angle.

9. Find the radius of a circle with a diameter of 18 in.

10. Find the diameter of a circle with a radius of 38 cm.

11. The radius of a sphere is 2.7 cm. Find the diameter.

12. The radius of sphere is $2\frac{1}{2}$ ft. Find the diameter.

13. The diameter of a sphere is 6 ft 2 in. Find the radius.

14. The diameter of a sphere is 8.4 m. Find the radius.

15. The radius of a sphere is $9\frac{1}{4}$ ft. Find the diameter.

16. The diameter of a sphere is 3.5 m. Find the radius.

1. _____
2. _____
3. _____
4. _____
5. _____
6. _____
7. _____
8. _____
9. _____
10. _____
11. _____
12. _____
13. _____
14. _____
15. _____
16. _____

Answers

Try it 4.1B
1. 43°
2. 45°
3. 16 cm
4. 7 in.

Quiz 4.1B
1. 67°
2. 61° and 90°
3. 16 cm
4. 8 ft
5. 2 cm

Solutions to Practice Sheet 4.1B

1. The sum of the three angles of a triangle is 180°.
$$\angle A + \angle B + \angle C = 180°$$
$$\angle A + 34° + 83° = 180°$$
$$\angle A + 117° = 180°$$
$$\angle A + 117° - 117° = 180° - 117°$$
$$\angle A = 63°$$
The measure of the other angle is 63°.

2. The sum of the three angles of a triangle is 180°.
$$\angle A + \angle B + \angle C = 180°$$
$$\angle A + 103° + 24° = 180°$$
$$\angle A + 127° = 180°$$
$$\angle A + 127° - 127° = 180° - 127°$$
$$\angle A = 53°$$
The measure of the other angle is 53°.

3. In a right triangle, one angle measures 90° and the two acute angles are complementary.
$$\angle A + \angle B = 90°$$
$$\angle A + 30° = 90°$$
$$\angle A + 30° - 30° = 90° - 30°$$
$$\angle A = 60°$$
The other angles measure 90° and 60°.

4. The sum of the three angles of a triangle is 180°.
$$\angle A + \angle B + \angle C = 180°$$
$$\angle A + 49° + 121° = 180°$$
$$\angle A + 170° = 180°$$
$$\angle A + 170° - 170° = 180° - 170°$$
$$\angle A = 10°$$
The measure of the other angle is 10°.

5. The sum of the three angles of a triangle is 180°.
$$\angle A + \angle B + \angle C = 180°$$
$$\angle A + 64° + 76° = 180°$$
$$\angle A + 140° = 180°$$
$$\angle A + 140° - 140° = 180° - 140°$$
$$\angle A = 40°$$
The measure of the other angle is 40°.

6. In a right triangle, one angle measures 90° and the two acute angles are complementary.
$$\angle A + \angle B = 90°$$
$$\angle A + 33° = 90°$$
$$\angle A + 33° - 33° = 90° - 33°$$
$$\angle A = 57°$$
The other angles measure 90° and 57°.

7. The sum of the three angles of a triangle is 180°.
$$\angle A + \angle B + \angle C = 180°$$
$$\angle A + 18° + 99° = 180°$$
$$\angle A + 117° = 180°$$
$$\angle A + 117° - 117° = 180° - 117°$$
$$\angle A = 63°$$
The measure of the other angle is 63°.

8. The sum of the three angles of a triangle is 180°.
$$\angle A + \angle B + \angle C = 180°$$
$$\angle A + 15° + 43° = 180°$$
$$\angle A + 58° = 180°$$
$$\angle A + 58° - 58° = 180° - 58°$$
$$\angle A = 122°$$
The measure of the other angle is 122°.

9. $r = \dfrac{1}{2}d$
$r = \dfrac{1}{2}(18 \text{ in.}) = 9 \text{ in.}$
The radius is 9 in.

10. $d = 2r$
$d = 2(38 \text{ cm}) = 76 \text{ cm}$
The diameter is 76 cm.

11. $d = 2r$
$d = 2(2.7 \text{ cm}) = 5.4 \text{ cm}$
The diameter is 5.4 cm.

12. $d = 2r$
$d = 2\left(2\dfrac{1}{2} \text{ ft}\right) = 2\left(\dfrac{5}{2} \text{ ft}\right) = 5 \text{ ft}$
The diameter is 5 ft.

13. $r = \dfrac{1}{2}d$

$r = \dfrac{1}{2}(6 \text{ ft } 2 \text{ in.}) = 3 \text{ ft } 1 \text{ in.}$

The radius is 3 ft 1 in.

14. $r = \dfrac{1}{2}d$

$r = \dfrac{1}{2}(8.4 \text{ m}) = 4.2 \text{ m}$

The radius is 4.2 m.

15. $d = 2r$

$d = 2\left(9\dfrac{1}{4} \text{ ft}\right) = 2\left(\dfrac{37}{4} \text{ ft}\right) = \dfrac{37}{2} \text{ ft} = 18\dfrac{1}{2} \text{ ft}$

The diameter is $18\dfrac{1}{2}$ ft .

16. $r = \dfrac{1}{2}d$

$r = \dfrac{1}{2}(3.5 \text{ m}) = 1.75 \text{ m}$

The radius is 1.75 m.

Objective 4.1C – Solve problems involving angles formed by intersecting lines

Key Words and Concepts

Two angles that are on opposite sides of the intersection of two lines are called _____ **angles**.

> *Note*   Vertical angles have the same measure.

In the diagram below,

$\angle w$ and $\angle y$ are vertical angles, and $\angle x$ and $\angle z$ are vertical angles.

$\angle w = \angle y$ and $\angle x = \angle z$.

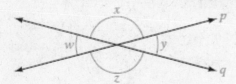

Two angles that share a common side are called _____ **angles**.

In the figure above,

$\angle x$ and $\angle y$ are adjacent angles, as are $\angle y$ and $\angle z$, $\angle z$ and $\angle w$, and $\angle w$ and $\angle x$.

Adjacent angles of intersecting lines are supplementary angles.

> *Note*   Each pair of adjacent angles listed here adds to 180°.

Example

In the diagram $\angle b = 105°$. Find the measures of the other angles.

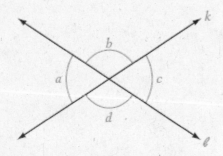

Try it

1. In the diagram to the left, $\angle a = 70°$. Find the measures of the other angles.

Solution

$\angle d = 105°$, since $\angle b$ and $\angle d$ are vertical angles.

Since $\angle a$ and $\angle b$ are adjacent angles of intersecting lines, they are supplementary.

$$\angle a + \angle b = 180°$$
$$\angle a + 105° = 180°$$
$$\angle a + 105° - 105° = 180° - 105°$$
$$\angle a = 75°$$

Thus, $\angle a = 75°$.

$\angle c = 75°$, since $\angle a$ and $\angle c$ are vertical angles.

A line intersecting two other lines at two different points is called a **transversal**. If the lines cut by a transversal are parallel lines and the transversal is perpendicular to the parallel lines, then all eight angles formed are right angles, as in the diagram below.

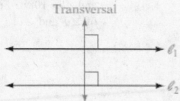

If the lines cut by a transversal are parallel lines and the transversal is **not** perpendicular to the parallel lines, then all four acute angles have the same measure and all four obtuse angles have the same measure.

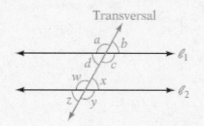

For the figure to the left, there are two groups of angles that have the same measure.

$$\angle a = \angle c = \angle w = \angle y$$

$$\angle b = \angle d = \angle x = \angle z$$

Alternate interior angles are two nonadjacent angles that are on opposite sides of the transversal and between the parallel lines.

    For the figure above,

        $\angle c$ and $\angle w$ are alternate interior angles.

        $\angle d$ and $\angle x$ are alternate interior angles.

    Note   Alternate interior angles have the same measure.

Alternate exterior angles are two nonadjacent angles that are on opposite sides of the transversal and outside the parallel lines.

    For the figure above,

        $\angle a$ and $\angle y$ are alternate exterior angles.

        $\angle b$ and $\angle z$ are alternate exterior angles.

    Note   Alternate exterior angles have the same measure.

Corresponding angles are two angles that are on the same side of the transversal and are both acute angles or are both obtuse angles.

    For the figure above, the following pairs of angles are corresponding angles:

        $\angle a$ and $\angle w$, $\angle d$ and $\angle z$, $\angle b$ and $\angle x$,   and $\angle c$ and $\angle y$.

    Note   Corresponding angles have the same measure.

Example

In the figure, $\ell_1 \parallel \ell_2$, and $\angle b = 119°$.
Find the measures of $\angle a$ and $\angle c$.

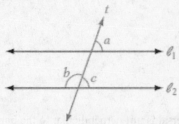

Try it

2. In the figure at the left, $\ell_1 \parallel \ell_2$, and $\angle b = 108°$. Find the measures of $\angle a$ and $\angle c$.

Solution

Since $\angle b$ and $\angle c$ form a straight angle, $\angle b + \angle c = 180°$.

$$\angle b + \angle c = 180°$$
$$119° + \angle c = 180°$$
$$119° - 119° + \angle c = 180° - 119°$$
$$\angle c = 61°$$

Since $\angle a$ and $\angle c$ are corresponding angles,
$$\angle a = \angle c = 61°.$$

Reflect on it

- Two angles that are on opposite sides of the intersection of two lines are called vertical angles.
     Vertical angles have the same measure
- Two angles that share a common side are called adjacent angles.
- Alternate interior angles are two nonadjacent angles that are on opposite sides of the transversal and between the parallel lines.
- Alternate exterior angles are two nonadjacent angles that are on opposite sides of the transversal and outside the parallel lines.
- **Corresponding angles** are two angles that are on the same side of the transversal and are both acute angles or are both obtuse angles.

Quiz Yourself 4.1C

1. Find the measures of $\angle a$, $\angle b$, and $\angle c$ in the diagram if $\angle d = 113°$.

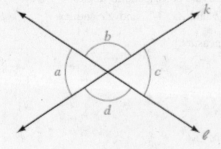

2. Find the measures of $\angle a$ and $\angle b$ in the diagram.

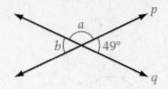

3. Find the measures of $\angle a$ and $\angle b$ in the diagram given that $\ell_1 \parallel \ell_2$.

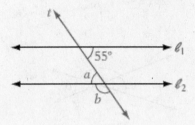

4. Find the measures of $\angle a$ and $\angle b$ in the diagram given that $\ell_1 \parallel \ell_2$.

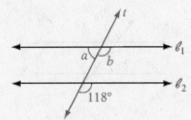

Practice Sheet 4.1C

Find the measures of angles *a* and *b*.

1.

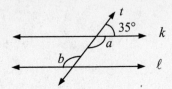

2.

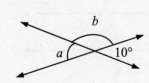

3.

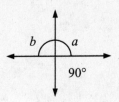

4.

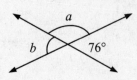

1. _____

2. _____

3. _____

4. _____

Find the measures of angles *a* and *b*.   *k* ∥ *ℓ*

5.

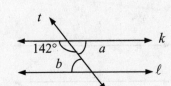

6.

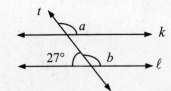

7.

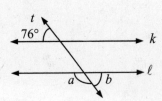

8.

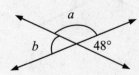

5. _____

6. _____

7. _____

8. _____

Answers

Try it 4.1C

1. $\angle b = 110°$, $\angle c = 70°$, $\angle d = 110°$
2. $\angle a = 72°$, $\angle b = 72°$

Quiz 4.1C

1. $\angle a = \angle c = 67°$; $\angle b = 113°$
2. $\angle a = 131°$; $\angle b = 49°$
3. $\angle a = 55°$; $\angle b = 125°$
4. $\angle a = 62°$; $\angle b = 118°$

Solutions to Practice Sheet 4.1C

1. $\angle a$ and $76°$ are supplementary angles.
$$\angle a + 76° = 180°$$
$$\angle a + 76° - 76° = 180° - 76°$$
$$\angle a = 104°$$
$\angle b = 76°$　Vertical angle

2. $\angle a$ and $48°$ are supplementary angles.
$$\angle a + 48° = 180°$$
$$\angle a + 48° - 48° = 180° - 48°$$
$$\angle a = 132°$$
$\angle b = 48°$　Vertical angle

3. $\angle a$ and $90°$ are supplementary angles.
$$\angle a + 90° = 180°$$
$$\angle a + 90° - 90° = 180° - 90°$$
$$\angle a = 90°$$
$\angle b = 90°$　Vertical angle

4. $\angle b$ and $10°$ are supplementary angles.
$$\angle b + 10° = 180°$$
$$\angle b + 10° - 10° = 180° - 10°$$
$$\angle b = 170°$$
$\angle a = 10°$　Vertical angle

5. $\angle a$ and $35°$ are supplementary angles.
$$\angle a + 35° = 180°$$
$$\angle a + 35° - 35° = 180° - 35°$$
$$\angle a = 145°$$
$\angle b = \angle a = 145°$　Alternate interior angle

6. $\angle a$ and $142°$ are supplementary angles.
$$\angle a + 142° = 180°$$
$$\angle a + 142° - 142° = 180° - 142°$$
$$\angle a = 38°$$
$\angle b = \angle a = 38°$　Alternate interior angle

7. $\angle b = 76°$　Alternate exterior angle
$\angle a$ and $\angle b$ are supplementary angles.
$$\angle a + \angle b = 180°$$
$$\angle a + 76° = 180°$$
$$\angle a + 76° - 76° = 180° - 76°$$
$$\angle a = 104°$$

8. $\angle b$ and $27°$ are supplementary angles.
$$\angle b + 27° = 180°$$
$$\angle b + 27° - 27° = 180° - 27°$$
$$\angle b = 153°$$
$\angle a = \angle b = 153°$　Corresponding angle

Objective 4.2A – Find the perimeter of a plane geometric figure

Key Words and Concepts

A **polygon** is a closed figure determined by three or more line segments that lie in a plane.

The **sides of a polygon** are the line segments that form the polygon.

A **regular polygon** is one in which each side has the _____ length and each angle has the _____ measure.
(same/different)   (same/different)

The name of a polygon is based on the number of its sides.

Number of Sides	Name of the Polygon
3	Triangle
4	Quadrilateral
5	Pentagon
6	Hexagon
7	Heptagon
8	Octagon
9	Nonagon
10	Decagon

Triangles are distinguished by the number of equal sides and also by the measures of their angles.

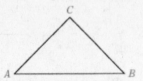      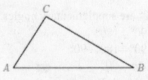

An **isosceles triangle** has two sides of equal length. The angles opposite each of the equal sides are of equal measure.

$AC = BC$
$\angle A = \angle B$

The three sides of an **equilateral triangle** are of equal length. The three angles are of equal measure.

$AB = BC = AC$
$\angle A = \angle B = \angle C$

A **scalene triangle** has no two sides of equal length. No two angles are of equal measure.

Label each triangle below as either right, acute, or obtuse.

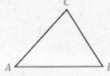      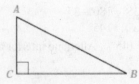

What is the difference between an acute triangle, an obtuse triangle, and a right triangle?

_____

_____

Quadrilaterals also are distinguished by their sides and angles, as shown below.

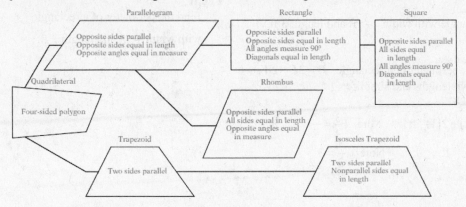

The **perimeter** of a plane geometric figure is a measure of the distance around the figure.

> *Note*   The perimeter of a polygon is the sum of the lengths of its sides.

Perimeter of a Triangle

Let a, b, and c be the lengths of the sides of a triangle.

The perimeter of the triangle is $P = a + b + c$.

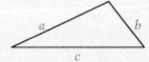

Example

The sides of a triangle have lengths 4 in., 5 in., and 8 in. Find the perimeter of the triangle.

Solution

Use the formula for perimeter of a triangle, $P = a + b + c$.
Substitute the known lengths and simplify.

$$P = a + b + c$$
$$= 4 \text{ in.} + 5 \text{ in.} + 8 \text{ in.}$$
$$= 17 \text{ in.}$$

The perimeter of the triangle is 17 in.

Try it

1. The sides of a triangle have lengths 8 cm, 10 cm, and 15 cm. Find the perimeter of the triangle.

Perimeter of a Rectangle

Let L be the length (usually the longer side) of a rectangle and let W be the width (usually the shorter side) of a rectangle.

The perimeter of the rectangle is $P = 2L + 2W$.

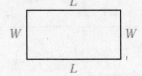

Example

Find the perimeter of a rectangle with width $1\frac{3}{4}$ ft and length 3 ft.

Try it

2. Find the perimeter of a rectangle with width $\frac{1}{3}$ ft and length 5 ft.

Solution

Use the formula for perimeter of a rectangle, $P = 2L + 2W$.

Substitute the known lengths and simplify.

$$P = 2L + 2W$$
$$= 2(3 \text{ ft}) + 2\left(1\frac{3}{4} \text{ ft}\right) \quad \text{Note } 1\frac{3}{4} = \frac{7}{4}$$
$$= 6 \text{ ft} + \frac{7}{2} \text{ ft} \quad \text{Note } \frac{7}{2} = 3\frac{1}{2}$$
$$= 9\frac{1}{2} \text{ ft}$$

The perimeter of the rectangle is $9\frac{1}{2}$ ft.

Perimeter of a Square

Let s be the length of a side of a square.

The perimeter of the square is $P = 4s$.

Example

Find the perimeter of a square with side length 9 yd.

Try it

3. Find the perimeter of a square with side length 5 in.

Solution

Use the formula for perimeter of a square, $P = 4s$.

Substitute the known length and simplify.

$$P = 4s$$
$$= 4(9) \text{ yd}$$
$$= 36 \text{ yd}$$

The perimeter of the square is 36 yd.

The perimeter of a circle is called its **circumference**.

The circumference of a circle is equal to the product of pi (π) and the diameter.

Name 2 ways that the value of π can be approximated. _____

Circumference of a Circle

Let d be the diameter of a circle.

Then the circumference of the circle is $C = \pi d$.

Note   Because the diameter is twice the radius, the circumference is also given by $C = 2\pi r$.

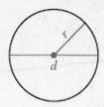

Example

Find the circumference of a circle with radius 4.5 cm. Use 3.14 for π.

Solution

Use the formula for circumference of a circle, $C = 2\pi r$.
Substitute the known length and simplify.

$$C = 2\pi r$$
$$\approx 2(3.14)(4.5 \text{ cm})$$
$$= 28.26 \text{ cm}$$

The circumference of the circle is approximately 28.26 cm.

Try it

4. Find the circumference of a circle with radius 6.5 in.. Use 3.14 for π.

Reflect on it

- An acute triangle has three acute sides.
- An obtuse triangle has one obtuse angle.
- A right triangle has a right angle.
- The perimeter of a triangle is $P = a + b + c$.
- The perimeter of a rectangle is $P = 2L + 2W$.
- The perimeter of a square is $P = 4s$.
- The circumference of a circle is $C = \pi d$ or $C = 2\pi r$.

Quiz Yourself 4.2A

1. Find the perimeter of the triangle shown.

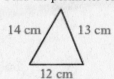

14 cm   13 cm

12 cm

2.   Find the perimeter of the rectangle shown.

5 ft

18 ft

3.   Find the perimeter of the square shown.

2 m

2 m

4.   A circle has radius 1.5 yd. Find its circumference. Use 3.14 for π.

5.   Suppose that you wish to enclose with fencing a plot of land that is 73 feet long and 48 feet wide.
     How much fencing will you need?

6.   The diameter of a circular region is 70 feet. What is the circumference of the region?
     Use 3.14 for π, and round to the nearest whole number.

Practice Sheet 4.2A

Simplify.

1. Find the perimeter of a triangle with sides 3 ft, 4 ft, and 6 ft.

2. Find the perimeter of a triangle with sides 4 ft, 5 ft, and 8 ft.

3. Find the perimeter of a triangle in which each side is 18 cm.

4. Find the perimeter of a triangle with sides 21.3 cm, 17.4 cm, and 14.8 cm.

5. Find the circumference of a circle with radius of 6 cm. Use 3.14 for π.

6. Find the circumference of a circle with a diameter of 28 in. Use $\frac{22}{7}$ for π.

7. Find the perimeter of a square in which the sides are equal to 13 m.

8. Find the perimeter of a triangle in which each side is $4\frac{1}{3}$ ft.

9. Find the perimeter of a rectangle with a width of 5 ft and a length of $8\frac{1}{2}$ ft.

10. Find the perimeter of a square in which the sides are equal to $8\frac{1}{4}$ ft.

11. Find the perimeter of a triangle with sides $10\frac{1}{2}$ ft, $15\frac{1}{4}$ ft, and $12\frac{3}{4}$ ft.

12. Find the circumference of a circle with radius of 15 cm. Use 3.14 for π.

13. Find the circumference of a circle with a diameter of 35 ft. Use $\frac{22}{7}$ for π.

14. Find the perimeter of a triangle in which each side is 9 ft.

15. Find the perimeter of a square in which the sides are equal to 15.5 m.

16. Find the perimeter of a triangle with sides 3 ft, 5 ft, and 6 ft.

17. Find the perimeter of a five-sided figure with sides of 19 cm, 36 cm, 25 cm, 39 cm, and 20 cm.

18. Find the perimeter of a six-sided figure with sides of 15 ft, 32 ft, 21 ft, 34 ft, 17 ft, and 26 ft.

1. _____

2. _____

3. _____

4. _____

5. _____

6. _____

7. _____

8. _____

9. _____

10. _____

11. _____

12. _____

13. _____

14. _____

15. _____

16. _____

17. _____

18. _____

Name:
Module 4: Geometry and Right Triangle Trigonometry

Class:
Section 4.2: Perimeter of a Plane Geometric Figure

Answers

Try it 4.2A
1. 33 cm
2. $10\frac{2}{3}$ ft
3. 20 in.
4. 40.82 in.

Quiz 4.2A
1. 39 cm
2. 46 ft
3. 8 m
4. 9.42 yd
5. 242 ft
6. 220 ft

Solutions to Practice Sheet 4.2A

1. $P = a + b + c$
   $P = 3\text{ ft} + 4\text{ ft} + 6\text{ ft}$
   $P = 13\text{ ft}$
   The perimeter of the triangle is 13 ft.

2. $P = a + b + c$
   $P = 4\text{ ft} + 5\text{ ft} + 8\text{ ft}$
   $P = 17\text{ ft}$
   The perimeter of the triangle is 17 ft.

3. $P = a + b + c$
   $P = 18\text{ cm} + 18\text{ cm} + 18\text{ cm}$
   $P = 54\text{ cm}$
   The perimeter of the triangle is 54 ft.

4. $P = a + b + c$
   $P = 21.3\text{ cm} + 17.4\text{ cm} + 14.8\text{ cm}$
   $P = 53.5\text{ cm}$
   The perimeter of the triangle is 53.5 cm.

5. $C = 2\pi r$
   $= 2\pi(6\text{ cm})$
   $\approx 2(3.14)(6\text{ cm})$
   $= 37.68\text{ cm}$
   The circumference of the circle is approximately 37.68 cm.

6. $C = \pi d$
   $= \pi(28\text{ in.})$
   $\approx \frac{22}{7}(28\text{ in.})$
   $= 88\text{ in.}$
   The circumference of the circle is approximately 88 in.

7. $P = 4s$
   $P = 4(13\text{ m})$
   $P = 52\text{ m}$
   The perimeter of the square is 52 m.

8. $P = a + b + c$
   $P = 4\frac{1}{3}\text{ ft} + 4\frac{1}{3}\text{ ft} + 4\frac{1}{3}\text{ ft}$
   $P = 13\text{ ft}$
   The perimeter of the triangle is 13 ft.

9. $P = 2L + 2W$
   $P = 2\left(8\frac{1}{2}\text{ ft}\right) + 2(5\text{ ft})$
   $P = 27\text{ ft}$
   The perimeter of the rectangle is 27 ft.

10. $P = 4s$
    $P = 4\left(8\frac{1}{4}\text{ ft}\right)$
    $P = 33\text{ ft}$
    The perimeter of the square is 33 ft.

11. $P = a + b + c$
    $P = 10\frac{1}{2}\text{ ft} + 15\frac{1}{4}\text{ ft} + 12\frac{3}{4}\text{ ft}$
    $P = 38\frac{1}{2}\text{ ft}$
    The perimeter of the triangle is $38\frac{1}{2}$ ft.

12. $C = 2\pi r$
    $= 2\pi(15\text{ cm})$
    $\approx 2(3.14)(15\text{ cm})$
    $= 94.2\text{ cm}$
    The circumference of the circle is approximately 94.2 cm.

13. $C = \pi d$
    $= \pi(35\text{ ft})$
    $\approx \frac{22}{7}(35\text{ ft})$
    $= 110\text{ ft}$
    The circumference of the circle is approximately 110 ft.

14. $P = a + b + c$
    $P = 9\text{ ft} + 9\text{ ft} + 9\text{ ft}$
    $P = 27\text{ ft}$
    The perimeter of the triangle is 27 ft.

15. $P = 4s$
    $P = 4(15.5\text{ m})$
    $P = 62\text{ m}$
    The perimeter of the square is 62 m.

16. $P = a + b + c$
$P = 3 \text{ ft} + 5 \text{ ft} + 6 \text{ ft}$
$P = 14 \text{ ft}$
The perimeter of the triangle is 14 ft.

17. $P = a + b + c + d + e$
$P = 19 \text{ cm} + 36 \text{ cm} + 25 \text{ cm} + 39 \text{ cm} + 20 \text{ cm}$
$P = 139 \text{ cm}$
The perimeter of the five-sided
figure is 139 cm.

18. $P = a + b + c + d + e + f$
$P = 15 \text{ ft} + 32 \text{ ft} + 21 \text{ ft} + 34 \text{ ft} + 17 \text{ ft} + 26 \text{ ft}$
$P = 145 \text{ ft}$
The perimeter of the six-sided
figure is 145 ft.

Objective 4.2B – Find the perimeter of a composite geometric figure

Key Words and Concepts

A **composite geometric figure** is a figure made from two or more geometric figures.

The perimeter of a composite geometric figure can be found by adding the perimeters or circumferences of the appropriate parts of the figures composing the composite figure.

Example

Find the perimeter of the figure shown here. Use 3.14 for π.

Solution
This figure is composed of a square and four half-circles. The perimeter is found by adding the circumferences of the **four** half circles.

Recall $C = \pi d$, so a half circle will have

circumference, $C = \frac{1}{2}\pi d$.

$$P = 4 \cdot \frac{1}{2}\pi d$$
$$\approx 2(3.14)(4 \text{ cm})$$
$$= 25.12 \text{ cm}$$

The perimeter of the figure is approximately 25.12 cm.

Try it

1. Find the perimeter of the figure shown here. Use 3.14 for π.

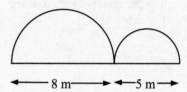

$\longleftarrow$ 8 m $\longrightarrow$ $\longleftarrow$ 5 m $\longrightarrow$

Reflect on it
- In the Example above, why was the formula for the perimeter of a square used?

Quiz Yourself 4.2B
1. Find the perimeter of the figure below. Use 3.14 for π.

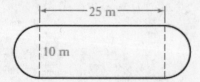

2. Find the perimeter of the figure below.

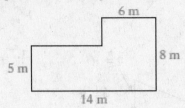

3. Find the perimeter of the figure below. Use 3.14 for π.

4. The figure below is formed by a square and one-half a circle with diameter $d = 14$ cm. Find the perimeter of the figure. Use 3.14 for π.

Practice Sheet 4.2B

Find the perimeter. Use 3.14 for π.

1.

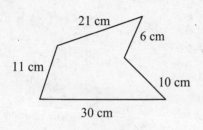

21 cm
6 cm
11 cm
10 cm
30 cm

2.

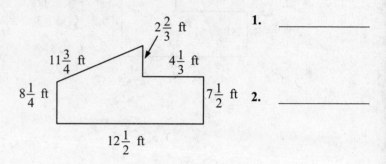

$2\frac{2}{3}$ ft
$11\frac{3}{4}$ ft
$4\frac{1}{3}$ ft
$8\frac{1}{4}$ ft
$7\frac{1}{2}$ ft
$12\frac{1}{2}$ ft

1. _____

2. _____

3.

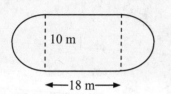

10 m
←—18 m—→

4.

14.6 m
13.6 m   13.5 m
12.4 m   14.1 m
23.6 m

3. _____

4. _____

5.

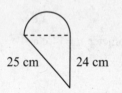

25 cm   24 cm

6.

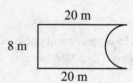

20 m
8 m
20 m

5. _____

6. _____

7.

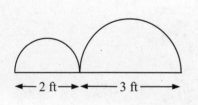

←2 ft→ ←—3 ft—→

8.

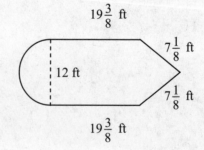

$19\frac{3}{8}$ ft
$7\frac{1}{8}$ ft
12 ft
$7\frac{1}{8}$ ft
$19\frac{3}{8}$ ft

7. _____

8. _____

Answers

Try it 4.2B
1. 33.41 m

Quiz 4.2B
1. 81.4 m
2. 44 m
3. 40.26 cm
4. 63.98 cm

Solutions to Practice Sheet 4.2B

1. Perimeter
   = sum of the lengths of the sides
   = 11 cm + 21 cm + 6 cm + 10 cm + 30 cm
   = 78 cm
   The perimeter is 78 cm.

2. Perimeter
   = sum of the lengths of the sides
   = $8\frac{1}{4}$ ft + $11\frac{3}{4}$ ft + $2\frac{2}{3}$ ft + $4\frac{1}{3}$ ft + $7\frac{1}{2}$ ft + $12\frac{1}{2}$ ft
   = 47 ft
   The perimeter is 47 ft.

3. Perimeter = $\begin{array}{c}\text{Circumference}\\\text{of a circle}\end{array}$ + $\begin{array}{c}\text{lengths of}\\\text{the two}\\\text{sides}\end{array}$
   = $\pi d + 2L$
   ≈ 3.14(10 m) + 2(18 m)
   = 67.4 m
   The perimeter is approximately 67.4 m.

4. Perimeter
   = sum of the lengths of the sides
   = 12.4 m + 13.6 m + 14.6 m + 13.5 m
   + 14.1 m + 23.6 m
   = 91.8 m
   The perimeter is 91.8 m.

5. Find the diameter using the Pythagorean theorem.
   $$c^2 = a^2 + b^2$$
   $$25^2 = a^2 + 24^2$$
   $$625 - 576 = a^2$$
   $$49 = a^2 \qquad \text{The negative solution}$$
   $$7 = a \qquad \text{is excluded.}$$

   Perimeter = $\begin{array}{c}\text{Circumference}\\\text{of }\frac{1}{2}\text{ circle}\end{array}$ + $\begin{array}{c}\text{lengths of}\\\text{the sides}\end{array}$
   = $\frac{1}{2}\pi d + c + b$
   ≈ $\frac{1}{2}$(3.14)(7 cm) + 25 cm + 24 cm
   = 59.99 cm
   The perimeter is approximately 59.99 cm.

6. Perimeter = $\begin{array}{c}\text{Circumference}\\\text{of }\frac{1}{2}\text{ circle}\end{array}$ + $\begin{array}{c}\text{lengths of}\\\text{the sides}\end{array}$
   = $\frac{1}{2}\pi d + a + b + c$
   ≈ $\frac{1}{2}$(3.14)(8 m) + 8 m
   + 20 m + 20 m
   = 60.56 m
   The perimeter is approximately 60.56 m.

7. Perimeter = $\begin{array}{c}\text{Circumference}\\\text{of }\frac{1}{2}\text{ circle}\end{array}$ + $\begin{array}{c}\text{Circumference}\\\text{of }\frac{1}{2}\text{ circle}\end{array}$
   = $\frac{1}{2}\pi d + \frac{1}{2}\pi d$
   ≈ $\frac{1}{2}$(3.14)(2 ft) + $\frac{1}{2}$(3.14)(3 ft)
   = 7.85 ft
   The perimeter is approximately 7.85 ft.

8. Perimeter = $\begin{array}{c}\text{Circumference}\\\text{of }\frac{1}{2}\text{ circle}\end{array}$ + $\begin{array}{c}\text{lengths of}\\\text{the sides}\end{array}$
   = $\frac{1}{2}\pi d + a + b + c + d$
   ≈ $\frac{1}{2}$(3.14)(12 ft) + $19\frac{3}{8}$ ft
   + $7\frac{1}{8}$ ft + $7\frac{1}{8}$ ft + $19\frac{3}{8}$ ft
   = 71.84 ft
   The perimeter is approximately 71.84 ft.

Objective 4.2C – Solve application problems

Example

The rectangular lot shown in the figure is being fenced. The fencing along the road costs $6.20 per foot. The rest of the fencing costs $5.85 per foot. Find the total cost to fence the lot.

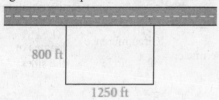

Try it

1. Binding is to be sewn around the edge of a rectangular carpet that measures 6 ft by 8 ft. If binding costs $3 per foot, find the total cost to bind the carpet.

Solution

To find the total cost:
- Multiply the length along the road by the per-foot cost of the fencing.
- Find the perimeter of the other three sides.
- Multiply the perimeter by the per-foot cost of the fencing.
- Determine the total of the costs.

The length of fencing along the road is 1250 ft.
The cost for this fencing is
$$1250(\$6.20) = \$7750.$$

The perimeter of the other three sides is
$$P = 800 \text{ ft} + 1250 \text{ ft} + 800 \text{ ft}$$
$$= 2850 \text{ ft}.$$

The cost for this fencing is
$$2850(\$5.85) = \$16,672.50.$$

Thus, the total cost to fence the lot is
$$\$7750 + \$16,672.50 = \$24,422.50$$

Quiz Yourself 4.2C

1. Find the length of weather stripping installed around the arched door shown below. Use 3.14 for π.

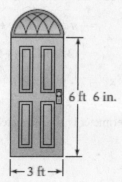

2. A circular track has radius 158 ft. If fencing for the track costs $7/ft, then how much will fencing for the entire track cost? Use 3.14 for π, and round the cost to the nearest $10.

3. How many feet of binding are required to bind the edge of a rectangular quilt that measures 4 ft by 9 ft?

4. A tire has diameter 14 in. How many feet does the tire travel when the wheel makes 10 revolutions? Use 3.14 for π, and round your answer to the nearest whole number of feet.

Practice Sheet 4.2C

Solve.

1. Find the number of feet of fence needed to fence a park that is $1\frac{3}{8}$ mi long and $\frac{5}{8}$ mi wide.

2. A horse trainer is planning to enclose a circular exercise area. How much fencing is needed if it has a diameter of 80 ft? Use 3.14 for π.

1. _____

2. _____

3. An irrigation system waters a circular field which has a 60-foot radius. Find the distance around the outside edge of the field. Use 3.14 for π.

4. Find the length of weather stripping needed to put around a rectangular window that is 9 ft high and 22 ft long.

3. _____

4. _____

5. Find the length of aluminum framing needed to frame a picture that is 6 ft by 4 ft.

6. Find the length of a rubber gasket needed to fit around a circular porthole that has a 16 in. diameter. Use 3.14 for π.

5. _____

6. _____

7. A rain gutter being installed on a home has the dimensions shown below. At $9.10 per meter, how much does it cost to install the rain gutter?

```
        20 m
   ┌──────────────┐
12 m│   10 m       │12 m
   │   ┌────┐      │
   └───┘6 m  6 m └──┘
```

8. A metal frame is being built to install around a rectangular sign which is 2 m wide and 3.5 m long. At $2.12 per meter, find the cost of the frame.

7. _____

8. _____

9. A security fence is to be built around a factory. The lot on which the factory is located is 1000 ft by 500 ft. If the fencing costs $3.25 per foot, what is the total cost of installing the fence?

10. The dimensions of a rectangular classroom are 30 ft by 24 ft. At $1.10 per foot, how much does it cost to purchase an outlet strip to go around the room? Subtract 3 ft for a doorway.

9. _____

10. _____

Answers

Try it 4.2C
1. $84

Quiz 4.2C
1. 20.71 ft
2. $6950
3. 26 ft
4. 440 ft

Solutions to Practice Sheet 4.2C

1. STRATEGY
   To find the amount of fencing, use the
   formula for the perimeter of a rectangle.

   SOLUTION
   $P = 2L + 2W$
   $= 2\left(1\frac{3}{8} \text{ mi}\right) + 2\left(\frac{5}{8} \text{ mi}\right)$
   $= 4 \text{ mi}$
   The amount of fencing needed is 4 mi.

2. STRATEGY
   To find the amount of fencing, use the
   formula for the perimeter of a circle.

   SOLUTION
   $P = \pi d$
   $\approx 3.14(80 \text{ ft})$
   $= 251.2 \text{ ft}$
   The amount of fencing needed is 251.2 ft.

3. STRATEGY
   To find the distance, use the
   formula for the perimeter of a circle.

   SOLUTION
   $P = 2\pi r$
   $\approx 2(3.14)(60 \text{ ft})$
   $= 376.8 \text{ ft}$
   The distance is 376.8 ft.

4. STRATEGY
   To find length, use the
   formula for the perimeter of a rectangle.

   SOLUTION
   $P = 2L + 2W$
   $= 2(22 \text{ ft}) + 2(9 \text{ ft})$
   $= 62 \text{ ft}$
   The length of weather stripping
   needed is 62 ft.

5. STRATEGY
   To find length, use the
   formula for the perimeter of a rectangle.

   SOLUTION
   $P = 2L + 2W$
   $= 2(6 \text{ ft}) + 2(4 \text{ ft})$
   $= 20 \text{ ft}$
   The length of framing needed is 20 ft.

6. STRATEGY
   To find the amount of rubber, use the
   formula for the perimeter of a circle.

   SOLUTION
   $P = \pi d$
   $\approx 3.14(16 \text{ in.})$
   $= 50.24 \text{ in.}$
   The amount of rubber needed is 50.24 in.

7. STRATEGY
   To find the cost of the installation:
   • Find the perimeter of the building.
   • Multiply the perimeter by the per-meter
   cost of the rain gutters.

   SOLUTION
   $P = 12 \text{ m} + 20 \text{ m} + 12 \text{ m} + 5 \text{ m}$
   $\quad + 6 \text{ m} + 10 \text{ m} + 6 \text{ m} + 5 \text{ m}$
   $P = 76 \text{ m}$
   $\text{Cost} = 76 \times 9.10 = 691.6$
   The cost for installation is $691.60.

8. STRATEGY
   To find the cost of the frame:
   • Find the perimeter of the sign.
   • Multiply the perimeter by the per-meter
   cost of the frame.

   SOLUTION
   $P = 2L + 2W$
   $P = 2(3.5 \text{ m}) + 2(2 \text{ m})$
   $P = 11 \text{ m}$
   $\text{Cost} = 11 \times 2.12 = 23.32$
   The cost is $23.32.

9. STRATEGY
   To find the cost of the fence:
   - Find the perimeter of the fence.
   - Multiply the perimeter by the per-foot
   cost of the fencing.

   SOLUTION
   $P = 2L + 2W$
   $P = 2(1000 \text{ ft}) + 2(500 \text{ ft})$
   $P = 3000 \text{ ft}$
   $\text{Cost} = 3000 \times 3.25 = 9750$
   The cost is $9750.

10. STRATEGY
    To find the cost of the outlet strip:
    - Find the perimeter of the classroom.
    - Subtract 3 ft from the perimeter.
    - Multiply the difference by the per-foot
    cost of the outlet strip.

    SOLUTION
    $P = 2L + 2W$
    $P = 2(30 \text{ ft}) + 2(24 \text{ ft})$
    $P = 108 \text{ ft}$
    $108 \text{ ft} - 3 \text{ ft} = 105 \text{ ft}$
    $\text{Cost} = 105 \times 1.10 = 115.5$
    The cost is $115.50.

Objective 4.3A – Find the area of a geometric figure

Key Words and Concepts

Define area. _____

_____

Area of a Rectangle

Let L be the length of a rectangle and W be the width of the rectangle.

The area of the rectangle is $A = LW$.

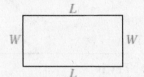

Example

A rectangle has length 13 in. and width 4 in. Find its area.

Solution

Use the formula for the area of a rectangle.
Substitute the known values and simplify.

$A = LW$

$\quad = (13 \text{ in.})(4 \text{ in.})$

$\quad = 52 \text{ in}^2$

The area of the rectangle is 52 in^2.

Try it

1. A rectangle has length 12 cm and width 9 cm. Find its area.

Area of a Square

Let s be the length of one side of a square.

The area of the square is $A = s^2$.

Example

The length of a side of a square is 3 m. Find its area.

Solution

Use the formula for the area of a square.
Substitute the known value and simplify.

$A = s^2$

$\quad = (3 \text{ m})^2$

$\quad = 9 \text{ m}^2$

The area of the square is 9 m^2.

Try it

2. The length of a side of a square is 5 in. Find its area.

Area of a Triangle

In the triangle below, $\overline{AB}$ is the base b of the triangle, and $\overline{CD}$, which is perpendicular to the base b, is the height h.

The area of a triangle is $A = \frac{1}{2}bh$.

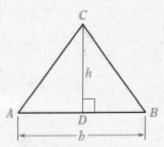

Example

Find the area of a triangle with base length 13 yd and height 5 yd.

Solution

Use the formula for the area of a triangle.
Substitute the known values and simplify.

$$A = \frac{1}{2}bh$$
$$= \frac{1}{2}(13 \text{ yd})(5 \text{ yd})$$
$$= 32.5 \text{ yd}^2$$

The area of the triangle is 32.5 yd^2.

Try it

3. Find the area of a triangle with base length 8 ft and height 3 ft.

Area of a Circle

Let r be the radius of a circle.

The area of the circle is $A = \pi r^2$.

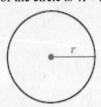

Example

Find the area of a circle with radius 5 m. Use 3.14 for π.

Solution

Use the formula for the area of a circle.
Substitute the known value and simplify.

$$A = \pi r^2$$
$$\approx (3.14)(5 \text{ m})^2$$
$$= 78.5 \text{ m}^2$$

The area of the circle is approximately 78.5 m^2.

Try it

4. Find the area of a circle with radius 15 in. Use 3.14 for π.

Reflect on it
- Area is a measure of the amount of surface in a region.
- The area of the rectangle is $A = LW$.
- The area of the square is $A = s^2$.
- The area of a triangle is $A = \frac{1}{2}bh$.
- The area of the circle is $A = \pi r^2$.

Quiz Yourself 4.3A
For Exercises 1 through 4, find the area of the given figure. Use 3.14 for π if necessary.

1.

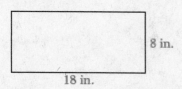

8 in.

18 in.

2.

4 in.

4 in.

3.

3 cm

4.

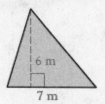

6 m

7 m

5. Find the area of a triangle with base length 9 cm and height 5 cm.

6. Find the area of a circle with diameter 11 m. Use 3.14 for π. Round to the nearest tenth.

Practice Sheet 4.3A

Solve.

1. Find the area of a triangle with a base of 8 ft and a height of $1\frac{3}{4}$ ft.

2. Find the area of a right triangle with a base of 11 cm and a height of 6.4 cm.

1. _____

2. _____

3. Find the area of a square with a side of 7 ft.

4. Find the area of a square with a side of 29 cm.

3. _____

4. _____

5. Find the area of a rectangle with a length of 52 cm and a width of 27 cm.

6. Find the area of a rectangle with a length of 25 in., and a width of 13 in.

5. _____

6. _____

7. Find the area of a circle with a radius of 11 in. Use 3.14 for π.

8. Find the area of a square with a side of 7.8 cm.

7. _____

8. _____

9. Find the area of a triangle with a base of 27 cm and a height of 30 cm.

10. Find the area of a rectangle with a length of 39 cm and a width of 25 cm.

9. _____

10. _____

Answers

Try it 4.3A
1. 108 cm^2
2. 25 in^2
3. 12 ft^2
4. 47.1 in^2

Quiz 4.3A
1. 144 in^2
2. 16 in^2
3. 28.26 cm^2
4. 21 m^2
5. 22.5 cm^2
6. 95.0 m^2

Solutions to Practice Sheet 4.3A

1. $A = \frac{1}{2}bh$

    $A = \frac{1}{2}(8 \text{ ft})\left(1\frac{3}{4} \text{ ft}\right) = \frac{1}{2}(8 \text{ ft})\left(\frac{7}{4} \text{ ft}\right)$

    $A = 7 \text{ ft}^2$

2. $A = \frac{1}{2}bh$

    $A = \frac{1}{2}(11 \text{ cm})(6.4 \text{ cm})$

    $A = 35.2 \text{ cm}^2$

3. $A = s^2 = (7 \text{ ft})^2 = 49 \text{ ft}^2$

4. $A = s^2 = (29 \text{ cm})^2 = 841 \text{ cm}^2$

5. $A = LW = (52 \text{ cm})(27 \text{ cm}) = 1404 \text{ cm}^2$

6. $A = LW = (25 \text{ in.})(13 \text{ in.}) = 325 \text{ in}^2$

7. $A = \pi r^2 \approx 3.14(11 \text{ in.})^2 \approx 379.94 \text{ in}^2$

8. $A = s^2 = (7.8 \text{ cm})^2 = 60.84 \text{ cm}^2$

9. $A = \frac{1}{2}bh$

    $A = \frac{1}{2}(27 \text{ cm})(30 \text{ cm})$

    $A = 405 \text{ cm}^2$

10. $A = LW = (39 \text{ cm})(25 \text{ cm}) = 975 \text{ cm}^2$

Objective 4.3B – Find the area of a composite geometric figure

Example
Find the area of the figure below. Use 3.14 for π.

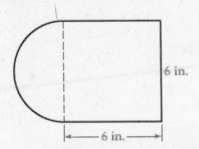

6 in.

6 in.

Try it
1. Find the area of the figure shown here. Use 3.14 for π.

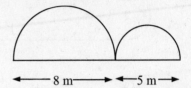

8 m · 5 m

Solution
This figure is composed of a square and half a circle.
- Find the area of each figure.
- Add the areas to find the area of the composite figure.

The area of the square is

$$A = s^2$$
$$= (6 \text{ in.})^2$$
$$= 36 \text{ in}^2$$

The area of the half-circle is

$$A = \frac{1}{2}\pi r^2$$
$$\approx \frac{1}{2}(3.14)(3 \text{ in.})^2$$
$$= 14.13 \text{ in}^2$$

The total area is

$$36 \text{ in}^2 + 14.13 \text{ in}^2 = 50.13 \text{ in}^2$$

Reflect on it
- In Quiz Yourself 2 (see next page), will it be easier to
  add the areas of two rectangles   *OR*   subtract the areas of two rectangles
  to determine the area of the composite figure? Why?

Quiz Yourself 4.3B
In Exercises 1 through 4, find the area. Use 3.14 for π if necessary.

1.

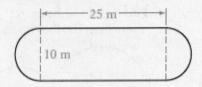

25 m

10 m

2.

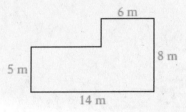

6 m

8 m

5 m

14 m

3.

Radius
= 6 cm

4.   The figure below is formed by a square and one-half of a circle with diameter $d = 14$ cm.

d

Practice Sheet 4.3B

Find the area. Use 3.14 for π.

1.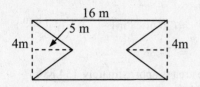

2.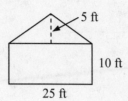

1. _____

2. _____

3.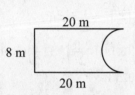

4.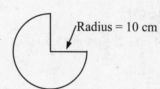

Radius = 10 cm

3. _____

4. _____

5.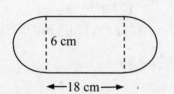

6.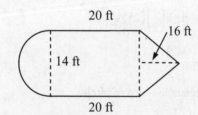

5. _____

6. _____

7.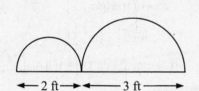

8.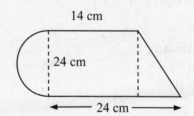

7. _____

8. _____

315

Answers

Try it 4.3B
   1.   139.73 m^2

Quiz 4.3B
   1.   328.5 m^2
   2.   88 m^2
   3.   84.78 cm^2
   4.   119.07 cm^2

Solutions to Practice Sheet 4.3B

1.   Area = area of rectangle $- 2 \times$ area of triangle

$$= LW - 2\left(\frac{1}{2}bh\right)$$

$$= (16 \text{ m} \times 4 \text{ m}) - 2\left(\frac{1}{2}\right)(4 \text{ m} \times 5 \text{ m})$$

$$= 44 \text{ m}^2$$

The area is 44 m^2.

2.   Area = area of rectangle + area of triangle

$$= LW + \frac{1}{2}bh$$

$$= (25 \text{ ft} \times 10 \text{ ft}) + \left(\frac{1}{2}\right)(25 \text{ ft} \times 5 \text{ ft})$$

$$= 312.5 \text{ ft}^2$$

The area is 312.5 ft^2.

3.   Area = area of rectangle + area of circle

$$= LW + \pi\left(\frac{d}{2}\right)^2$$

$$\approx (18 \text{ cm} \times 6 \text{ cm}) + 3.14\left(\frac{6 \text{ cm}}{2}\right)^2$$

$$= 136.26 \text{ cm}^2$$

The area is approximately 136.26 cm^2.

4.   Area $= \frac{3}{4} \times$ area of circle

$$= \frac{3}{4}\left(\pi r^2\right)$$

$$\approx \frac{3}{4}(3.14)(10 \text{ cm})^2$$

$$= 235.5 \text{ cm}^2$$

The area is approximately 235.5 cm^2.

5.   Area = area of rectangle $- \frac{1}{2} \times$ area of circle

$$= LW - \frac{1}{2}\pi\left(\frac{d}{2}\right)^2$$

$$\approx (20 \text{ m} \times 8 \text{ m}) - \left(\frac{1}{2}\right)(3.14)\left(\frac{8 \text{ m}}{2}\right)^2$$

$$= 134.88 \text{ m}^2$$

The area is approximately 134.88 m^2.

6.   Area = area of rectangle $+ \frac{1}{2} \times$ area of circle
   + area of triangle

$$= LW + \frac{1}{2}\pi\left(\frac{d}{2}\right)^2 + \frac{1}{2}bh$$

$$\approx (20 \text{ ft} \times 14 \text{ ft}) + \left(\frac{1}{2}\right)(3.14)\left(\frac{14 \text{ ft}}{2}\right)^2$$

$$+ \frac{1}{2}(14 \text{ ft})(16 \text{ ft})$$

$$= 468.93 \text{ ft}^2$$

The area is approximately 468.93 ft^2.

7.   Area $= \frac{1}{2} \times$ area of circle $+ \frac{1}{2} \times$ area of circle

$$= \frac{1}{2}\pi\left(\frac{d}{2}\right)^2 + \frac{1}{2}\pi\left(\frac{d}{2}\right)^2$$

$$\approx \left(\frac{1}{2}\right)(3.14)\left(\frac{2 \text{ ft}}{2}\right)^2 + \left(\frac{1}{2}\right)(3.14)\left(\frac{3 \text{ ft}}{2}\right)^2$$

$$= 5.1025 \text{ ft}^2$$

The area is approximately 5.1025 ft^2.

8.   Base of the triangle is 24 cm $-$ 14 cm = 10 cm

Area = area of rectangle $+ \frac{1}{2} \times$ area of circle
   + area of triangle

$$= LW + \frac{1}{2}\pi\left(\frac{d}{2}\right)^2 + \frac{1}{2}bh$$

$$\approx (14 \text{ cm} \times 24 \text{ cm}) + \left(\frac{1}{2}\right)(3.14)\left(\frac{24 \text{ cm}}{2}\right)^2$$

$$+ \frac{1}{2}(10 \text{ cm})(24 \text{ cm})$$

$$= 682.08 \text{ cm}^2$$

The area is approximately 682.08 cm^2.

Objective 4.3C – Solve application problems

Example

Carpet is to be installed in the room diagrammed below. If the carpeting will cost $5/square meter to install, then what will the installation cost be?

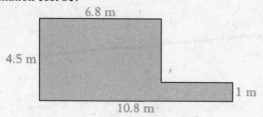

Try it

1. Carpet is to be installed in the room diagrammed below. If the carpeting will cost $0.50/square foot to install, then what will the installation cost be?

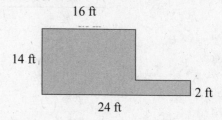

Solution

To find the installation cost,
- Find the area of the composite figure.
- Multiply the area by the $5/square meter charge.

The area of the above room may be computed as either
the sum of the areas of two rectangles, *OR*
the difference of the areas of two rectangles.

Here, we show the solution using the sum of the areas.
The "big rectangle" (on the left) has length 6.8 m and width 4.5 m, so its area is

$$(6.8\text{ m})(4.5\text{ m}) = 30.6\text{ m}^2.$$

The "small rectangle" (on the right) has length $10.8\text{m} - 6.8\text{m},$ or 4 m. Its area is

$$(4\text{ m})(1\text{ m}) = 4\text{ m}^2.$$

The total area of the room is
$$30.6\text{ m}^2 + 4\text{ m}^2, \text{ or } 34.6\text{ m}^2$$

Here, we show the solution using the difference of the areas.
Notice also that the area of the room may be computed by viewing the room as being 10.8 m by 4.5 m, with an area of 4 m by 3.5 m removed. Computing the area this way gives
$$(10.8\text{ m})(4.5\text{ m}) - (4\text{ m})(3.5\text{ m})$$
$$= 48.6\text{ m}^2 - 14\text{ m}^2$$
$$= 34.6\text{ m}^2$$
This is the same total area.

The cost of installation is then $5(34.6), or $173.

Quiz Yourself 4.3C

1. Carpet is to be installed in the room diagrammed below. If the carpeting will cost $0.40/square foot to install, then what will the installation cost be?

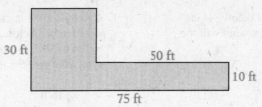

2. A room has dimensions 24 ft. by 14 ft. How many 2 ft. by 2 ft. tiles should be purchased to tile the floor of the room?

3. A park in the shape of a square is to be seeded with bags of grass seed. Each bag of grass seed covers $1000 \, \text{ft}^2$ and costs $6. If each side of the park has length 150 ft, then how many bags of seed should be purchased if only whole bags may be purchased?

Practice Sheet 4.3C

Solve.

1. Find the area of a circular rug that is 18 ft in diameter.

2. Find the area of a circular portrait that has a 24 in. diameter.

1. _____

2. _____

3. A circular skating rink has a 144-foot radius. Find the area of the surface of the ice rink.

4. Find the area of a rectangular park that is 300 m long and 180 m wide.

3. _____

4. _____

5. Find the area of a concrete driveway with the following measurements.

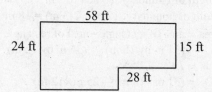

6. A carpet is to be placed in two rooms as shown in the diagram. At $17.50 per square meter, how much will it cost to carpet the area?

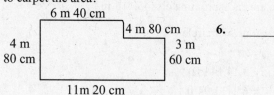

5. _____

6. _____

7. How many square yards of carpet are needed to cover the indoor arena?

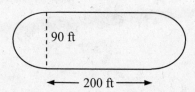

8. Find the area of the 3-meter boundary around the swimming pool.

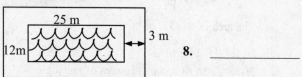

7. _____

8. _____

Answers

Try it 4.3C
1. $120

Quiz 4.3C
1. $500
2. 84
3. 23

Solutions to Practice Sheet 4.3C

1. Area = area of circle

$$= \pi \left(\frac{d}{2}\right)^2$$

$$\approx 3.14 \left(\frac{18 \text{ ft}}{2}\right)^2$$

$$= 254.34 \text{ ft}^2$$

The area is approximately 254.34 ft^2.

2. Area = area of circle

$$= \pi \left(\frac{d}{2}\right)^2$$

$$\approx 3.14 \left(\frac{24 \text{ in.}}{2}\right)^2$$

$$= 452.16 \text{ in}^2$$

The area is approximately 452.16 in^2.

3. Area = area of circle

$$= \pi r^2$$

$$\approx 3.14 (144 \text{ ft})^2$$

$$= 65,111.04 \text{ ft}^2$$

The area is approximately $65,111.04 \text{ ft}^2$.

4. Area = area of rectangle

$$= LW$$

$$= 300 \text{ m} \times 180 \text{ m}$$

$$= 54,000 \text{ m}^2$$

The area is $54,000 \text{ m}^2$.

5. $24 \text{ ft} - 15 \text{ ft} = 9 \text{ ft}$

Area = area of rectangle − area of rectangle
   (58 ft by 24 ft)   (28 ft by 9 ft)

$$= LW - LW$$

$$= (58 \text{ ft} \times 24 \text{ ft}) - (28 \text{ ft} \times 9 \text{ ft})$$

$$= 1140 \text{ ft}^2$$

The area is 1140 ft^2.

6. STRATEGY
   - Convert lengths to meters.
   - Find the length of the missing side.
   - Find the area.
   - Find the total cost.

SOLUTION

$4.8 \text{ m} - 3.6 \text{ m} = 1.2 \text{ m}$

Area = area of rectangle − area of rectangle
   11.2 m by 4.8 m   4.8 m by 1.2 m

$$= LW - LW$$

$$= (11.2 \text{ m} \times 4.8 \text{ m}) - (4.8 \text{ m} \times 1.2 \text{ m})$$

$$= 48 \text{ m}^2$$

$\text{Cost} = 48 \times 17.50 = 840$

The cost is $840.

7. Area = area of rectangle + area of circle

$$= LW + \pi \left(\frac{d}{2}\right)^2$$

$$\approx (200 \text{ ft} \times 90 \text{ ft}) + 3.14 \left(\frac{90 \text{ ft}}{2}\right)^2$$

$$= 24,358.5 \text{ ft}^2$$

The area is approximately $24,358.5 \text{ ft}^2$.

8. Length of boundary: $25 \text{ m} + 2(3 \text{ m}) = 31 \text{ m}$
   Width of boundary: $12 \text{ m} + 2(3 \text{ m}) = 18 \text{ m}$

Area = area of rectangle − area of rectangle
   (31 m by 18 m)   (25 m by 12 m)

$$= LW - LW$$

$$= (31 \text{ m} \times 18 \text{ m}) - (25 \text{ m} \times 12 \text{ m})$$

$$= 258 \text{ m}^2$$

The area is 258 m^2.

Objective 4.4A – Find the volume of a geometric solid

Key Words and Concepts

Define volume. _____

_____

Volume of a Rectangular Solid

Let L be the length, W be the width, and H be the height of a rectangular solid.

    The volume of the solid is $V = LWH$.

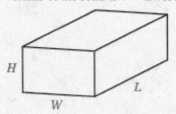

Example

Find the volume of a rectangular solid with length 3.5 m, width 2 m, and height 5 m.

  Solution

    Use the formula for the volume of a rectangular solid.
    Substitute the known values and simplify.

$$V = LWH$$
$$= (3.5 \text{ m})(2 \text{ m})(5 \text{ m})$$
$$= 35 \text{ m}^3$$

    The volume of the rectangular solid is 35 m³.

Try it

1. Find the volume of a rectangular solid with length 6 in, width 4 in, and height 3 in.

A **cube** is a rectangular solid for which the length, width, and height are all equal.

Volume of a Cube

Let s be the length of one side of a cube.

    The volume of the cube is $V = s^3$.

Example

Find the volume of a cube with side length $s = 8$ cm.

Solution

Use the formula for the volume of a cube.
Substitute the known value and simplify.

$$V = s^3$$
$$= (8 \text{ cm})^3$$
$$= 512 \text{ cm}^3$$

The volume of the cube is 512 cm^3.

Try it

2. Find the volume of a cube with side length $s = 6$ yd.

Volume of a Sphere

Let r be the radius of a sphere.

The volume V of the sphere is $V = \frac{4}{3}\pi r^3$.

Example

Find the volume of a sphere with radius 12 in. Use 3.14 for π. Round to the nearest whole number.

Solution

Use the formula for the volume of a sphere.
Substitute the known value and simplify.

$$V = \frac{4}{3}\pi r^3$$
$$\approx \frac{4}{3}(3.14)(12 \text{ in.})^3$$
$$= 7234.56 \text{ in}^3$$
$$\approx 7235 \text{ in}^3$$

The volume of the sphere is approximately 7235 in^3.

Try it

3. Find the volume of a sphere with radius 4 in. Use 3.14 for π. Round to the nearest whole number.

Volume of a Cylinder

Let r be the radius of a cylinder and h be the height of the cylinder.

The volume of the cylinder is $V = \pi r^2 h$.

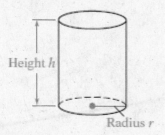

Height h

Radius r

Example

A cylinder has radius 8 cm and height 9 cm. What is its volume? Use 3.14 for π.

Solution

Use the formula for the volume of a sphere.
Substitute the known values and simplify.

$$V = \pi r^2 h$$
$$\approx (3.14)(8 \text{ cm})^2 (9 \text{ cm})$$
$$= 1808.64 \text{ cm}^3$$

The volume of the cylinder is approximately 1808.64 cm³.

Try it

4. A cylinder has radius 3.5 in. and height 7 in. What is its volume? Use 3.14 for π.

Reflect on it

- Volume is a measure of the amount of space inside a closed surface, or figure in space.
- The volume of a rectangular solid is $V = LWH$.
- The volume of the cube is $V = s^3$.
- The volume V of the sphere is $V = \dfrac{4}{3}\pi r^3$.
- The volume of the cylinder is $V = \pi r^2 h$.

Quiz Yourself 4.4A

In Exercises 1 through 4, find the volume. Use 3.14 for π when necessary.
Round to the nearest hundredth when necessary.

1.

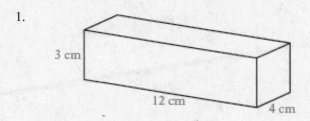

3 cm

12 cm

4 cm

2.
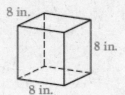

3.
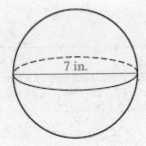

4.
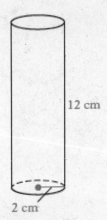

Practice Sheet 4.4A

Solve.

1. Find the volume of a cylinder with a radius of 10 cm and a height of 21 cm. Use $\frac{22}{7}$ for π.

2. Find the volume of a cylinder with a radius of 22 cm and a height of 35 cm. Use $\frac{22}{7}$ for π.

1. _____

2. _____

3. Find the volume of a cube with a side of 5 ft 3 in. Round to the nearest hundredth.

4. Find the volume of a cube with a side of 22 cm.

3. _____

4. _____

5. Find the volume of a cube with a side of 6 ft 4 in. Round to the nearest hundredth.

6. Find the volume of a sphere with a radius of 6 mm. Use 3.14 for π. Round to the nearest hundredth.

5. _____

6. _____

7. Find the volume of a sphere with a radius of 12 cm. Use 3.14 for π. Round to the nearest hundredth.

8. Find the volume of a sphere with a radius of 1.4 mm. Use 3.14 for π. Round to the nearest hundredth.

7. _____

8. _____

9. Find the volume of a rectangular solid with a length of 2 m 24 cm, and a width of 75 cm, and a height of 50 cm.

10. Find the volume of a rectangular solid with a length of 5 m, and width of 250 cm, and a height of 150 cm.

9. _____

10. _____

Answers

Try it 4.4A
1. 72 in^3
2. 216 yd^3
3. 268 in^3
4. 269.255 in^3

Quiz 4.4A
1. 144 cm^3
2. 512 in^3
3. 179.50 in^3
4. 150.72 cm^3

Solutions to Practice Sheet 4.4A

1. $V = \pi r^2 h$
   $V \approx \dfrac{22}{7}(10 \text{ cm})^2(21 \text{ cm}) \approx 6600 \text{ cm}^3$

2. $V = \pi r^2 h$
   $V \approx \dfrac{22}{7}(22 \text{ cm})^2(35 \text{ cm}) \approx 53,240 \text{ cm}^3$

3. $V = s^3$
   $V = (5 \text{ ft } 3 \text{ in.})^3 = \left(5\dfrac{3}{12} \text{ ft}\right)^3 = (5.25 \text{ ft})^3$
   $V \approx 144.70 \text{ ft}^3$

4. $V = s^3 = (22 \text{ cm})^3 = 10,648 \text{ cm}^3$

5. $V = s^3$
   $V = (6 \text{ ft } 4 \text{ in.})^3 = \left(6\dfrac{4}{12} \text{ ft}\right)^3 = \left(\dfrac{76}{12} \text{ ft}\right)^3$
   $V = \dfrac{6859}{27} \text{ ft}^3$
   $V \approx 254.04 \text{ ft}^3$

6. $V = \dfrac{4}{3}\pi r^3$
   $V \approx \dfrac{4}{3}(3.14)(6 \text{ mm})^3 \approx 904.32 \text{ mm}^3$

7. $V = \dfrac{4}{3}\pi r^3$
   $V \approx \dfrac{4}{3}(3.14)(12 \text{ cm})^3 \approx 7234.56 \text{ cm}^3$

8. $V = \dfrac{4}{3}\pi r^3$
   $V \approx \dfrac{4}{3}(3.14)(1.4 \text{ mm})^3 \approx 11.49 \text{ mm}^3$

9. $V = LWH$
   $V = (2 \text{ m } 24 \text{ cm})(75 \text{ cm})(50 \text{ cm})$
   $V = (224 \text{ cm})(75 \text{ cm})(50 \text{ cm})$
   $V = 840,000 \text{ cm}^3$

10. $V = LWH$
    $V = (5 \text{ m})(250 \text{ cm})(150 \text{ cm})$
    $V = (500 \text{ cm})(250 \text{ cm})(150 \text{ cm})$
    $V = 18,750,000 \text{ cm}^3$

Objective 4.4B – Find the volume of a composite geometric solid

Key Words and Concepts

A **composite geometric solid** is a solid made from two or more geometric solids.

Example

Find the volume of the solid shown here. The solid consists of a cylinder capped by a hemisphere. Use 3.14 for π.

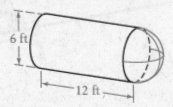

Solution

Find the volume of each solid figure.
Add the volumes.

The radius of the cylinder is

$$\frac{1}{2}(6 \text{ ft}) = 3 \text{ ft.}$$

The volume of the cylinder is

$$V = \pi r^2 h$$
$$\approx (3.14)(3 \text{ ft})^2 (12 \text{ ft})$$
$$= 339.12 \text{ ft}^3$$

The volume of the hemisphere (half of a sphere) is

$$V = \frac{1}{2} \cdot \frac{4}{3} \pi r^3$$
$$\approx \frac{2}{3}(3.14)(3 \text{ ft})^3$$
$$= 56.52 \text{ ft}^3$$

Thus, the total volume is approximately

$$339.12 \text{ ft}^3 + 56.52 \text{ ft}^3, \text{ or } 395.64 \text{ ft}^3.$$

Try it

1. Find the volume of the solid shown in the Example (to the left) replace the diameter of 6 ft with 4 in, and replace 12 ft with 9 in. Use 3.14 for π.

Reflect on it

- A composite geometric solid is a solid made from two or more geometric solids.
- When finding the volume of a composite geometric solid, under what circumstances will you need to

    add the volumes of the geometric solids?

    subtract the volumes of the geometric solids?

Quiz Yourself 4.4B

In Exercises 1 through 3, find the volume of the given figure. Use 3.14 for π, and round to the nearest hundredth, if necessary.

1.

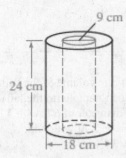

2.

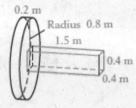

3.

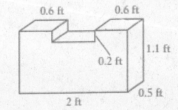

Practice Sheet 4.4B

Find the volume. Write measurements involving two units in terms of the larger unit before working the problem. Use 3.14 for π.

1.

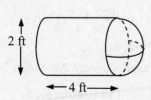

2 ft

←— 4 ft —→

2.

12 in.   4 in.

←— 6 in. —→

1. _____

2. _____

3.

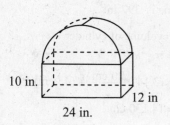

10 in.

12 in

24 in.

4.

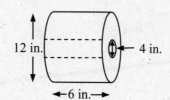

8 cm

←20 cm

60 cm

2 m 30 cm

3. _____

4. _____

5.

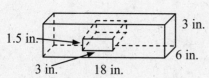

1.5 in.

3 in.

3 in.   18 in.

6 in.

6.

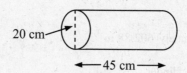

20 cm

←— 45 cm —→

5. _____

6. _____

Answers

Try it 4.4B

1. 129.79 in^3

Quiz 4.4B

1. 4578.12 cm^3
2. 0.64 m^3
3. 1.02 ft^3

Solutions to Practice Sheet 4.4B

1. Volume = volume of cylinder
$+ \dfrac{1}{2}$ volume of sphere

$$= \pi r^2 h + \dfrac{1}{2}\left(\dfrac{4}{3}\pi r^3\right)$$

$$\approx (3.14)\left(\dfrac{2 \text{ ft}}{2}\right)^2 (4 \text{ ft})$$

$$+ \dfrac{1}{2}\left(\dfrac{4}{3}\right)(3.14)\left(\dfrac{2 \text{ ft}}{2}\right)^3$$

$$\approx 14.65 \text{ ft}^3$$

The volume is approximately 14.65 ft^3.

2. Volume = volume of cylinder
$-$ volume of cylinder

$$= \pi r^2 h - \pi r^2 h$$

$$\approx (3.14)\left(\dfrac{12 \text{ in.}}{2}\right)^2 (6 \text{ in.})$$

$$- (3.14)\left(\dfrac{4 \text{ in.}}{2}\right)^2 (6 \text{ in.})$$

$$= 602.88 \text{ in}^3$$

The volume is approximately 602.88 in^3.

3. Volume = $\dfrac{1}{2}$ volume of cylinder
$+$ volume of rectangular solid

$$= \dfrac{1}{2}(\pi r^2 h) + LWH$$

$$\approx \dfrac{1}{2}(3.14)\left(\dfrac{24 \text{ in.}}{2}\right)^2 (12 \text{ in.})$$

$$+ (24 \text{ in.})(12 \text{ in.})(10 \text{ in.})$$

$$= 5592.96 \text{ in}^3$$

The volume is approximately 5592.96 in^3.

4. Volume = volume of rectangular solid
$-$ volume of cylinder

$$= LWH - \pi r^2 h$$

$$\approx (2.30 \text{ m})(0.08 \text{ m})(0.60 \text{ m})$$

$$- (3.14)\left(\dfrac{0.20 \text{ m}}{2}\right)^2 (2.30 \text{ m})$$

$$= 0.03818 \text{ m}^3$$

The volume is approximately 0.03818 m^3.

5. Volume = volume of rectangular solid
$-$ volume of rectangular solid

$$= LWH - LWH$$

$$= (18 \text{ in.})(6 \text{ in.})(3 \text{ in.})$$

$$- (3 \text{ in.})(1.5 \text{ in.})(6 \text{ in.})$$

$$= 297 \text{ in}^3$$

The volume is 297 in^3.

6. Volume = volume of cylinder

$$= \pi r^2 h$$

$$\approx (3.14)\left(\dfrac{20 \text{ cm}}{2}\right)^2 (45 \text{ cm})$$

$$= 14{,}130 \text{ cm}^3$$

The volume is approximately $14{,}130 \text{ cm}^3$.

Objective 4.4C – Solve application problems

Example

The tanker pictured below is half full of oil. What volume of oil is in the tanker? Use 3.14 for π, and round the final answer to the nearest whole number.

Try it

1. Find the volume of a railroad box car that is 50 ft long, 9.5 ft wide, and 13 ft high.

Solution

The tanker is composed of a cylinder and a full sphere (two hemispheres). Notice that the radius of the tanker is

$$\frac{1}{2}(8 \text{ ft}) = 4 \text{ ft}.$$

The volume of the cylindrical part is

$$V = \pi r^2 h$$
$$\approx (3.14)(4 \text{ ft})^2 (30 \text{ ft})$$
$$= 1507.20 \text{ ft}^3$$

The volume of the spherical part is

$$V = \frac{4}{3}\pi r^3$$
$$\approx \frac{4}{3}(3.14)(4 \text{ ft})^3$$
$$\approx 267.95 \text{ ft}^3$$

The volume of oil is half of the total tanker volume. So the volume of oil is approximately

$$\frac{1}{2}\left(1507.20 \text{ ft}^3 + 267.95 \text{ ft}^3\right), \text{ or}$$

approximately 888 ft^3.

Quiz Yourself 4.4C
In Exercises 1 through 3, use 3.14 for π when necessary.

1. The diagram shows a concrete floor. What volume of concrete will be needed to form the floor? Round to the nearest whole number.

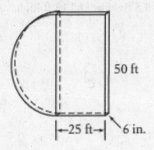

2. Find the volume of the bushing shown below.

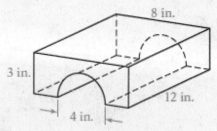

3. A rectangular tank at a fish hatchery is 8 yd long, 4 yd wide, and 1.5 yd deep. Find the volume of water in the tank when the tank is full.

Practice Sheet 4.4C

Solve.

1. A propane gas storage tank which is in the shape of a cylinder, is 10 m high and has a 6-meter diameter. Find the volume of the gas storage tank. Use 3.14 for π.

2. A silo, which is the shape of a cylinder, is 27 ft in diameter and has a height of 48 ft. Find the volume of the silo. Use 3.14 for π.

1. _____

2. _____

3. Find the volume of a spherical oxygen tank that is 9 m in diameter. Use 3.14 for π. Round to the nearest tenth.

4. Find the volume of a spherical water tank that is 18 ft in diameter. Use 3.14 for π. Round to the nearest hundredth.

3. _____

4. _____

5. Find the volume of a railroad car that is 16 m long, 5 m wide, and 5.2 m high.

6. A hole is being dug for installing a swimming pool. The hole is 24 ft long, 12 ft wide, and 8 ft deep. Find the volume of the hole.

5. _____

6. _____

7. How many gallons of water will fill a fish tank that is 24 in. long, 12 in. wide and 10 in. high? Round to the nearest tenth. (1 gal = 231 in.^{3})

8. How many gallons of water will fill an aquarium that is 10 in. wide, 18 in. high and 8 in. long? (1 gal = 231 in.^{3})

7. _____

8. _____

9. Find the volume of the concrete needed to build the steps.

10. The floor of a building is shown below. The concrete floor is 6 in. thick. Find the cost of the floor at $2.50 per cubic foot.

9. _____

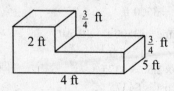

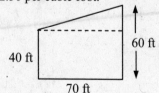

10. _____

Answers

Try it 4.4C
1. 6175 ft^3

Quiz 4.4C
1. 1116 ft^3
2. 212.64 in^3
3. 48 yd^3

Solutions to Practice Sheet 4.4C

1. Volume = volume of cylinder
$$= \pi r^2 h$$
$$\approx (3.14)\left(\frac{6 \text{ m}}{2}\right)^2 (10 \text{ m})$$
$$= 282.6 \text{ m}^3$$
The volume is approximately 282.6 m^3.

2. Volume = volume of cylinder
$$= \pi r^2 h$$
$$\approx (3.14)\left(\frac{27 \text{ ft}}{2}\right)^2 (48 \text{ ft})$$
$$= 27,468.72 \text{ ft}^3$$
The volume is approximately $27,468.72 \text{ ft}^3$.

3. Volume = volume of sphere
$$= \frac{4}{3}\pi r^3$$
$$\approx \frac{4}{3}(3.14)\left(\frac{9 \text{ m}}{2}\right)^3$$
$$\approx 381.5 \text{ m}^3$$
The volume is approximately 381.5 m^3.

4. Volume = volume of sphere
$$= \frac{4}{3}\pi r^3$$
$$\approx \frac{4}{3}(3.14)\left(\frac{18 \text{ ft}}{2}\right)^3$$
$$\approx 3052.08 \text{ ft}^3$$
The volume is approximately 3052.08 ft^3.

5. Volume = volume of rectangular solid
$$= LWH$$
$$= (16 \text{ m})(5 \text{ m})(5.2 \text{ m})$$
$$= 416 \text{ m}^3$$
The volume is 416 m^3.

6. Volume = volume of rectangular solid
$$= LWH$$
$$= (24 \text{ ft})(12 \text{ ft})(8 \text{ ft})$$
$$= 2304 \text{ ft}^3$$
The volume is 2304 ft^3.

7. Volume = volume of rectangular solid
$$= LWH$$
$$= (24 \text{ in.})(12 \text{ in.})(10 \text{ in.})$$
$$= 2880 \text{ in.}^3$$
Convert from cubic inches to gallons.
$$2880 \text{ in.}^3 \cdot \frac{1 \text{ gal}}{231 \text{ in.}^3} \approx 12.5 \text{ gal}$$
The fish tank will need approximately 12.5 gal of water.

8. Volume = volume of rectangular solid
$$= LWH$$
$$= (8 \text{ in.})(10 \text{ in.})(18 \text{ in.})$$
$$= 1440 \text{ in.}^3$$
Convert from cubic inches to gallons.
$$1440 \text{ in.}^3 \cdot \frac{1 \text{ gal}}{231 \text{ in.}^3} \approx 6.2 \text{ gal}$$
The aquarium will need approximately 6.2 gal of water.

9. Volume = volume of rectangular solid
$$+ \text{ volume of rectangular solid}$$
$$= LWH + LWH$$
$$= (4 \text{ ft})(5 \text{ ft})\left(\frac{3}{4}\text{ft}\right) + (2 \text{ ft})(5 \text{ ft})\left(\frac{3}{4}\text{ft}\right)$$
$$= 22.5 \text{ ft}^3$$
The volume is 22.5 ft^3.

10. Find the area of the figure.
To find the volume, multiply the area by the thickness, 6 in., or 0.5 ft.
Find the cost.
Area = area of rectangle + area of triangle
$$= LW + \frac{1}{2}bh$$
$$= (70 \text{ ft})(40 \text{ ft}) + \frac{1}{2}(70 \text{ ft})(60 - 40 \text{ ft})$$
$$= 3500 \text{ ft}^2$$
Volume $= (3500 \text{ ft}^2)(0.5 \text{ ft})$
$$= 1750 \text{ ft}^3$$
$1750 \times 2.50 = 4375$
The cost is $4375.

Objective 4.5A – Apply the Pythagorean Theorem

Key Words and Concepts

The Greek mathematician Pythagoras is generally credited with the discovery that the square of the hypotenuse of a right triangle is equal to the sum of the squares of the two legs.

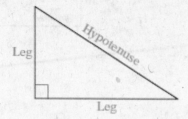

Pythagorean Theorem

If a and b are the lengths of the legs of a right triangle and c is the length of the hypotenuse, then $c^2 = a^2 + b^2$.

If the length of the hypotenuse is unknown, use

$$\text{Hypotenuse} = \sqrt{(\text{leg})^2 + (\text{leg})^2}$$

If the length of a leg is unknown, use

$$\text{Leg} = \sqrt{(\text{hypotenuse})^2 - (\text{leg})^2}$$

Example

Find the length of the hypotenuse in the right triangle shown in the figure. Round to the nearest thousandth.

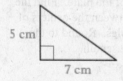

Solution
Use the formula for unknown hypotenuse.
Substitute the known values and simplify.

$$\begin{aligned}
\text{Hypotenuse} &= \sqrt{(\text{leg})^2 + (\text{leg})^2} \\
&= \sqrt{5^2 + 7^2} \\
&= \sqrt{74} \\
&\approx 8.602
\end{aligned}$$

The length of the hypotenuse is approximately 8.602 cm.

Try it

1. Find the length of the hypotenuse in the right triangle shown in the figure. Round to the nearest thousandth.

Example

Find the length of the leg in the right triangle shown in the figure.

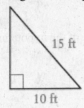

15 ft

10 ft

Solution

Use the formula for unknown hypotenuse.
Substitute the known values and simplify.

$$\text{Leg} = \sqrt{(\text{hypotenuse})^2 - (\text{leg})^2}$$
$$= \sqrt{15^2 - 10^2}$$
$$= \sqrt{125}$$
$$\approx 11.180$$

The length of the leg is approximately 11.180 ft.

Try it

2. Find the length of the leg in the right triangle shown in the figure.

8 cm   16 cm

Example

Find the distance between the centers of the holes in the plate shown in the figure. Round to the nearest hundredth.

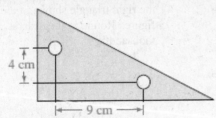

4 cm

9 cm

Solution

The Pythagorean Theorem can be used to find the distance. The lengths 4 cm and 9 cm are thought of as the legs of a right triangle, and the missing distance is the hypotenuse.

$$\text{Hypotenuse} = \sqrt{(\text{leg})^2 + (\text{leg})^2}$$
$$= \sqrt{4^2 + 9^2}$$
$$= \sqrt{97}$$
$$\approx 9.85$$

The distance between the centers of the holes is approximately 9.85 cm.

Try it

3. Four holes are drilled in the circular plate in the figure below. The centers of the holes are 3 in. from the center of the plate. Find the distance between the centers of adjacent holes. Round to the nearest thousandth.

Reflect on it

- Pythagorean Theorem put simply is $c^2 = a^2 + b^2$.

Quiz Yourself 4.5A

In Exercises 1 through 3, find the unknown side of the triangle. Round to the nearest thousandth.

1.

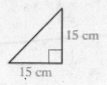

15 cm

15 cm

2.

9 m   12 m

3.

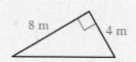

8 m   4 m

4.   An L-shaped sidewalk from a parking lot to a memorial is shown in the figure below. If the distances are as labeled, find the distance from the corner to the memorial.

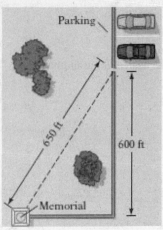

5.   A vinyl fence is to be built around the plot shown in the figure below. At $12.90 per meter, how much will it cost to build the fence?

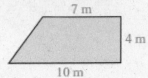

Practice Sheet 4.5A

Find the unknown side of the triangle. Round to the nearest hundredth.

1.
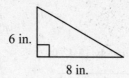

2.

1. _____

2. _____

3.

4.

3. _____

4. _____

5.
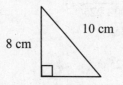

6.

5. _____

6. _____

Solve. Round to the nearest thousandth.

7. A guy wire holds a television antenna in place. Find the distance along the ground from the base of the antenna to the guy wire.

8. Find the length of a ramp used to move a fork lift from the ground to the loading ramp.

7. _____

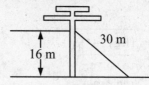

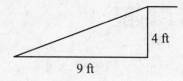

8. _____

9. A car is driven 24 mi west and then 10 mi north. How far is the car from the starting point?

10. Find the distance around the hiking trail.

9. _____

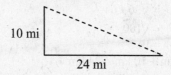

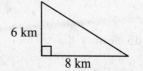

10. _____

Answers

Try it 4.5A

1. 9.615 cm
2. 13.856 cm
3. 4.243 in.

Quiz 4.5A

1. 21.213 cm
2. 7.937 m
3. 8.944 m
4. 250 ft
5. $335.40

Solutions to Practice Sheet 4.5A

1. $\text{Hypotenuse} = \sqrt{(\text{leg})^2 + (\text{leg})^2}$
   $= \sqrt{6^2 + 8^2}$
   $= \sqrt{36 + 64}$
   $= \sqrt{100}$
   $= 10$
   The hypotenuse is 10 in.

2. $\text{Leg} = \sqrt{(\text{hypotenuse})^2 - (\text{leg})^2}$
   $= \sqrt{14^2 - 12^2}$
   $= \sqrt{196 - 144}$
   $= \sqrt{52}$
   ≈ 7.21
   The leg is approximately 7.21 cm.

3. $\text{Leg} = \sqrt{(\text{hypotenuse})^2 - (\text{leg})^2}$
   $= \sqrt{10^2 - 8^2}$
   $= \sqrt{100 - 64}$
   $= \sqrt{36}$
   $= 6$
   The leg is 6 cm.

4. $\text{Leg} = \sqrt{(\text{hypotenuse})^2 - (\text{leg})^2}$
   $= \sqrt{30^2 - 27^2}$
   $= \sqrt{900 - 729}$
   $= \sqrt{171}$
   ≈ 13.08
   The leg is approximately 13.08 cm.

5. $\text{Leg} = \sqrt{(\text{hypotenuse})^2 - (\text{leg})^2}$
   $= \sqrt{18^2 - 12^2}$
   $= \sqrt{324 - 144}$
   $= \sqrt{180}$
   ≈ 13.42
   The leg is approximately 13.42 ft.

6. $\text{Hypotenuse} = \sqrt{(\text{leg})^2 + (\text{leg})^2}$
   $= \sqrt{20^2 + 15^2}$
   $= \sqrt{400 + 225}$
   $= \sqrt{625}$
   $= 25$
   The hypotenuse is 25 cm.

7. $\text{Leg} = \sqrt{(\text{hypotenuse})^2 - (\text{leg})^2}$
   $= \sqrt{30^2 - 16^2}$
   $= \sqrt{900 - 256}$
   $= \sqrt{644}$
   ≈ 25.377
   The distance is approximately 25.377 m.

8. $\text{Hypotenuse} = \sqrt{(\text{leg})^2 + (\text{leg})^2}$
   $= \sqrt{9^2 + 4^2}$
   $= \sqrt{81 + 16}$
   $= \sqrt{97}$
   ≈ 9.849
   The length of the ramp is approximately 9.849 ft.

9. $\text{Hypotenuse} = \sqrt{(\text{leg})^2 + (\text{leg})^2}$
   $= \sqrt{10^2 + 24^2}$
   $= \sqrt{100 + 576}$
   $= \sqrt{676}$
   $= 26$
   The car is 26 mi from the starting point.

10. $\text{Hypotenuse} = \sqrt{(\text{leg})^2 + (\text{leg})^2}$
    $= \sqrt{6^2 + 8^2}$
    $= \sqrt{36 + 64}$
    $= \sqrt{100}$
    $= 10$
    6 km + 8 km + 10 km = 24 km
    The distance around the hiking trail is 24 km.

Objective 4.6A – Find the values of trigonometric ratios of a right triangle

Key Words and Concepts

In this section, we will examine **right triangle trigonometry**—that is, trigonometry that applies only to right triangles.

When working with right triangles, it is convenient to refer to the side *opposite* an angle and the side *adjacent to* (next to) an angle.

Note   The hypotenuse of a right triangle is defined to be neither adjacent nor opposite either of the acute angles of the triangle.

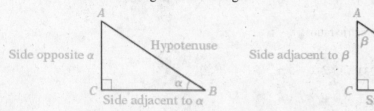

Note   It is common to use the Greek letters α (alpha), β (beta), and θ (theta).

The Trigonometric Ratios of an Acute Angle of a Right Triangle

If θ is an acute angle of a right triangle *ABC*, then

$$\sin\theta = \frac{\text{length of opposite side}}{\text{length of hypotenuse}} \qquad \csc\theta = \frac{\text{length of hypotenuse}}{\text{length of opposite side}}$$

$$\cos\theta = \frac{\text{length of adjacent side}}{\text{length of hypotenuse}} \qquad \sec\theta = \frac{\text{length of hypotenuse}}{\text{length of adjacent side}}$$

$$\tan\theta = \frac{\text{length of opposite side}}{\text{length of adjacent side}} \qquad \cot\theta = \frac{\text{length of adjacent side}}{\text{length of opposite side}}$$

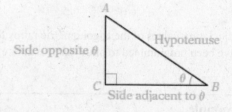

For convenience, we will write **opp**, **adj**, and **hyp** as the abbreviation for the length of the **opp**osite side, **adj**acent side, and **hyp**otenuse, respectively. Using this convention, the definitions of the trigonometric ratios are written as:

$$\sin\theta = \frac{\text{opp}}{\text{hyp}} \qquad \csc\theta = \frac{\text{hyp}}{\text{opp}}$$

$$\cos\theta = \frac{\text{adj}}{\text{hyp}} \qquad \sec\theta = \frac{\text{hyp}}{\text{adj}}$$

$$\tan\theta = \frac{\text{opp}}{\text{adj}} \qquad \cot\theta = \frac{\text{adj}}{\text{opp}}$$

When working with trigonometric ratios, be sure to draw a diagram and label the adjacent and opposites of an angle.

$$\sin\theta = \frac{\text{opp}}{\text{hyp}}$$

$$\cos\theta = \frac{\text{adj}}{\text{hyp}}$$

$$\tan\theta = \frac{\text{opp}}{\text{adj}}$$

Example

For a right triangle with hypotenuse 26, adjacent 10, and opposite 24, find the values of $\sin\theta$, $\cos\theta$, and $\tan\theta$. Round to the nearest ten-thousandth.

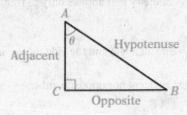

Solution

Use the definitions of the trigonometric ratios.

$$\sin\theta = \frac{\text{opp}}{\text{hyp}} = \frac{24}{26} \approx 0.9231$$

$$\cos\theta = \frac{\text{adj}}{\text{hyp}} = \frac{10}{26} \approx 0.3846$$

$$\tan\theta = \frac{\text{opp}}{\text{adj}} = \frac{24}{10} = 2.4$$

Try it

1. For a right triangle with hypotenuse 10, adjacent 8, and opposite 6, find the values of $\sin\theta$, $\cos\theta$, and $\tan\theta$. Round to the nearest ten-thousandth.

Calculating values of the trigonometric ratios for most angles would be quite difficult. Fortunately, many calculator have been programmed to allow us to estimate these values.

Example

Given $\cos\theta = 0.2473$, find the value of θ. Use a calculator. Round to the nearest tenth of a degree.

Solution

This is equivalent to finding $\cos^{-1}(0.2473)$.

$$\cos^{-1}(0.2473) \approx 75.7°$$
$$\theta \approx 75.7°$$

Try it

2. Given $\sin\theta = 0.8926$, find the value of θ. Use a calculator. Round to the nearest tenth of a degree.

Example

For the right triangle with adjacent side 36 mi and opposite side 24 mi, find the measure of angle A. Round to the nearest tenth of a degree.

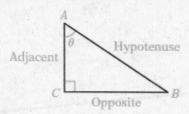

Try it

3. For the right triangle with hypotenuse 50 ft and opposite side 15 ft, find the measure of angle A. Round to the nearest tenth of a degree.

Solution

We are given the adjacent and opposite sides of angle A.
Use the tangent ratio to find the measure of angle A.

$$\tan A = \frac{\text{opp}}{\text{adj}}$$
$$\tan A = \frac{24}{36}$$
$$A = \tan^{-1}\left(\frac{24}{36}\right)$$
$$A \approx 33.7^\circ$$

The measure of angle A is approximately 33.7°.

Reflect on it

- What do you think SOHCAHTOA represents?
- For estimating angle measurements, calculators have two settings, Radian and Degree. Choose Degree setting.

Quiz Yourself 4.6A

For Exercises 1 to 4, use the following diagram.

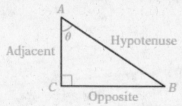

1. If the hypotenuse is 50, adjacent side is 48, and opposite side is 14, find the values of $\sin\theta$, $\cos\theta$, and $\tan\theta$. Round to the nearest ten-thousandth.

2. If the adjacent side is 9 and opposite side is 10, find the values of $\sin\theta$, $\cos\theta$, and $\tan\theta$. Round to the nearest ten-thousandth.

3. Angle A is 53° and the length of side AC is 29 ft. Find the length of side BC.
    Round to the nearest tenth of a foot.

4. The adjacent side is 42 in. and the hypotenuse is 68 in. Find the value of θ.
    Round to the nearest tenth of an inch.

5. Given $\cos\theta = 0.7361$, find the value of θ. Round to the nearest tenth of a degree.

Practice Sheet 4.6A

Solve.

1. For a right triangle with adjacent side 6 and opposite side 9, find the values of $\sin\theta$, $\cos\theta$, and $\tan\theta$. Round to the nearest ten-thousandth.

2. For a right triangle with adjacent side 12 and opposite side 20, find the values of $\sin\theta$, $\cos\theta$, and $\tan\theta$. Round to the nearest ten-thousandth.

1. _____

2. _____

3. Given $\cos\theta = 0.3298$, find the value of θ.

4. Given $\sin\theta = 0.9327$, find the value of θ.

3. _____

4. _____

5. Given $\tan\theta = 8.9271$, find the value of θ.

6. Given $\cos\theta = 0.1209$, find the value of θ.

5. _____

6. _____

7. For the right triangle with opposite side 19 in. and $\theta = 64°$, find the length of the adjacent side. Round to the nearest tenth of an inch.

8. For the right triangle with opposite side 34 m and hypotenuse 85 m, find the measure of angle θ. Round to the nearest tenth of a degree.

7. _____

8. _____

Answers

Try it 4.6A

1. $\sin\theta = 0.6$, $\cos\theta = 0.8$, $\tan\theta = 0.75$
2. $\theta \approx 63.2°$
3. $17.5°$

Quiz 4.6A

1. $\sin\theta = 0.28$, $\cos\theta = 0.96$, $\tan\theta \approx 0.2917$
2. $\sin\theta \approx 0.7433$, $\cos\theta = 0.6690$, $\tan\theta \approx 1.1111$
3. 38.5 ft
4. $\theta \approx 51.9°$
5. $\theta = 42.6°$

Solutions to Practice Sheet 4.6A

1. First find the hypotenuse.

   $$hyp = \sqrt{6^2 + 9^2} = \sqrt{36+81} = \sqrt{117} \approx 10.8167$$

   $$\sin\theta = \frac{opp}{hyp} = \frac{9}{\sqrt{117}} \approx 0.8321$$

   $$\cos\theta = \frac{adj}{hyp} = \frac{6}{\sqrt{117}} \approx 0.5547$$

   $$\tan\theta = \frac{opp}{adj} = \frac{9}{6} = 1.5$$

2. First find the hypotenuse.

   $$hyp = \sqrt{20^2 + 12^2} = \sqrt{400+144} = \sqrt{544}$$
   $$\approx 23.3238$$

   $$\sin\theta = \frac{opp}{hyp} = \frac{20}{\sqrt{544}} \approx 0.8575$$

   $$\cos\theta = \frac{adj}{hyp} = \frac{12}{\sqrt{544}} \approx 0.5145$$

   $$\tan\theta = \frac{opp}{adj} = \frac{20}{12} \approx 1.6667$$

3. $\cos\theta = 0.3298$
   $$\theta = \cos^{-1}(0.3298) \approx 70.7°$$

4. $\sin\theta = 0.9327$
   $$\theta = \sin^{-1}(0.9327) \approx 68.9°$$

5. $\tan\theta = 8.9271$
   $$\theta = \tan^{-1}(8.9271) \approx 83.6°$$

6. $\cos\theta = 0.1209$
   $$\theta = \cos^{-1}(0.1209) \approx 83.1°$$

7. $$\tan\theta = \frac{opp}{adj}$$
   $$\tan 64° = \frac{19}{adj}$$
   $$adj = \frac{19}{\tan 64}$$
   $$adj \approx 9.3 \text{ in.}$$

8. $$\sin\theta = \frac{opp}{hyp}$$
   $$\sin\theta = \frac{34}{85}$$
   $$\theta = \sin^{-1}\left(\frac{34}{85}\right)$$
   $$\theta \approx 23.6°$$

Objective 4.6B – Solve applications involving trigonometric ratios

Key Words and Concepts

Example

Suppose engineers measure angle A as 31.3° and AC as 150 ft. Find the length BC across the ravine. Round to the nearest tenth.

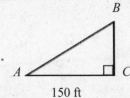

150 ft

Solution

Use the tangent function.

$$\tan A = \frac{\text{opp}}{\text{adj}}$$

$$\tan 31.3° = \frac{BC}{150}$$

$$BC = 150(\tan 31.3°)$$

$$BC \approx 91.2 \text{ ft}$$

The distance BC is approximately 91.2 ft.

Try it

1. A surveyor wants to determine the distance across a lake. The surveyor measures distance BC as 47 m and angle A as 64°. Find the distance AB across the lake. Round to the nearest tenth.

Angles of elevation and depression are measured with respect to a horizontal line.

If the object being observed is *above* the observer, the acute angle formed by the line of sight and the horizontal line is an **angle of elevation**.

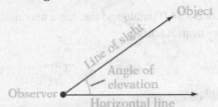

If the objective being observed is *below* the observer, the acute angle formed by the line of sight and the horizontal line is an **angle of depression**.

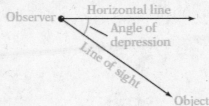

Example

The angle of elevation from a point 150 ft away from the base of a flagpole to the top of the flagpole is 28°. Find the height of the flagpole. Round ot the nearest tenth of a foot.

Solution
Draw a diagram.

h

$28°$

150 ft

To find the height h, write a trigonometric ratio that relates the given information and the unknown side of the triangle.

$$\tan 28° = \frac{h}{150} \qquad \text{Use tangent}$$

$$150(\tan 28°) = 150\left(\frac{h}{150}\right) \qquad \text{Multiply each side by 150}$$

$$79.8 \approx h \qquad \text{Use a calculator to find the approximation}$$

The height of the flagpole is approximately 79.8 ft.

Try it

2. The angle of depression from the top of a church tower that is 120 ft high to a historic marker on the ground is 31°. How far is the historic marker from the base of the church tower? Round to the nearest tenth.

Reflect on it

- If the object being observed is *above* the observer, the acute angle formed by the line of sight and the horizontal line is an angle of elevation.
- If the objective being observed is *below* the observer, the acute angle formed by the line of sight and the horizontal line is an angle of depression.

Quiz Yourself 4.6B

1. A skateboarder wishes to build a jump ramp that is inclined at a 14.0° angle and that has a maximum height of 25.0 inches. Find the horizontal width x of the ramp. Round to the nearest tenth.

2. A 15-foot ladder is resting against a wall and makes an angle of 63° with the ground. Find the height which the ladder reaches on the wall. Round to the nearest tenth.

3. Two totem poles are 25 ft apart. The angle of depression from the top of one totem pole to the top of the second totem pole is 9°. How many feet shorter than the first totem pole is the second totem pole? Round to the nearest tenth.

4. A surveyor determines that the angle of elevation from a base of a statue to the top of a tower is 35.2°. The statue is 47 ft from the tower. How tall is the tower? Round to the nearest tenth of a foot.

Practice Sheet 4.6B

Solve.

1. The angle of elevation from a point 40 ft from the base of the light pole to the top of the pole is 55°. Find the height of the light pole. Round to the nearest tenth of a foot.

2. An observer 6 feet tall stands h feet from the base of a pole that is 30 feet high. The angle of elevation from the observer to the top of the pole is 28°. Find the distance h. Round to the nearest tenth of a foot.

1. _____

2. _____

3. The angle of depression from the top of a monument that is 50 m high to a marker is 34°. How far is the marker from the base of the monument? Round to the nearest tenth of a meter.

4. A ladder rests against a wall at a point 15 ft up on the wall and makes an angle of 47° with the horizontal. Find the length of the ladder. Round to the nearest tenth of a foot.

3. _____

4. _____

5. A 10-foot ladder rests against a wall and makes an angle of 65° with a horizontal. Find the horizontal distance from the base of the ladder to the wall. Round to the nearest tenth of a foot.

6. A ramp is inclined at a 9° angle with a rise of 10 feet. Find the horizontal width of the ramp. Round to the nearest tenth of a foot.

5. _____

6. _____

Answers

Try it 4.6B
1. 96.4 m
2. 199.7 ft

Quiz 4.6B
1. 100.3 in.
2. 13.4 ft
3. 4.0 ft
4. 208.7 ft

Solutions to Practice Sheet 4.6B

1. $\tan 55^\circ = \dfrac{h}{40}$

    $h = 40(\tan 55^\circ)$

    $h \approx 57.1 \text{ ft}$

2. $\tan 28^\circ = \dfrac{30-6}{h}$

    $h = \dfrac{24}{\tan 28^\circ}$

    $h \approx 45.1 \text{ ft}$

3. $\tan 34^\circ = \dfrac{50}{x}$

    $x = \dfrac{50}{\tan 34^\circ}$

    $x \approx 74.1 \text{ m}$

4. $\sin 47^\circ = \dfrac{15}{L}$

    $L = \dfrac{15}{\sin 47^\circ}$

    $L \approx 20.5 \text{ ft}$

5. $\cos 65^\circ = \dfrac{x}{10}$

    $x = 10(\cos 65^\circ)$

    $x \approx 4.2 \text{ ft}$

6. $\tan 9^\circ = \dfrac{10}{x}$

    $x = \dfrac{10}{\tan 9^\circ}$

    $x \approx 63.1 \text{ ft}$

Application and Activities

Discussion and reflection questions

1.   Does the container for products you purchase matter to you?

2.   Have you ever picked one product over another because you liked the packaging more?

3.   Discuss the process you believe companies go through to decide new product packaging.

Group Activity (2-3 people)

Objectives for the lesson:

In this lesson you will use and enhance your understanding of the following:

- Defining geometric figures
- Finding the values of trigonometric ratios of a right triangle
- Applying the Pythagorean Theorem
- Finding the perimeter of a geometric figure
- Finding the area of a geometric figure
- Finding the volume of a geometric figure
- Describing geometric figures
- Finding the area of a composite geometric figure
- Solving application problems

1.   Carol is designing a new container for her shampoo line. She is starting to design the base first, and wants to use a right triangle with one angle having a measure of 32°. She also wants the longest side of her right triangle to be 7 inches long.

Find the length of the side opposite the 32° angle, y, using a trigonometric ratio.

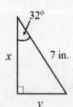

Find the third side, x, using the Pythagorean Theorem.

2.   Find the perimeter and the area of the right triangle Carol is going to use for the base of her shampoo container. Round your answer to the nearest tenth.

3. Carol's new shampoo container is going to be in the shape of a right triangular prism.   The base of her
    bottle will be the right triangle she just created.

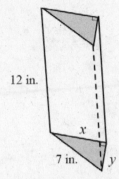

12 in.

7 in.

x

y

The volume of a right prism is given by
$$V = (\text{Area of base}) \cdot (\text{length})$$

Find the volume of Carol's new shampoo container if the length of the container is
going to be 12 inches.

4. To create the bottle, the manufacturer uses a net of the shape, as shown.   Describe which geometric shapes
    are used in the net.

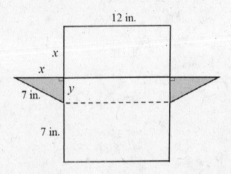

12 in.

x

x

7 in.

y

7 in.

5. Find the total area of the composite figure represented by the net in question #4, and round to the nearest
    whole number. The total area will give us the surface area of the right rectangular prism, which represents
    the amount of material needed to create the new shampoo bottle.

6. The company creating the bottles for Carol will charge her $0.03 per square inch of material, and $0.95 for
    labor to create each bottle. How much will it cost Carol to create each bottle?

7.   Carol's shampoo costs \$0.07 to mix per cubic inch. She wants to leave 10% of the volume of the bottle empty. How much will it cost her to make and put shampoo in the bottle?

8.   If Carol sells each bottle for \$25, how much profit will she earn per bottle?

Objective 5.1A – Graph points in a rectangular coordinate system

Key Words and Concepts

A **rectangular coordinate system** is formed by two number lines, one horizontal and one vertical, that intersect at the zero point of each line. The point of intersection is called the **origin**. The two lines are called **coordinate axes**, or simply **axes**.

The axes determine a **plane**. The two axes divide the plane into four regions called **quadrants**, numbered counterclockwise from I to IV.

Each point in the plane can be identified by a pair of numbers called an **ordered pair**.

The first number of the pair measures horizontal distance and is called the _____ .

The second number of the pair measures vertical distance and is called the _____ .

The **coordinates** of a point are the numbers in the ordered pair associated with the point. The **graph of an ordered pair** is the dot drawn at the coordinates of the point in the plane.

When drawing a rectangular coordinate system, we often label the horizontal axis x and the vertical axis y. In this case, the coordinate system is called an ***xy*-coordinate system**. The coordinates of the points are given by ordered pairs (x, y), where the first number is the ***x*-coordinate** and the second number is the ***y*-coordinate**.

Graphing, or plotting, an ordered pair in the plane means placing a _____ at the location given by the ordered pair.

Example
Graph the ordered pairs $(1, 1)$, $(1,-4)$, $(-2, 1)$, and $(1,-2)$.

    Solution
        Plot each point on the graph.

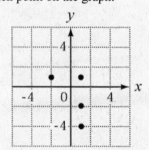

Try it
  1. Graph the ordered pairs $(-3, 4)$, $(-2, 2)$, $(2, -3)$, and $(4, 0)$.

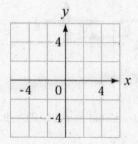

Example
Find the coordinates of each of the points.

Solution

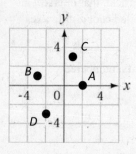

A is (2, 0), *B* is (–3, 1), *C* is (1, 3), *D* is (–2, -3)

Try it
2. Find the coordinates of each of the points.

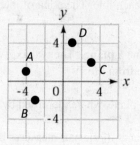

Reflect on it
- The first number of the pair measures horizontal distance and is called the abscissa.
- The second number of the pair measures horizontal distance and is called the ordinate.
- Graphing, or plotting, an ordered pair in the plane means placing a dot at the location given by the ordered pair.
- Each point in the plane is associated with an ordered pair, and each ordered pair is associated with a point on the plane.

Quiz Yourself 5.1A

1. Graph the ordered pairs (–1, 2), (2,–4), (–1, 3), and (1, 0).

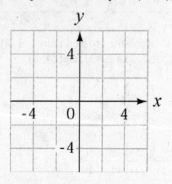

2. Graph the ordered pairs (4,–2), (–2,–2), (–2, 0), and (2, –1).

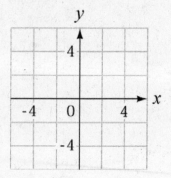

3. Graph the ordered pairs (2,–3), (–1,–2), (0, 1), and (4, 0).

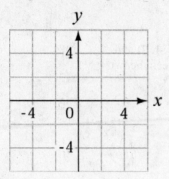

4. Graph the ordered pairs (4, 1), (–3,–2), (2, 2), and (0,–1).

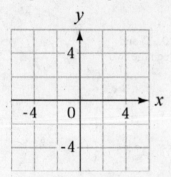

Practice Sheet 5.1A

Solve.

1. Graph the ordered pairs (2, 1) and (1, 3). Draw a line between the two points.

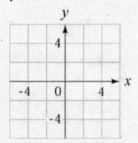

2. Graph the ordered pairs (–1, 2) and (2, –1). Draw a line between the two points.

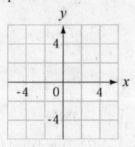

3. Graph the ordered pairs (–2,–3) and (3, 1). Draw a line between the two points.

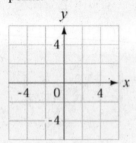

4. Graph the ordered pairs (0,–3) and (–2, 0). Draw a line between the two points.

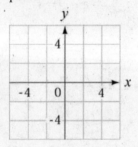

5. Find the coordinates of each of the points.

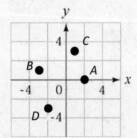

6. Find the coordinates of each of the points.

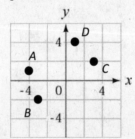

5. _____

6. _____

Answers

Try it 5.1A

1.

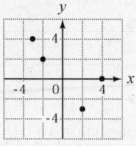

2. A is $(-4, 1)$, B is $(-3, -2)$, C is $(3, 2)$, D is $(1, 4)$

Quiz 5.1A

1.

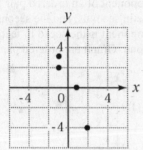

2.

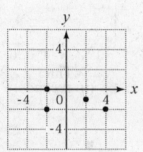

3.

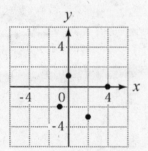

4.

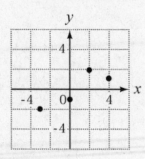

Solutions to Practice Sheet 5.1A

1.

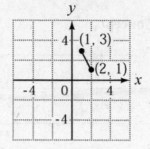

2.

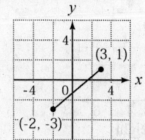

3.

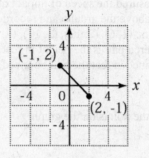

4.

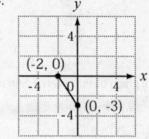

5. A is $(2, 0)$, B is $(-3, 1)$, C is $(1, 3)$,
    D is $(-2, -3)$

6. A is $(-4, 1)$, B is $(-3, -2)$, C is $(3, 2)$,
    D is $(1, 4)$

Objective 5.1B – Graph a scatter diagram

Key Words and Concepts

Define scatter diagram. _____

_____

Example

A physics student measured the speed of impact of a solid ball dropped from various heights above the ground and recorded the results as the following ordered pairs: (1, 6), (2, 12), (3, 17), (4, 22). The first component of the ordered pair is the height in feet, and the second component is the distance in feet the ball bounced. Graph the ordered pairs.

Solution
Plot each of the points on the graph.

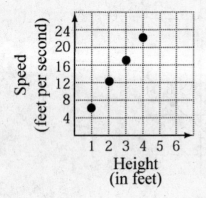

Try it

1. The speed of a steel ball as it rolls down a ramp is recorded as the following ordered pairs: (0, 0), (1, 3), (2, 6), (3, 9). The first component of an ordered pair is the time in seconds, and the second component is the speed in feet per second. Graph the ordered pairs.

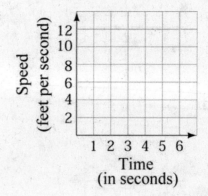

Reflect on it

• A scatter diagram is a graph of the ordered pairs of the known data.

Quiz Yourself 5.1B

1. A biochemist recorded the number of bacteria in a culture at different times as the following ordered pairs: (1, 6), (2, 10), (3, 16), (4, 24). The first component of the ordered pair is the time in hours, and the second component is the number of bacteria. Graph the ordered pairs.

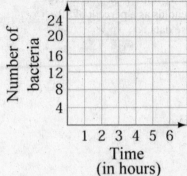

2. A chemistry student recorded the temperature of a chemical reaction at different times as the following ordered pairs: (3, 6), (6, 9), (9, 12), (12, 15). The first component of the ordered pair is the time measured in minutes, and the second component is temperature measured in degrees. Graph the ordered pairs.

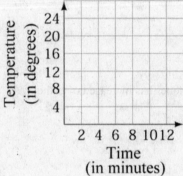

Practice Sheet 5.1B

Solve.

1. The table shows the number of U.S. households with a net worth of $1 million or more every three years from 1989 to 2001. Also shown is the millionaire households as a percent of all households in the United States. Graph the scatter diagram for these data.

Millionaire households (in millions), x	3.0	3.1	3.4	4.1	5.0
Percent of all households, y	3.2	3.2	3.4	4.0	4.7

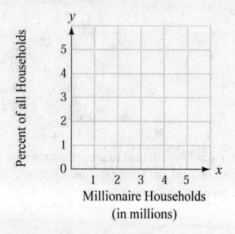

2. As shown in the table below, the percent of households with telephones has increased over the years. Draw a scatter diagram for these data.

Year, x	'20	'30	'40	'50	'60	'70	'80	'90	'00
Percent, y	35	41	37	62	78	90	93	95	97

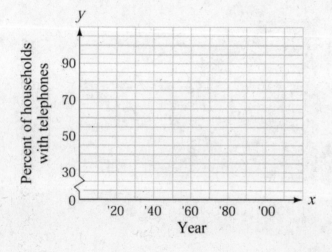

Answers

Try it 5.1B

1.

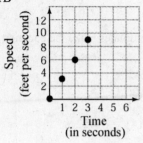

2.

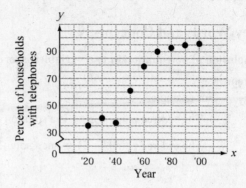

Quiz 5.1B

1.

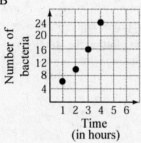

2.

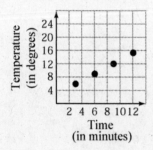

Solutions to Practice Sheet 5.1B

1.

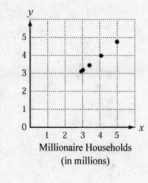

Objective 5.1C – Graph an equation in two variables

Key Words and Concepts

A **solution of an equation in two variables** is an ordered pair (x, y) whose coordinates make the equation a true statement.

For example,

The ordered pair $(-3, 4)$ is a solution of $2x^2 + y = 22$ because $2(-3)^2 + 4 = 22$.

The ordered pair $(4, -3)$ is not a solution of $2x^2 + y = 22$ because $2(4)^2 - 3 \neq 22$.

Example

Graph the equation $y = -\dfrac{1}{2}x - 2$ by first plotting the solutions of

the equation when $x = -4, -2, 0, 2, 4$.

Solution

Substitute each value of x into the equation and solve for y.

x	$y = -\dfrac{1}{2}x - 2$	y	(x, y)
-4	$y = -\dfrac{1}{2}(-4) - 2$	0	$(-4, 0)$
-2	$y = -\dfrac{1}{2}(-2) - 2$	-1	$(-2, -1)$
0	$y = -\dfrac{1}{2}(0) - 2$	-2	$(0, -2)$
2	$y = -\dfrac{1}{2}(2) - 2$	-3	$(2, -3)$
4	$y = -\dfrac{1}{2}(4) - 2$	-4	$(4, -4)$

Then graph the resulting ordered pairs by placing a dot at the coordinates of each point.

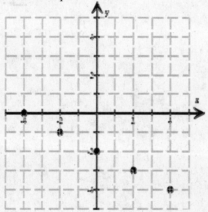

Try it

1. Graph the equation $y = 2x - 4$ by first plotting the solutions of the equation when $x = 0, 1, 2, 3, 4$.

Connect the points to form the graph.

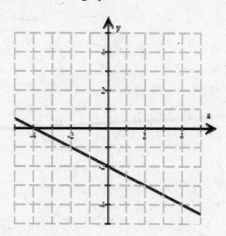

Reflect on it
- The graph of an equation is the graph of all of the ordered-pair solutions of the equation.

Quiz Yourself 5.1C
In Exercises 1 through 5, graph the equation. First plot the solutions of the equation for the given values of x, and then connect the points with a smooth graph.

1. $y = x + 1$
     $x = -4, -2, 0, 2, 4$

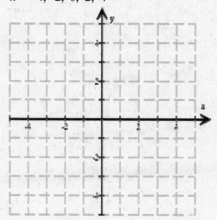

2. $y = -x - 1$

   $x = -4, -2, 0, 2, 4$

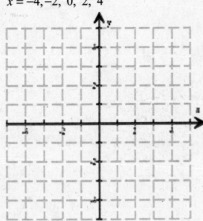

3. $y = |x - 2|$

   $x = -4, -2, 0, 2, 4$

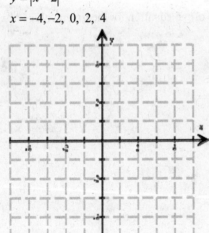

4. $y = \dfrac{1}{3}x + 1$

    $x = -6, -3, 0, 3, 6$

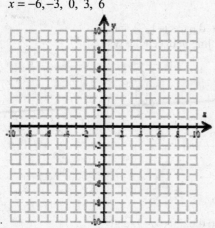

5. $y = -x^2 + 2$

    $x = -2, -1, 0, 1, 2$

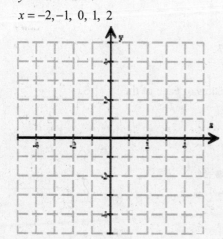

Practice Sheet 5.1C

Solve.

1. Graph the ordered pairs (2, 1) and (1, 3). Draw a line between the two points.

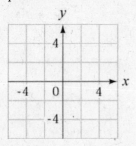

2. Graph the ordered pairs (–1, 2) and (2, –1). Draw a line between the two points.

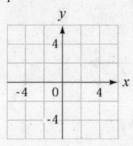

3. Graph the ordered pairs (–2, –3) and (3, 1). Draw a line between the two points.

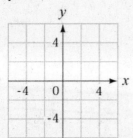

4. Graph the ordered pairs (0, –3) and (–2, 0). Draw a line between the two points.

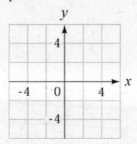

5. Find the coordinates of each of the points.

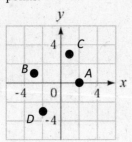

6. Find the coordinates of each of the points.

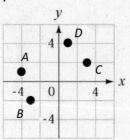

5. _____

6. _____

7. Draw a line through all points with an abscissa of –2.

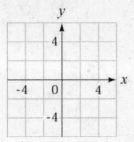

8. Draw a line through all points with an ordinate of 3.

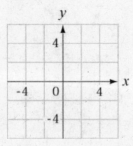

Answers

Try it 5.1C

1.
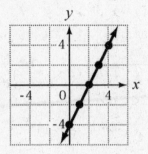

Quiz 5.1C

1.
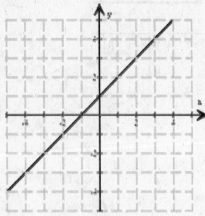

2.
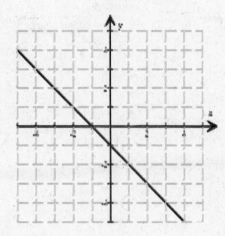

3.
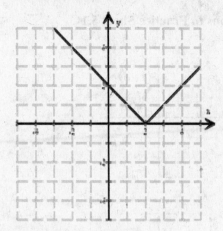

4.
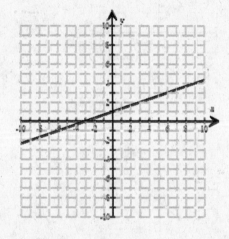

5.
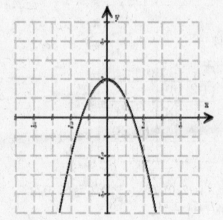

Solutions to Practice Sheet 5.1C

1.

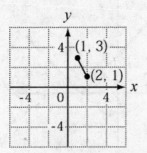

7.

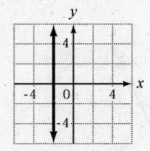

2.

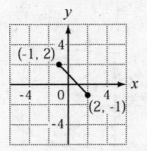

8.

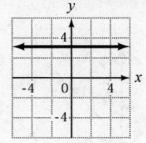

3.

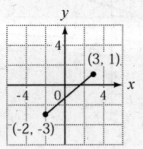

4.

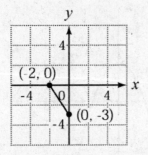

5. *A* is (2, 0), *B* is (–3, 1), *C* is (1, 3),
   D is (–2,–3)

6. *A* is (–4, 1), *B* is (–3,–2), *C* is (3, 2),
   D is (1, 4)

Objective 5.2A – Evaluate a function

Key Words and Concepts

A **function** is a set of ordered pairs in which no two ordered pairs have the same first coordinate.

The **domain** of a function is the set of _____ coordinates of the ordered pairs.
                                            (first/second)

The **range** of a function is the set of _____ coordinates of the ordered pairs.
                                            (first/second)

A **relation** is any set of ordered pairs. A function is a special type of relation. The concepts of domain and range apply to relations as well as to functions.

Example
Determine whether $\{(-2, 4), (-1, 1), (0, 0), (1, 1), (2, 4)\}$ is a function. State the domain and range.

Solution
No two ordered pairs have the same first element.
        Thus, the set of ordered pairs is a function.

The domain is $\{-2, -1, 0, 1, 2\}$.   The set of first coordinates

The range is $\{0, 1, 4\}$.   The set of second coordinates

Try it
1. Determine whether $\{(5, 3), (3, 1), (1, -1), (-1, -3), (-3, -5)\}$ is a function. State the domain and range.

Example
Determine whether $\{(4, -2), (1, -1), (0, 0), (1, 1), (4, 2)\}$ is a function. State the domain and range.

Solution
The ordered pairs $(4, -2)$ and $(4, 2)$ have the same first coordinate. (It is also the case that $(1, -1)$ and $(1, 1)$ have the same first coordinate.)
        Thus, the set of ordered pairs is not a function.

The domain is $\{0, 1, 4\}$.   The set of first coordinates

The range is $\{-2, -1, 0, 1, 2\}$.   The set of second coordinates

Try it
2. Determine whether $\{(5, 3), (5, 1), (2, 1), (0, 0), (2, -1)\}$ is a function. State the domain and range.

371

Let's look at the equation $y = 5x$. Because the value of y *depends* on the value of x, y is called the **dependent variable** and x is called the **independent variable**. We can say that y is a function of x.

To emphasize that the equation represents a function, **function notation** is used. Just as the variable x is commonly used to represent a number, the letter f is commonly used to name a function.

For example, $f(x) = 5x$, which is read "f of x is equal to $5x$."

Note $f(x)$ does *not* mean f times x.

The symbol $f(x)$ is the **value of the function** and represents the value of the dependent variable for a given value of the independent variable. We often write $y = f(x)$ to emphasize the relationship between the independent variable x and the dependent variable y.

Note y and $f(x)$ are different symbols for the same number.

What does it mean to evaluate a function? _____

_____

Example

Evaluate $f(x) = 7 - x^2$, when $x = -2$.

Solution
    Replace x by -2 and then simplify.
$$f(x) = 7 - x^2$$
$$f(-2) = 7 - (-2)^2$$
$$= 7 - 4$$
$$= 3$$
When x is -2, the value of the function is 3.

Try it

3. Evaluate $f(x) = 2x^2 - 3x + 5$, when $x = -1$.

When a function is represented by an equation, the domain of the function is all real numbers for which the value of the function is a real number.

Example

What is the domain of $P(x) = (x-1)^2$?

Solution
    Because the value of $(x-1)^2$ is a real number for any value of x, *no values* are excluded from the domain of $P(x) = (x-1)^2$.

    The domain of the function is all real numbers,
        or $\{x \mid -\infty < x < \infty\}$.

Try it

4. What is the domain of $f(x) = 5x^2 - 3x + 1$?

Example

What is the domain of $R(x) = \dfrac{1}{x-1}$?

Solution

For $x = 1$,

$$R(1) = \frac{1}{x-1} = \frac{1}{1-1} = \frac{1}{0}, \text{ which is undefined.}$$

So, 1 is excluded from the domain of R.

The domain is $\{x \mid x \neq 1\}$.

Try it

5. What is the domain of

$$f(x) = \frac{1}{x+6}?$$

Reflect on it

- A relation is any set of ordered pairs.
- A function is a set of ordered pairs in which *no two* ordered pairs have the *same first* coordinate.
- The domain of a function is the set of *first coordinates* of the ordered pairs.
- The range of a function is the set of *second coordinates* of the ordered pairs.
- The process of determining $f(x)$ for a given value of x is called **evaluating a function**.

Quiz Yourself 5.2A

1. Determine whether $\{(1, 0), (2, 0), (3, 1), (4, 1), (5, 2)\}$ is a function. State the domain and range.

2. Given $f(x) = -x + 5$, evaluate $f(-2)$.

3. Given $g(x) = 5 - x^2$, evaluate $g(-2)$.

4. Given $h(x) = x^3 - 2$, evaluate $h(-1)$.

5. What is the domain of $r(x) = \dfrac{5}{x^2}$? Write the answer in set-builder notation.

6. What is the domain of $H(x) = 5x^2$? Write the answer in set-builder notation.

Practice Sheet 5.2A

State whether the relation is a function.

1. $\{(1, 1), (2, 3), (4, 3), (5, 7)\}$ 2. $\{(-3,-1), (-1,-2), (1, 5), (2, 6), (1,-7)\}$ 1. _____

2. _____

Given $f(x) = 2x + 1$, evaluate:

3. $f(-1)$ 4. $f(2)$ 5. $f(0)$ 3. _____

4. _____

5. _____

Given $g(x) = x^2 + x - 1$, evaluate:

6. $g(0)$ 7. $g(1)$ 8. $g(-1)$ 6. _____

7. _____

8. _____

9. $g(2)$ 10. $g(-2)$ 11. $g(t)$ 9. _____

10. _____

11. _____

Find the domain and range of the function.

12. $\{(1, 1), (2, 3), (4, 3), (5, 7)\}$ 13. $\{(1, 4), (6, 2), (3, 1), (5, 4)\}$ 12. _____

13. _____

What values are excluded from the domain of the function?

14. $f(x) = \dfrac{3x + 5}{6}$ 15. $g(x) = \dfrac{1}{x + 5}$ 16. $h(x) = \dfrac{4 - x}{3 - x}$ 14. _____

15. _____

16. _____

Find the range of the function defined by the equation and the given domain.

17. $f(x) = 4x + 3$ 18. $g(x) = \dfrac{3}{4 - x}$ 17. _____
         domain = $\{-1, 1, 3\}$ domain = $\{-1, 0, 3\}$

18. _____

Answers

Try it 5.2A

1. Yes; Domain = {5, 3, 1,–1,–3};
   Range = {3, 1,–1,–3,–5}
2. No; Domain = {5, 2, 0}; Range: {3, 1, 0,–1}
3. 10
4. $\{x \mid -\infty < x < \infty\}$
5. $\{x \mid x \neq -6\}$

Quiz 5.2A

1. Yes; Domain = {1, 2, 3, 4, 5};
   Range = {0, 1, 2}
2. 7
3. 1
4. –3
5. $\{x \mid x \neq 0\}$
6. $\{x \mid -\infty < x < \infty\}$

Solutions to Practice Sheet 5.2A

1. No two ordered pairs have the same first element.
   Yes, the set of ordered pairs is a function.

2. The ordered pairs (1, 5) and (1,–7) have the same first coordinate.
   No, the set of ordered pairs is not a function.

3. $f(x) = 2x + 1$
   $f(-1) = 2(-1) + 1 = -1$

4. $f(x) = 2x + 1$
   $f(2) = 2(2) + 1 = 5$

5. $f(x) = 2x + 1$
   $f(0) = 2(0) + 1 = 1$

6. $g(x) = x^2 + x - 1$
   $g(0) = 0^2 + 0 - 1 = -1$

7. $g(x) = x^2 + x - 1$
   $g(1) = 1^2 + 1 - 1 = 1$

8. $g(x) = x^2 + x - 1$
   $g(-1) = (-1)^2 + (-1) - 1 = -1$

9. $g(x) = x^2 + x - 1$
   $g(2) = 2^2 + 2 - 1 = 5$

10. $g(x) = x^2 + x - 1$
    $g(-2) = (-2)^2 + (-2) - 1 = 1$

11. $g(x) = x^2 + x - 1$
    $g(t) = t^2 + t - 1$

12. The domain is {1, 2, 4, 5}.
    The range is {1, 3, 7}.

13. The domain is {1, 3, 5, 6}.
    The range is {1, 2, 4}.

14. None

15. For $x = -5$, $g(-5) = \dfrac{1}{-5+5} = \dfrac{1}{0}$, which is undefined. So –5 is excluded from the domain of the function.

16. For $x = 3$, $h(3) = \dfrac{4-3}{3-3} = \dfrac{1}{0}$, which is undefined. So 3 is excluded from the domain of the function.

17. $f(x) = 4x + 3$ domain = {–1, 1, 3}
    $f(-1) = 4(-1) + 3 = -1$
    $f(1) = 4(1) + 3 = 7$
    $f(3) = 4(3) + 3 = 15$
    For the given domain, the range is {–1, 7, 15}.

18. $g(x) = \dfrac{3}{4-x}$ domain = {–1, 0, 3}
    $g(-1) = \dfrac{3}{4-(-1)} = \dfrac{3}{5}$
    $g(0) = \dfrac{3}{4-0} = \dfrac{3}{4}$
    $g(3) = \dfrac{3}{4-3} = 3$
    For the given domain, the range is $\left\{\dfrac{3}{5}, \dfrac{3}{4}, 3\right\}$.

Objective 5.2B – Graph a function

Key Words and Concepts

The **graph** of a function is a graph of all of the ordered pairs of the function.

Example

Graph the function $f(x) = x^2 + 2x - 2$. First plot the ordered pairs of the function for $x = -3, -2, -1, 0, 1, 2, 3$. Then connect the points to form the graph. Estimate the domain and range from the graph. Write the domain and range in set-builder notation and in interval notation.

> Solution
>
> Evaluate the function at each point.

x	$f(x) = x^2 + 2x - 2$	y	(x, y)
−3	$f(-3) = (-3)^2 + 2(-3) - 2$	1	(−3, 1)
−2	$f(-2) = (-2)^2 + 2(-2) - 2$	−2	(−2, −2)
−1	$f(-1) = (-1)^2 + 2(-1) - 2$	−3	(−1, −3)
0	$f(0) = (0)^2 + 2(0) - 2$	−2	(0, −2)
1	$f(1) = (1)^2 + 2(1) - 2$	1	(1, 1)
2	$f(2) = (2)^2 + 2(2) - 2$	6	(2, 6)
3	$f(3) = (3)^2 + 2(3) - 2$	13	(3, 13)

> Plot the points.

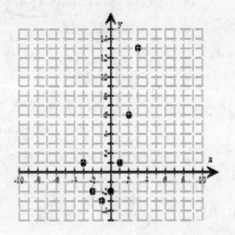

Try it

1. Graph the function $f(x) = -x^2 + 1$.

   First plot the ordered pairs of the function for $x = -3, -2, -1, 0, 1, 2, 3$. Then connect the points to form the graph. Estimate the domain and range from the graph. Write the domain and range in set-builder notation.

Connect the points with a smooth curve.

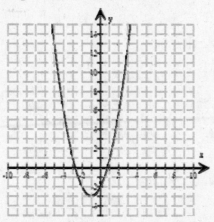

Because $x^2 + 2x - 2$ is a real number for all values of x, we can write the domain in set-builder notation as

$$\{x \mid -\infty < x < \infty\}.$$

In interval notation this is

$$(-\infty, \infty).$$

Because the values of y are all greater than or equal to -3, we can write the range in set-builder notation as

$$\{y \mid y \geq -3\}.$$

In interval notation this is

$$[-3, \infty).$$

Reflect on it
- Every ordered pair (x, y) on the graph of a function f satisfies $y = f(x)$, and every ordered pair that satisfies $y = f(x)$ is a point on the graph of the function.

Quiz Yourself 5.2B
In Exercises 1 through 4, graph the function. First plot the ordered pairs of the function for the given values of the independent variable. Then connect the points to form a graph. Estimate the domain and range from the graph. Write the domain and range in set-builder notation.

1. $f(x) = -x^2 + 4x - 1$, for $x = -1, 0, 1, 2, 3, 4, 5$

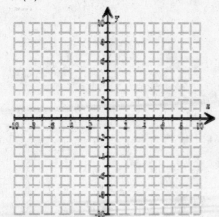

2. $f(x) = -|x-1| + 2$ for $x = -3, -2, -1, 0, 1, 2, 3$

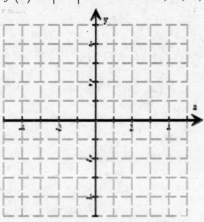

3. $g(m) = |3m| + 3$ for $m = -3, -2, -1, 0, 1, 2, 3$

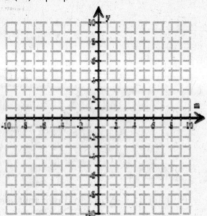

4. $f(t) = |3t+3|$, for $t = -3, -2, -1, 0, 1, 2, 3$

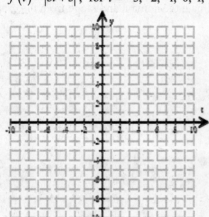

Practice Sheet 5.2B

Graph each function by evaluating the function at the given values of *x*. Plot the resulting ordered pairs. Then connect the points to form the graph.

1. $f(x) = 2x - 4$,
$x = 0, 1, 2, 3, 4$

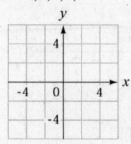

2. $g(x) = -3x - 3$
$x = -2, -1, 0$

3. $h(x) = -x^2 + 3$
$x = -2, -1, 0, 1, 2$

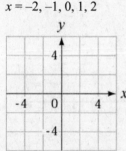

4. $P(x) = x^2 + 2x + 1$
$x = -3, -2, -1, 0, 1$

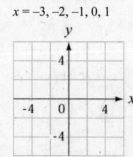

5. $f(x) = -|x - 1| + 2$
$x = -3, -2, -1, 0, 1, 2, 3, 4$

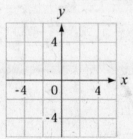

6. $g(x) = |x| + 2$
$x = -3, -2, -1, 0, 1, 2, 3$

Answers

Try it 5.2B

1. Domain $= \{x \mid -\infty < x < \infty\}$

   Range $= \{y \mid y \leq 1\}$

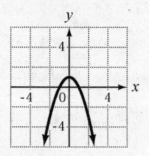

Quiz 5.2B

1. Domain $= \{x \mid -\infty < x < \infty\}$

   Range $= \{y \mid y \leq 3\}$

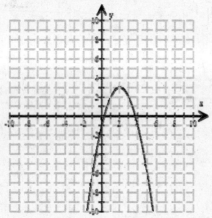

2. Domain $= \{x \mid -\infty < x < \infty\}$

   Range $= \{y \mid y \leq 2\}$

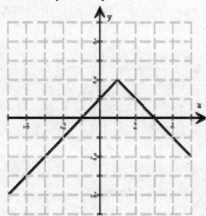

3. Domain $= \{x \mid -\infty < x < \infty\}$

   Range $= \{y \mid y \geq 3\}$

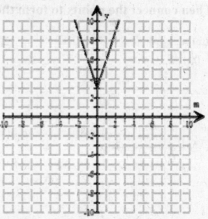

4. Domain $= \{x \mid -\infty < x < \infty\}$

   Range $= \{y \mid y \geq 0\}$

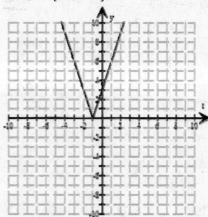

Solutions to Practice Sheet 5.2B

1. $f(x) = 2x - 4,$
   $x = 0, \ 1, \ 2, \ 3, \ 4$

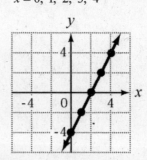

2. $g(x) = -3x - 3,$
    $x = -2, \ -1, \ 0$

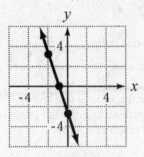

6. $g(x) = |x| + 2,$
    $x = -3, -2, -1, \ 0, \ 1, \ 2, \ 3$

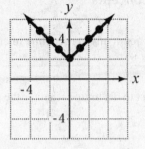

3. $h(x) = -x^2 + 3,$
    $x = -2, -1, \ 0, \ 1, \ 2$

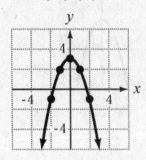

4. $P(x) = x^2 + 2x + 1,$
    $x = -3, -2, -1, \ 0, \ 1$

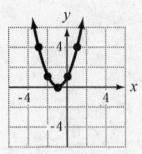

5. $f(x) = -|x - 1| + 2,$
    $x = -3, -2, -1, \ 0, \ 1, \ 2, \ 3, \ 4$

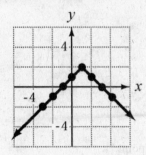

Name:
Module 5: Applications of Functions

Class:
Section 5.2: Introduction to Functions

Objective 5.2C – Apply the vertical line test

Key Words and Concepts

Vertical Line Test
A graph defines a function if any vertical line intersects the graph at no more than one point.

Example
Use the vertical line test to determine whether the graph is the graph of a function.

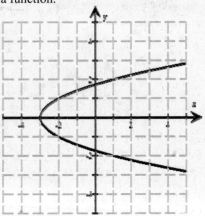

Try it
1. Use the vertical line test to determine whether the graph is the graph of a function.

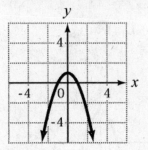

Solution

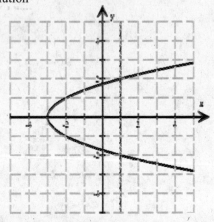

The dashed vertical line is one of infinitely many vertical lines that could be drawn passing through two points of the graph. Thus the graph is not the graph of a function.

Reflect on it
• A graph is *not* a function if a vertical line intersects the graph at more than one point.

382
© 2017 Cengage Learning. All Rights Reserved. May not be scanned, copied or duplicated, or posted to a publicly accessible website, in whole or in part.

Quiz Yourself 5.2C

For Exercises 1 through 4, use the vertical line test to determine whether the graph is the graph of a function.

1.

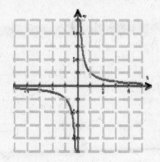

2.

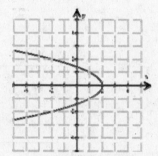

3.

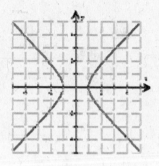

4.

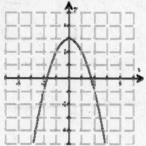

Name: Class:
Module 5: Applications of Functions   Section 5.2: Introduction to Functions

Practice Sheet 5.2C

Use the vertical line test to see if the graph represents function.

1.

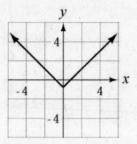

2.

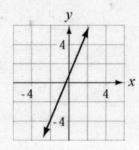

1. _____

2. _____

3.

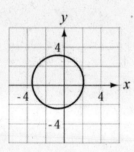

4.

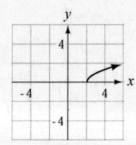

3. _____

4. _____

5.

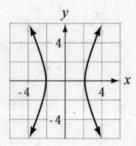

6.

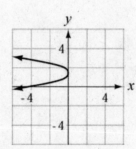

5. _____

6. _____

7.

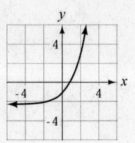

8.

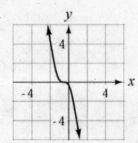

7. _____

8. _____

384

Answers

Try it 5.2C
1. Yes

Quiz 5.2C
1. Yes
2. No
3. No
4. Yes

Solutions to Practice Sheet 5.2C

1. Every vertical line intersects the graph at most once. Yes, the graph is the graph of a function.

2. Every vertical line intersects the graph at most once. Yes, the graph is the graph of a function.

3. There are vertical lines that intersect the graph at more than once point. No, the graph is not the graph of a function.

4. Every vertical line intersects the graph at most once. Yes, the graph is the graph of a function.

5. There are vertical lines that intersect the graph at more than once point. No, the graph is not the graph of a function.

6. There are vertical lines that intersect the graph at more than once point. No, the graph is not the graph of a function.

7. Every vertical line intersects the graph at most once. Yes, the graph is the graph of a function.

8. Every vertical line intersects the graph at most once. Yes, the graph is the graph of a function.

Objective 5.3A – Linear Functions

Key Words and Concepts

A function that can be written in the form $f(x) = mx + b$ (or $y = mx + b$) is called a **linear function** because its graph is a straight line.

Example

Graph: $f(x) = -\dfrac{2}{3}x + 2$

Solution

Find at least three ordered pairs.

Hint: Because the coefficient of x is a fraction with denominator 3, choosing values of x that are divisible by 3 simplifies the calculations.

x	$y = f(x)$
-3	4
0	2
3	0

Graph the ordered pairs and draw a line through the points.

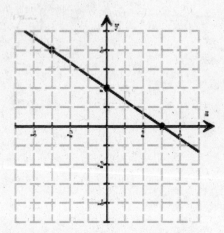

There are a variety of applications of linear functions.

Try it

1. Graph: $f(x) = \dfrac{1}{4}x - 1$

Example

An international data-roaming plan for a device costs $5 for the connection plus $0.15/megabyte (MB). The equation $C = 5 + 0.15d$, where C is the cost in dollars and d is the amount of data used (measured in MB), describes this relationship. How much will it cost to use 150 MB of roaming data on this device? How much data has been used if the roaming bill is $22.55? Graph the equation $C = 5 + 0.15d$ for $0 \le d \le 200$. The point (187, 33.05) is on the graph. Write a sentence that describes the meaning of this ordered pair.

Solution

If 150 MB of roaming data is used, the cost is:

$C = 5 + 0.15d$ Copy the formula

$C = 5 + 0.15(150)$ Substitute 150 for d

$C = 27.50$

The cost for 150 MB of roaming data is $27.50.

If the bill is $22.55, find the amount of data used by replacing C in the equation with 22.55:

$C = 5 + 0.15d$ Copy the formula

$22.55 = 5 + 0.15d$ Substitute 22.55 for C

$17.55 = 0.15d$

$117 = d$

If the bill is $22.55, 117 MB of data was used.

To graph the equation, substitute 0 and 200 for d to determine two points on the graph.

These two points may be more convenient to plot than the two points given in the solution above.

$C = 5 + 0.15d$ $C = 5 + 0.15d$

$C = 5 + 0.15(0)$ $C = 5 + 0.15(200)$

$C = 5$ $C = 35$

Plot the points (0, 5) and (200, 35) and connect with a line.

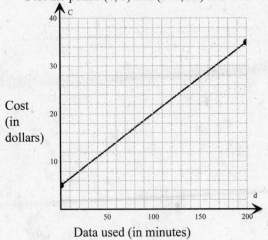

Try it

2. A roller coaster has a maximum speed of 99 ft/s. The equation that describes the total number of feet traveled by the roller coaster in t seconds at this speed is given by $D = 99t$. Graph this equation for $0 \le t \le 10$. The point (6, 594) is on this graph. Write a sentence that describes the meaning of this ordered pair.

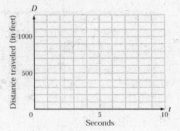

Reflect on it
- Choose values of x that make calculations simple.

Quiz Yourself 5.3A

1. Graph: $f(x) = -x$

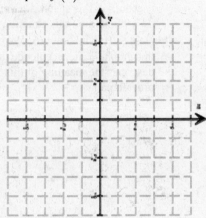

2. Graph: $y = -2x + 1$

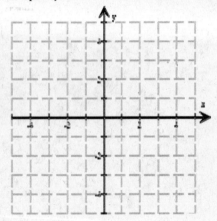

3. Graph: $y = -\dfrac{1}{2}x$

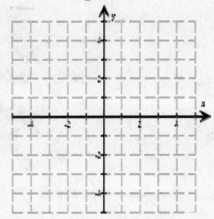

4. Graph: $f(x) = \dfrac{2}{5}x - 2$

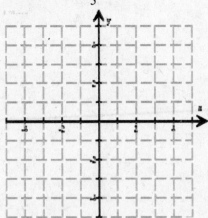

5. A truck costs $50 to rent, plus $0.20/mile driven. The equation $C = 50 + 0.20m$ gives the cost C in dollars for renting a truck and driving it m miles. Graph this equation for $0 \le m \le 500$. The point with coordinates (312, 112.40) is on the graph. Write a sentence that describes the meaning of this ordered pair.

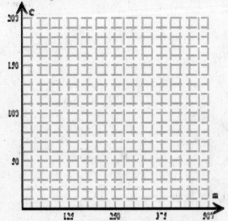

Practice Sheet 5.3A

Graph.

1. $y = 2x - 1$

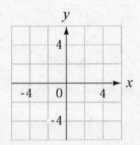

2. $y = 3x - 3$

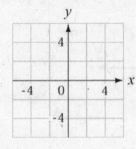

3. $y = \frac{1}{3}x$

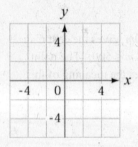

4. $y = -\frac{1}{2}x + 3$

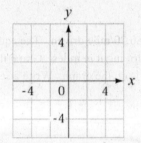

Solve.

5. Loren receives $10 per hour as a mathematics tutor. The equation that describes her wages is $w = 10t$, where t is the number of hours she spends tutoring. Graph this equation for $0 \le t \le 20$. The ordered pair (16, 160) is on the graph. Write a sentence that describes the meaning of this ordered pair.

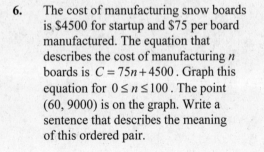

5. _____

6. The cost of manufacturing snow boards is $4500 for startup and $75 per board manufactured. The equation that describes the cost of manufacturing n boards is $C = 75n + 4500$. Graph this equation for $0 \le n \le 100$. The point (60, 9000) is on the graph. Write a sentence that describes the meaning of this ordered pair.

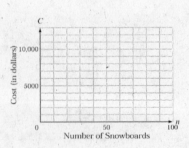

6. _____

390

Answers

Try it 5.3A

1.

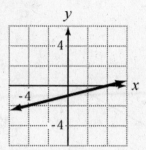

2.

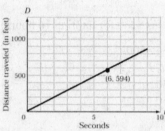

The roller coaster travels 594 ft in 6 s.

Quiz 5.3A

1.

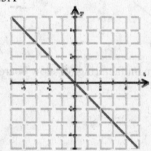

2.

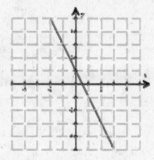

3.

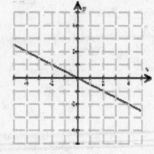

4.

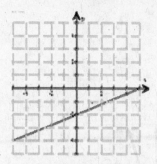

5. The cost for renting the truck and driving 312 miles is $112.40.

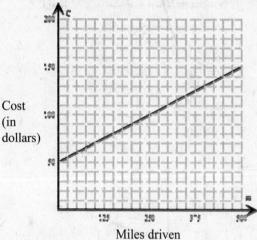

Miles driven

Solutions to Practice Sheet 5.3A

1. $y = 2x - 1$

x	y
-2	-5
0	-1
2	3

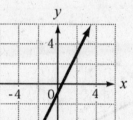

2. $y = 3x - 3$

x	y
0	−3
1	0
2	3

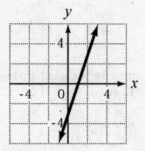

3. $y = \dfrac{1}{3}x$

x	y
−3	−1
0	0
3	1

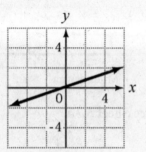

4. $y = -\dfrac{1}{2}x + 3$

x	y
−2	4
0	3
2	2

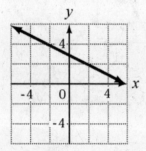

5. $w = 10t$

t	w
5	50
10	100
16	160

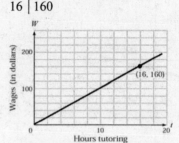

Loren earns \$160 for tutoring 16 hours.

6. $C = 75n + 4500$

n	C
30	6750
50	8250
60	9000

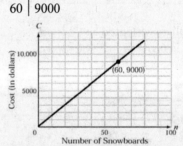

The cost of manufacturing 60 snow boards is \$9000.

Objective 5.3B – Graph an equation of the form $Ax + By = C$

Key Words and Concepts

The equation $Ax + By = C$, where A and B are coefficients and C is a constant, is also a *linear equation in two variables*.

To graph an equation of the form $Ax + By = C$,

First solve the equation for y.

Then follow the same procedure used to graph an equation of the form $y = mx + b$.

Example
Graph: $4x - 2y = 8$

Try it
1. Graph: $3x + y = -3$

Solution

Solve for y.
$$4x - 2y = 8$$
$$-2y = -4x + 8$$
$$y = 2x - 4$$

Find at least 3 solutions.

x	$y = f(x)$
−1	−6
0	−4
1	−2

Graph the ordered pairs and draw a line through the points.

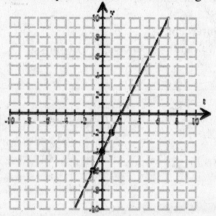

Graph of $y = b$

The graph of $y = b$ is a horizontal line passing through the point with coordinates $(0, b)$.

Constant Function

A function given by $f(x) = b$, where b is a constant, is a **constant function**.

The graph of a constant function is a horizontal line passing through the point $P(0, b)$.

Graph of $x = a$

The graph of $x = a$ is a vertical line passing through the point with coordinates $(a, 0)$.

Example
Graph: $y = f(x) = -1$

Try it
2. Graph: $x - 3 = 0$

Solution
The graph of $y = -1$ is a horizontal line passing through the point with coordinates $(0, -1)$.

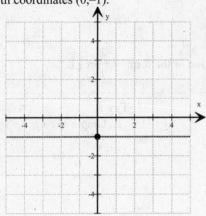

The point at which a graph crosses the x-axis is called an **x-intercept**.
Since any point on the x-axis has y-coordinate 0, we can find x-intercepts by letting $y = 0$.

The point at which a graph crosses the y-axis is called a **y-intercept**.
Since any point on the y-axis has x-coordinate 0, we can find y-intercepts by letting $x = 0$.

Example
Graph $2x - 4y = 8$ by using the x- and y-intercepts.

Try it
3. Graph $5y = 2x - 10$ by using the x- and y-intercepts.

Solution
To find the x-intercept, let $y = 0$. Then solve for x.
$$2x - 4y = 8$$
$$2x - 4(0) = 8$$
$$2x = 8$$
$$x = 4$$
The x-intercept has coordinates $(4, 0)$.

To find the y-intercept, let $x = 0$. Then solve for y.
$$2x - 4y = 8$$
$$2(0) - 4y = 8$$
$$-4y = 8$$
$$y = -2$$
The y-intercept has coordinates $(0, -2)$.

Graph the x- and y-intercepts.
Then draw a line through the two points.

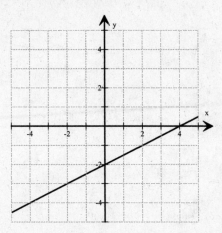

Reflect on it

- The graph of $y = b$ is a horizontal line.
- The graph of $x = a$ is a vertical line.
- To find the x-intercept, set $y = 0$ and solve for x.
- To find the y-intercept, set $x = 0$ and solve for y.

Quiz Yourself 5.3B

1. Graph: $2x + y = 4$

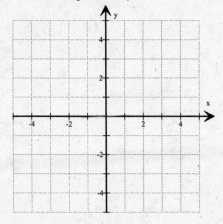

2.   Graph: $x = -1$

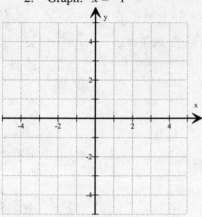

3.   Graph: $y - 4 = 0$

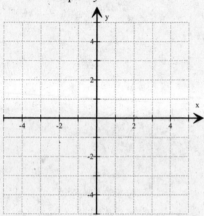

For Exercises 4 through 6, find the x- and y-intercepts and graph.

4. $x - y = -3$

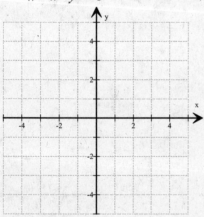

5. $2x - 5y = -10$

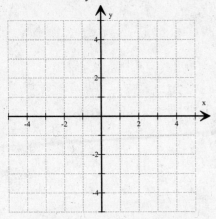

6. $2x + y = 5$

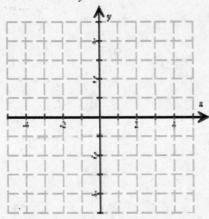

Practice Sheet 5.3B

Graph.

1. $x - 2y = -4$

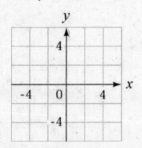

2. $2x - 5y = -10$

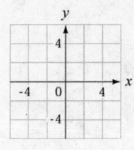

3. $3x - 2y = 6$

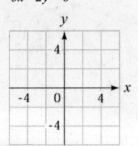

4. $y = 2$

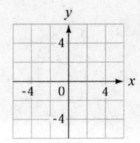

Find the x– and y–intercepts and graph.

5. $3x - y = 6$

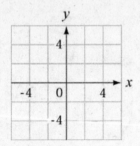

6. $y = -2x + 2$

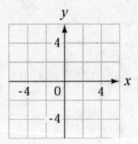

7. $y = -\frac{1}{3}x - 2$

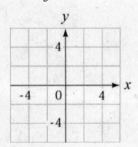

8. $3x + y = -3$

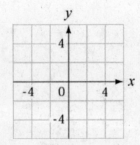

Answers

Try it 5.3B

1.
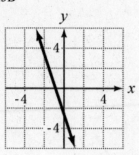

2.
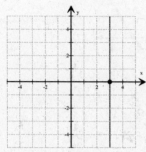

3. *x*-intercept: (5, 0); *y*-intercept: (0,–2)

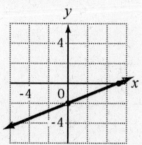

Quiz 5.3B

1.
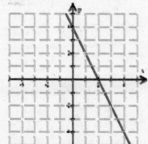

2.
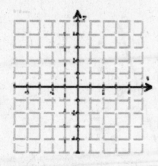

3.
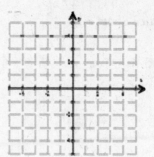

4. x-intercept: $(-3,0)$; y-intercept: $(0,3)$

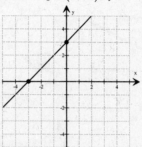

5. x-intercept: $(-5,0)$; y-intercept: $(0,2)$

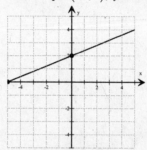

6. x-intercept: $\left(\dfrac{5}{2},0\right)$; y-intercept: $(0,5)$

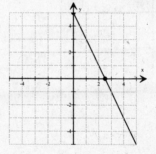

Solutions to Practice Sheet 5.3B

1. $x - 2y = -4$

$\quad -2y = -x - 4$

$\quad\quad y = \dfrac{1}{2}x + 2$

x	y
-2	1
0	2
2	3

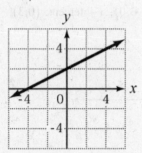

2. $2x - 5y = -10$

$\quad -5y = -2x - 10$

$\quad\quad y = \dfrac{2}{5}x + 2$

x	y
-5	0
0	2
5	4

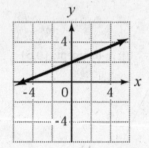

3. $3x - 2y = 6$

$\quad -2y = -3x + 6$

$\quad\quad y = \dfrac{3}{2}x - 3$

x	y
0	-3
2	0
4	3

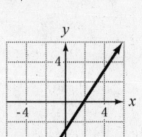

4. $y = 2$

x	y
-2	2
0	2
2	2

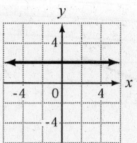

5. x-intercept: $\quad 3x - y = 6$

$\quad\quad\quad\quad\quad\quad 3x - 0 = 6$

$\quad\quad\quad\quad\quad\quad\quad\quad x = 2$

x-intercept: $(2, 0)$

y-intercept: $\quad 3x - y = 6$

$\quad\quad\quad\quad\quad\quad 3(0) - y = 6$

$\quad\quad\quad\quad\quad\quad\quad -y = 6$

$\quad\quad\quad\quad\quad\quad\quad\quad y = -6$

y-intercept: $(0, -6)$

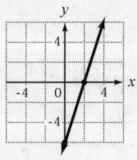

6. x-intercept:
$$y = -2x + 2$$
$$0 = -2x + 2$$
$$2x = 2$$
$$x = 1$$

    x-intercept: (1, 0)

    y-intercept:
$$y = -2x + 2$$
$$y = -2(0) + 2$$
$$y = 2$$

    y-intercept: (0, 2)

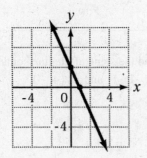

8. x-intercept:
$$3x + y = -3$$
$$3x + 0 = -3$$
$$3x = -3$$
$$x = -1$$

    x-intercept: (−1, 0)

    y-intercept:
$$3x + y = -3$$
$$3(0) + y = -3$$
$$y = -3$$

    y-intercept: (0, −3)

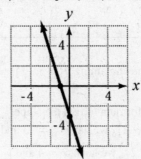

7. x-intercept:
$$y = -\frac{1}{3}x - 2$$
$$0 = -\frac{1}{3}x - 2$$
$$\frac{1}{3}x = -2$$
$$x = -6$$

    x-intercept: (−6, 0)

    y-intercept:
$$y = -\frac{1}{3}x - 2$$
$$y = -\frac{1}{3}(0) - 2$$
$$y = -2$$

    y-intercept: (0, −2)

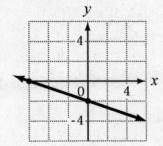

Objective 5.3C – Find the slope of a line given two points

Key Words and Concepts

The **slope** of a line is a measure of the slant of the line. The symbol for slope is m.

Slope Formula

The slope of the line containing the two points $P_1(x_1, y_1)$ and $P_2(x_2, y_2)$ is given by

$$m = \frac{\text{change in } y}{\text{change in } x} = \frac{\Delta y}{\Delta x} = \frac{y_2 - y_1}{x_2 - x_1}, \; x_1 \neq x_2$$

Example

Find the slope of the line containing $P_1(-3, -2)$ and $P_2(4, 2)$.

Solution

Use the slope formula to find the slope.

From $P_1(-3, -2)$, we have $x_1 = -3$ and $y_1 = -2$.

From $P_2(4, 2)$, we have $x_1 = 4$ and $y_1 = 2$.

Substitute the values into the slope formula and simplify.

$$m = \frac{y_2 - y_1}{x_2 - x_1} = \frac{2 - (-2)}{4 - (-3)} = \frac{4}{7}$$

The slope of the line is $\frac{4}{7}$.

Note   A line that slants upward to the right has a **positive slope**.

Try it

1. Find the slope of the line containing $P_1(-1, -1)$ and $P_2(4, 3)$.

Example

Find the slope of the line containing $P_1(3, -2)$ and $P_2(-4, -1)$.

Solution

Use the slope formula to find the slope.

From $P_1(3, -2)$, we have $x_1 = 3$ and $y_1 = -2$.

From $P_2(-4, -1)$, we have $x_1 = -4$ and $y_1 = -1$.

Substitute the values into the slope formula and simplify.

$$m = \frac{y_2 - y_1}{x_2 - x_1} = \frac{-1 - (-2)}{-4 - 3} = \frac{1}{-7} = -\frac{1}{7}$$

The slope of the line is $-\frac{1}{7}$.

Note   A line that slants downward to the right has a **negative slope**.

Try it

2. Find the slope of the line containing $P_1(-2, 5)$ and $P_2(3, 1)$.

Example

Find the slope of the line containing $P_1(-3, 1)$ and $P_2(5, 1)$.

Solution

Use the slope formula to find the slope.

From $P_1(-3, 1)$, we have $x_1 = -3$ and $y_1 = 1$.

From $P_2(5, 1)$, we have $x_1 = 5$ and $y_1 = 1$.

Substitute the values into the slope formula and simplify.

$$m = \frac{y_2 - y_1}{x_2 - x_1} = \frac{1-1}{5-(-3)} = \frac{0}{8} = 0$$

The slope of the line is 0.

Note   A horizontal line has **zero slope**.

Try it

3. Find the slope of the line containing $P_1(2,-3)$ and $P_2(5,-3)$.

Example

Find the slope of the line containing $P_1(-3,-1)$ and $P_2(-3, 4)$.

Solution

Use the slope formula to find the slope.

From $P_1(-3,-1)$, we have $x_1 = -3$ and $y_1 = -1$.

From $P_2(-3, 4)$, we have $x_1 = -3$ and $y_1 = 4$.

Substitute the values into the slope formula and simplify.

$$m = \frac{y_2 - y_1}{x_2 - x_1} = \frac{4-(-1)}{-3-(-3)} = \frac{5}{0}$$

Division by zero is undefined.

The slope of the line is undefined.

Note   A vertical line has **undefined slope**. A vertical line is also said to have **no slope**.

Try it

4. Find the slope of the line containing $P_1(5, 0)$ and $P_2(5, 5)$.

Reflect on it
- A line that has positive slope slants upward to the right.
- A line that has negative slope slants downward to the right.
- A horizontal line has zero slope.
- A vertical line has undefined slope, or no slope.

Quiz Yourself 5.3C

1. Find the slope of the line containing $P_1(-5, 1)$ and $P_2(-1, 5)$.

2. Find the slope of the line containing $P_1(-5, 1)$ and $P_2(-5, -1)$.

3. Find the slope of the line containing $P_1(-5, 1)$ and $P_2(5, 1)$.

4. Find the slope of the line containing $P_1(-5, 1)$ and $P_2(5, -1)$.

5. Find the slope of the line containing $P_1\left(-\frac{1}{5}, 1\right)$ and $P_2\left(\frac{1}{5}, 5\right)$.

Practice Sheet 5.3C

Find the slope of the line containing the points.

1. $P_1(1,-3), P_2(5,-1)$ 2. $P_1(-6, 7), P_2(2, 3)$ 3. $P_1(0,-3), P_2(-1, 0)$

4. $P_1(1, 2), P_2(3, 4)$ 5. $P_1(3, 2), P_2(2, 1)$ 6. $P_1(-2, 1), P_2(-6, 3)$

7. $P_1(2, 3), P_2(3, 2)$ 8. $P_1(4,-5), P_2(1,-4)$ 9. $P_1(-1,-6), P_2(-4,-1)$

10. $P_1(-4,-5), P_2(-2, 0)$ 11. $P_1(4, 3), P_2(2, 1)$ 12. $P_1(1, 2), P_2(4, 8)$

13. $P_1(-4, 4), P_2(2,-2)$ 14. $P_1(-3, 4), P_2(-1,-6)$ 15. $P_1(0,-4), P_2(3,-10)$

16. $P_1(-1, 3), P_2(2, 4)$ 17. $P_1(2,-2), P_2(0, 4)$ 18. $P_1(-1, 2), P_2(-3, 4)$

19. $P_1(1, 2), P_2(-1, 2)$ 20. $P_1(1,-2), P_2(5,-1)$ 21. $P_1(2, 0), P_2(1,-1)$

1. _____

2. _____

3. _____

4. _____

5. _____

6. _____

7. _____

8. _____

9. _____

10. _____

11. _____

12. _____

13. _____

14. _____

15. _____

16. _____

17. _____

18. _____

19. _____

20. _____

21. _____

Answers

Try it 5.3C

1. $\dfrac{4}{5}$

2. $-\dfrac{4}{5}$

3. 0

4. undefined

Quiz 5.3C

1. 1

2. undefined

3. 0

4. $-\dfrac{1}{5}$

5. 10

Solutions to Practice Sheet 5.3C

1. $P_1(1,-3),\ P_2(5,-1)$

$m = \dfrac{y_2 - y_1}{x_2 - x_1} = \dfrac{-1-(-3)}{5-1} = \dfrac{2}{4} = \dfrac{1}{2}$

The slope is $\dfrac{1}{2}$.

2. $P_1(-6,\ 7),\ P_2(2,\ 3)$

$m = \dfrac{y_2 - y_1}{x_2 - x_1} = \dfrac{3-7}{2-(-6)} = \dfrac{-4}{8} = -\dfrac{1}{2}$

The slope is $-\dfrac{1}{2}$.

3. $P_1(0,-3),\ P_2(-1,\ 0)$

$m = \dfrac{y_2 - y_1}{x_2 - x_1} = \dfrac{0-(-3)}{-1-0} = -3$

The slope is -3.

4. $P_1(1,\ 2),\ P_2(3,\ 4)$

$m = \dfrac{y_2 - y_1}{x_2 - x_1} = \dfrac{4-2}{3-1} = \dfrac{2}{2} = 1$

The slope is 1.

5. $P_1(3,\ 2),\ P_2(2,\ 1)$

$m = \dfrac{y_2 - y_1}{x_2 - x_1} = \dfrac{1-2}{2-3} = 1$

The slope is 1.

6. $P_1(-2,\ 1),\ P_2(-6,\ 3)$

$m = \dfrac{y_2 - y_1}{x_2 - x_1} = \dfrac{3-1}{-6-(-2)} = \dfrac{2}{-4} = -\dfrac{1}{2}$

The slope is $-\dfrac{1}{2}$.

7. $P_1(2,\ 3),\ P_2(3,\ 2)$

$m = \dfrac{y_2 - y_1}{x_2 - x_1} = \dfrac{2-3}{3-2} = -1$

The slope is -1.

8. $P_1(4,-5),\ P_2(1,-4)$

$m = \dfrac{y_2 - y_1}{x_2 - x_1} = \dfrac{-4-(-5)}{1-4} = -\dfrac{1}{3}$

The slope is $-\dfrac{1}{3}$.

9. $P_1(-1,-6),\ P_2(-4,-1)$

$m = \dfrac{y_2 - y_1}{x_2 - x_1} = \dfrac{-1-(-6)}{-4-(-1)} = -\dfrac{5}{3}$

The slope is $-\dfrac{5}{3}$.

10. $P_1(-4,-5),\ P_2(-2,\ 0)$

$m = \dfrac{y_2 - y_1}{x_2 - x_1} = \dfrac{0-(-5)}{-2-(-4)} = \dfrac{5}{2}$

The slope is $\dfrac{5}{2}$.

11. $P_1(4,\ 3),\ P_2(2,\ 1)$

$m = \dfrac{y_2 - y_1}{x_2 - x_1} = \dfrac{1-3}{2-4} = \dfrac{-2}{-2} = 1$

The slope is 1.

12. $P_1(1,\ 2),\ P_2(4,\ 8)$

$m = \dfrac{y_2 - y_1}{x_2 - x_1} = \dfrac{8-2}{4-1} = \dfrac{6}{3} = 2$

The slope is 2.

13. $P_1(-4,\ 4),\ P_2(2,-2)$

$m = \dfrac{y_2 - y_1}{x_2 - x_1} = \dfrac{-2-4}{2-(-4)} = \dfrac{-6}{6} = -1$

The slope is -1.

14. $P_1(-3,\ 4),\ P_2(-1,-6)$

$m = \dfrac{y_2 - y_1}{x_2 - x_1} = \dfrac{-6-4}{-1-(-3)} = \dfrac{-10}{2} = -5$

The slope is -5.

15. $P_1(0,-4),\ P_2(3,-10)$

$m = \dfrac{y_2 - y_1}{x_2 - x_1} = \dfrac{-10-(-4)}{3-0} = \dfrac{-6}{3} = -2$

The slope is -2.

16. $P_1(-1, 3),\ P_2(2, 4)$

$$m = \frac{y_2 - y_1}{x_2 - x_1} = \frac{4-3}{2-(-1)} = \frac{1}{3}$$

The slope is $\frac{1}{3}$.

17. $P_1(2, -2),\ P_2(0, 4)$

$$m = \frac{y_2 - y_1}{x_2 - x_1} = \frac{4-(-2)}{0-2} = \frac{6}{-2} = -3$$

The slope is -3.

18. $P_1(-1, 2),\ P_2(-3, 4)$

$$m = \frac{y_2 - y_1}{x_2 - x_1} = \frac{4-2}{-3-(-1)} = \frac{2}{-2} = -1$$

The slope is -1.

19. $P_1(1, 2),\ P_2(-1, 2)$

$$m = \frac{y_2 - y_1}{x_2 - x_1} = \frac{2-2}{-1-1} = 0$$

The slope is 0.

20. $P_1(1, -2),\ P_2(5, -1)$

$$m = \frac{y_2 - y_1}{x_2 - x_1} = \frac{-1-(-2)}{5-1} = \frac{1}{4}$$

The slope is $\frac{1}{4}$.

21. $P_1(2, 0),\ P_2(1, -1)$

$$m = \frac{y_2 - y_1}{x_2 - x_1} = \frac{-1-0}{1-2} = 1$$

The slope is 1.

Objective 5.3D – Graph a line given a point and the slope

Key Words and Concepts

The equation $y = mx + b$ is called the **slope-intercept form** of a straight line. The slope of the line is m, the coefficient of x. The coordinates of the y-intercept are $(0, b)$.

When the equation of a straight line is in the form $y = mx + b$, its graph can be drawn by using the slope and y-intercept. First locate the y-intercept. Use the slope to find a second point on the line. Then draw a line through the two points.

Example

Graph $y = \dfrac{4}{3}x + 1$ by using the slope and y-intercept.

> Solution
> The coordinates of the y-intercept are $(0, 1)$.
>
> $$\text{The slope is } \frac{4}{3} = \frac{\text{change in } y}{\text{change in } x}.$$
>
> Beginning at the y-intercept, move up 4 units (change in y) and then right 3 units (change in x).
>
> The point $(3, 5)$ is a second point on the graph. Draw a line through the points with coordinates $(0, 1)$ and $(3, 5)$.

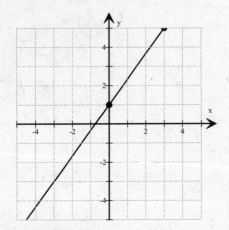

Try it

1. Graph $y = 2x - 5$ by using the slope and y-intercept.

Reflect on it

- The graph of a line can be drawn when *any* point on the line and the slope of the line are given.

Quiz Yourself 5.3D

1. Graph $y = \frac{1}{2}x - 2$ by using the slope and y-intercept.

2. Graph $y = -\frac{2}{3}x + 2$ by using the slope and y-intercept.

3. Graph $y = -\frac{3}{5}x$ by using the slope and y-intercept.

4. Graph the line that passes through the point $P(1,\ 2)$ and has slope $-\dfrac{3}{4}$.

5. Graph the line that passes through the point $P(-3,-3)$ and has slope 4.

Name:
Module 5: Applications of Functions

Class:
Section 5.3: Linear Functions

Practice Sheet 5.3D

Graph by using the slope and the *y*–intercept.

1. $y = \frac{1}{2}x + 3$

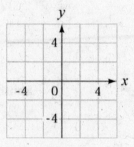

2. $y = \frac{3}{2}x + 2$

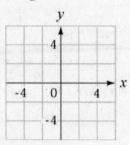

3. $y = -\frac{4}{3}x$

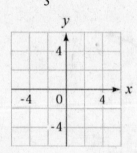

4. $y = 2x - 4$

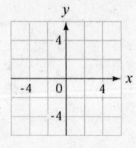

5. $3x - y = 3$

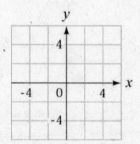

6. $2x + y = 4$

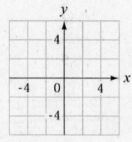

7. Graph the line that passes through point (1, 2) and has slope $\frac{1}{3}$.

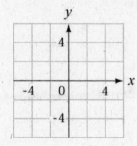

8. Graph the line that passes through point (–2, 1) and has slope –1.

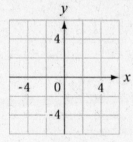

411

Answers

Try it 5.3D

1.

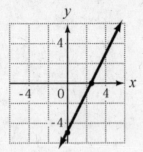

Quiz 5.3D

1.

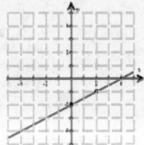

2.

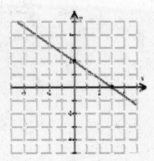

3.

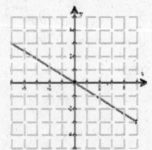

4.

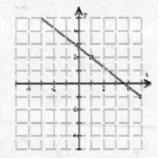

5.

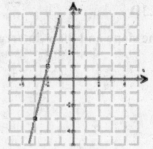

Solutions to Practice Sheet 5.3D

1. $y = \frac{1}{2}x + 3$

   $m = \frac{1}{2}$

   y-intercept: $(0, 3)$

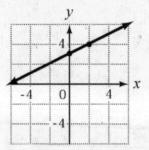

2. $y = \frac{3}{2}x + 2$

   $m = \frac{3}{2}$

   y-intercept: $(0, 2)$

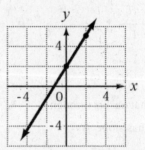

3. $y = -\dfrac{4}{3}x$

    $m = -\dfrac{4}{3}$

    y-intercept: $(0,\ 0)$

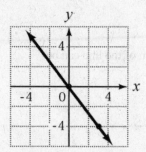

4. $y = 2x - 4$

    $m = 2$

    y-intercept: $(0, -4)$

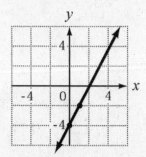

5. $y = 3x - 3$

    $m = 3$

    y-intercept: $(0, -3)$

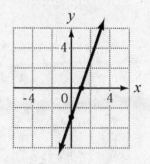

6. $y = -2x + 4$

    $m = -2$

    y-intercept: $(0,\ 4)$

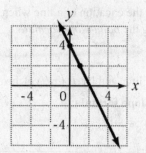

7. $(1,\ 2)$

    $m = \dfrac{1}{3}$

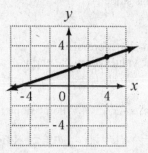

8. $(-2,\ 1)$

    $m = -1$

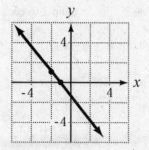

Objective 5.4A – Find the equation of a line given a point and the slope

Key Words and Concepts

One method of finding the equation of a line when the slope and *any* point on the line are known involves using the *point-slope formula*.

Point-Slope Formula

Let m be the slope of a line, and let $P_1(x_1, y_1)$ be a point on the line. The equation of the line can be found from the **point-slope formula**:

$$y - y_1 = m(x - x_1)$$

Example
Find the equation of the line that contains the point $P(-9,\ 2)$ and has slope $\frac{2}{3}$.

Solution
Use the point-slope formula.
Substitute the values for m, x_1, and y_1.
Solve for y.

$$y - y_1 = m(x - x_1) \qquad \text{Point-slope formula}$$

$$y - 2 = \frac{2}{3}\left[x - (-9)\right] \qquad m = \frac{2}{3},\ (x_1, y_1) = (-9,\ 2)$$

$$y - 2 = \frac{2}{3}x + 6 \qquad \text{Distribute } \frac{2}{3}$$

$$y = \frac{2}{3}x + 8 \qquad \text{Add 2 to each side}$$

The equation of the line is $y = \frac{2}{3}x + 8$.

Try it
1. Find the equation of the line that contains the point $P(6, -17)$ and has slope $-\frac{5}{3}$.

Reflect on it
• When the slope of a line is undefined, the point-slope formula cannot be used. Recall that when the slope of a line is undefined, the line is vertical. The equation of a vertical line is $x = a$, where a is the x-coordinate of the x-intercept.

Quiz Yourself 5.4A

In Exercises 1 through 5, find the equation of the line that contains the given point and has the given slope.

1. $P(4, -13)$, $m = -4$

2. $P(4, 0)$, $m = \dfrac{1}{2}$

3. $P(-10, -3)$, $m = \dfrac{2}{5}$

4. $P(-4, -1)$, slope is undefined

5. $P(-4, -1)$, $m = 0$

Practice Sheet 5.4A

Find the equation of the line that contains the given point and has the given slope.

1. Point $(0, 5)$. $m = -3$

2. Point $(-2, 0)$. $m = 4$

3. Point $(1, 2)$. $m = \frac{1}{3}$

4. Point $(3, -2)$. $m = -\frac{1}{3}$

5. Point $(-3, 2)$. $m = -1$

6. Point $(-3, 2)$. $m = \frac{1}{2}$

7. Point $(-1, 1)$. $m = \frac{2}{3}$

8. Point $(0, 3)$. $m = -\frac{1}{3}$

9. Point $(-1, 5)$. $m = -2$

10. Point $(-1, 3)$. $m = -\frac{2}{5}$

11. Point $(1, -2)$. $m = \frac{3}{4}$

12. Point $(-1, -3)$. $m = \frac{1}{3}$

13. Point $(-1, -1)$. $m = -\frac{1}{2}$

14. Point $(0, 0)$. $m = -\frac{1}{4}$

15. Point $(3, -2)$. $m = 2$

16. Point $(3, -4)$. $m = -2$

17. Point $(2, 4)$. $m = -\frac{1}{3}$

18. Point $(4, 1)$. $m = -\frac{3}{5}$

1. _____

2. _____

3. _____

4. _____

5. _____

6. _____

7. _____

8. _____

9. _____

10. _____

11. _____

12. _____

13. _____

14. _____

15. _____

16. _____

17. _____

18. _____

Answers

Try it 5.4A

1. $y = -\dfrac{5}{3}x - 7$

Quiz 5.4A

1. $y = -4x + 3$

2. $y = \dfrac{1}{2}x - 2$

3. $y = \dfrac{2}{5}x + 1$

4. $x = -4$

5. $y = -1$

Solutions to Practice Sheet 5.4A

1. $m = -3$, $(x_1, y_1) = (0, 5)$

$y - y_1 = m(x - x_1)$
$y - 5 = -3(x - 0)$
$y - 5 = -3x$
$y = -3x + 5$

The equation of the line is $y = -3x + 5$.

2. $m = 4$, $(x_1, y_1) = (-2, 0)$

$y - y_1 = m(x - x_1)$
$y - 0 = 4[x - (-2)]$
$y = 4x + 8$

The equation of the line is $y = 4x + 8$.

3. $m = \dfrac{1}{3}$, $(x_1, y_1) = (1, 2)$

$y - y_1 = m(x - x_1)$
$y - 2 = \dfrac{1}{3}(x - 1)$
$y - 2 = \dfrac{1}{3}x - \dfrac{1}{3}$
$y = \dfrac{1}{3}x + \dfrac{5}{3}$

The equation of the line is $y = \dfrac{1}{3}x + \dfrac{5}{3}$.

4. $m = -\dfrac{1}{3}$, $(x_1, y_1) = (3, -2)$

$y - y_1 = m(x - x_1)$
$y - (-2) = -\dfrac{1}{3}(x - 3)$
$y + 2 = -\dfrac{1}{3}x + 1$
$y = -\dfrac{1}{3}x - 1$

The equation of the line is $y = -\dfrac{1}{3}x - 1$.

5. $m = -1$, $(x_1, y_1) = (-3, 2)$

$y - y_1 = m(x - x_1)$
$y - 2 = -1[x - (-3)]$
$y - 2 = -x - 3$
$y = -x - 1$

The equation of the line is $y = -x - 1$.

6. $m = \dfrac{1}{2}$, $(x_1, y_1) = (-3, 2)$

$y - y_1 = m(x - x_1)$
$y - 2 = \dfrac{1}{2}[x - (-3)]$
$y - 2 = \dfrac{1}{2}x + \dfrac{3}{2}$
$y = \dfrac{1}{2}x + \dfrac{7}{2}$

The equation of the line is $y = \dfrac{1}{2}x + \dfrac{7}{2}$.

7. $m = \dfrac{2}{3}$, $(x_1, y_1) = (-1, 1)$

$y - y_1 = m(x - x_1)$
$y - 1 = \dfrac{2}{3}[x - (-1)]$
$y - 1 = \dfrac{2}{3}x + \dfrac{2}{3}$
$y = \dfrac{2}{3}x + \dfrac{5}{3}$

The equation of the line is $y = \dfrac{2}{3}x + \dfrac{5}{3}$.

8. $m = -\dfrac{1}{3}$, $(x_1, y_1) = (0, 3)$

$y - y_1 = m(x - x_1)$
$y - 3 = -\dfrac{1}{3}(x - 0)$
$y - 3 = -\dfrac{1}{3}x$
$y = -\dfrac{1}{3}x + 3$

The equation of the line is $y = -\dfrac{1}{3}x + 3$.

9. $m = -2$, $(x_1, y_1) = (-1, 5)$

$y - y_1 = m(x - x_1)$
$y - 5 = -2[x - (-1)]$
$y - 5 = -2x - 2$
$y = -2x + 3$

The equation of the line is $y = -2x + 3$.

10. $m = -\dfrac{2}{5}$, $(x_1, y_1) = (-1, 3)$

$$y - y_1 = m(x - x_1)$$
$$y - 3 = -\dfrac{2}{5}[x - (-1)]$$
$$y - 3 = -\dfrac{2}{5}x - \dfrac{2}{5}$$
$$y = -\dfrac{2}{5}x + \dfrac{13}{5}$$

The equation of the line is $y = -\dfrac{2}{5}x + \dfrac{13}{5}$.

11. $m = \dfrac{3}{4}$, $(x_1, y_1) = (1, -2)$

$$y - y_1 = m(x - x_1)$$
$$y - (-2) = \dfrac{3}{4}(x - 1)$$
$$y + 2 = \dfrac{3}{4}x - \dfrac{3}{4}$$
$$y = \dfrac{3}{4}x - \dfrac{11}{4}$$

The equation of the line is $y = \dfrac{3}{4}x - \dfrac{11}{4}$.

12. $m = \dfrac{1}{3}$, $(x_1, y_1) = (-1, -3)$

$$y - y_1 = m(x - x_1)$$
$$y - (-3) = \dfrac{1}{3}[x - (-1)]$$
$$y + 3 = \dfrac{1}{3}x + \dfrac{1}{3}$$
$$y = \dfrac{1}{3}x - \dfrac{8}{3}$$

The equation of the line is $y = \dfrac{1}{3}x - \dfrac{8}{3}$.

13. $m = -\dfrac{1}{2}$, $(x_1, y_1) = (-1, -1)$

$$y - y_1 = m(x - x_1)$$
$$y - (-1) = -\dfrac{1}{2}[x - (-1)]$$
$$y + 1 = -\dfrac{1}{2}x - \dfrac{1}{2}$$
$$y = -\dfrac{1}{2}x - \dfrac{3}{2}$$

The equation of the line is $y = -\dfrac{1}{2}x - \dfrac{3}{2}$.

14. $m = -\dfrac{1}{4}$, $(x_1, y_1) = (0, 0)$

$$y - y_1 = m(x - x_1)$$
$$y - 0 = -\dfrac{1}{4}(x - 0)$$
$$y = -\dfrac{1}{4}x$$

The equation of the line is $y = -\dfrac{1}{4}x$.

15. $m = 2$, $(x_1, y_1) = (3, -2)$

$$y - y_1 = m(x - x_1)$$
$$y - (-2) = 2(x - 3)$$
$$y + 2 = 2x - 6$$
$$y = 2x - 8$$

The equation of the line is $y = 2x - 8$.

16. $m = -2$, $(x_1, y_1) = (3, -4)$

$$y - y_1 = m(x - x_1)$$
$$y - (-4) = -2(x - 3)$$
$$y + 4 = -2x + 6$$
$$y = -2x + 2$$

The equation of the line is $y = -2x + 2$.

17. $m = -\dfrac{1}{3}$, $(x_1, y_1) = (2, 4)$

$$y - y_1 = m(x - x_1)$$
$$y - 4 = -\dfrac{1}{3}(x - 2)$$
$$y - 4 = -\dfrac{1}{3}x + \dfrac{2}{3}$$
$$y = -\dfrac{1}{3}x + \dfrac{14}{3}$$

The equation of the line is $y = -\dfrac{1}{3}x + \dfrac{14}{3}$.

18. $m = -\dfrac{3}{5}$, $(x_1, y_1) = (4, 1)$

$$y - y_1 = m(x - x_1)$$
$$y - 1 = -\dfrac{3}{5}(x - 4)$$
$$y - 1 = -\dfrac{3}{5}x + \dfrac{12}{5}$$
$$y = -\dfrac{3}{5}x + \dfrac{17}{5}$$

The equation of the line is $y = -\dfrac{3}{5}x + \dfrac{17}{5}$.

Objective 5.4B – Find the equation of a line given two points

Key Words and Concepts

The point-slope formula and the formula for slope are used to find the equation of a line when two points are known.

Example
Find the equation of the line containing $P_1(7,-4)$ and $P_2(2, 3)$.

Solution
Find the slope.

$$m = \frac{3-(-4)}{2-7} = -\frac{7}{5}$$

Use the point-slope formula.
Substitute the values for m, x_1, and y_1.
Solve for y.

$$y - y_1 = m(x - x_1) \qquad \text{Point-slope formula}$$

$$y-(-4) = -\frac{7}{5}(x-7) \qquad m = -\frac{7}{5}, \ (x_1, y_1) = (7,-4).$$

$$y + 4 = -\frac{7}{5}x + \frac{49}{5} \qquad \text{Distribute}$$

$$y = -\frac{7}{5}x + \frac{29}{5} \qquad \text{Subtract 4 from each side}$$

The equation of the line is $y = -\frac{7}{5}x + \frac{29}{5}$.

Try it
1. Find the equation of the line containing $P_1(-2,-9)$ and $P_2(3, 1)$.

Reflect on it
- Given two points, to find the equation of the line containing those points
  Find the slope.
  Use the point-slope formula to find the equation.
  Write the equation in the form $y = mx + b$.
- Does it matter which point is used in the point-slope formula?

Quiz Yourself 5.4B
In Exercises 1 through 5, find the equation of the line that contains the given points.
1. $P_1(-3, 5), \ P_2(-1,-1)$

2. $P_1(-3,-2),\ P_2(-3,-7)$

3. $P_1(4,-1),\ P_2(3,-5)$

4. $P_1(5,\ 2),\ P_2(0,\ 2)$

5. $P_1(-5,-1),\ P_2(7,\ 6)$

Practice Sheet 5.4B

Find the equation of the line through the given points.

1. $P_1(1, 4)$, $P_2(2, 3)$ **2.** $P_1(1, 3)$, $P_2(0, 4)$ 1. _____

  2. _____

3. $P_1(1, 2)$, $P_2(4, 4)$ **4.** $P_1(3, 1)$, $P_2(5, 2)$ 3. _____

  4. _____

5. $P_1(-2, 2)$, $P_2(1, 3)$ **6.** $P_1(-2, -2)$, $P_2(1, 1)$ 5. _____

  6. _____

7. $P_1(-2, -3)$, $P_2(2, 3)$ **8.** $P_1(-2, 1)$, $P_2(1, 5)$ 7. _____

  8. _____

9. $P_1(1, 0)$, $P_2(0, -2)$ **10.** $P_1(0, 3)$, $P_2(-1, 0)$ 9. _____

  10. _____

11. $P_1(0, 0)$, $P_2(2, 1)$ **12.** $P_1(1, -3)$, $P_2(0, 0)$ 11. _____

  12. _____

13. $P_1(1, 0)$, $P_2(-2, 2)$ **14.** $P_1(2, -4)$, $P_2(-1, 0)$ 13. _____

  14. _____

15. $P_1(3, -2)$, $P_2(-2, 4)$ **16.** $P_1(4, -5)$, $P_2(-3, 1)$ 15. _____

  16. _____

17. $P_1(-1, 3)$, $P_2(2, 0)$ **18.** $P_1(2, -2)$, $P_2(3, -2)$ 17. _____

  18. _____

Answers

Try it 5.4B

1. $y = 2x - 5$

Quiz 5.4B

1. $y = -3x - 4$
2. $x = -3$
3. $y = 4x - 17$
4. $y = 2$
5. $y = \frac{7}{12}x + \frac{23}{12}$

Solutions to Practice Sheet 5.4B

1. $P_1(1, 4), \ P_2(2, 3)$

$$m = \frac{y_2 - y_1}{x_2 - x_1} = \frac{3 - 4}{2 - 1} = -1$$

$$y - y_1 = m(x - x_1)$$
$$y - 4 = -1(x - 1)$$
$$y - 4 = -x + 1$$
$$y = -x + 5$$

The equation of the line is $y = -x + 5$.

2. $P_1(1, 3), \ P_2(0, 4)$

$$m = \frac{y_2 - y_1}{x_2 - x_1} = \frac{4 - 3}{0 - 1} = -1$$

$$y - y_1 = m(x - x_1)$$
$$y - 4 = -1(x - 0)$$
$$y - 4 = -x$$
$$y = -x + 4$$

The equation of the line is $y = -x + 4$.

3. $P_1(1, 2), \ P_2(4, 4)$

$$m = \frac{y_2 - y_1}{x_2 - x_1} = \frac{4 - 2}{4 - 1} = \frac{2}{3}$$

$$y - y_1 = m(x - x_1)$$
$$y - 2 = \frac{2}{3}(x - 1)$$
$$y - 2 = \frac{2}{3}x - \frac{2}{3}$$
$$y = \frac{2}{3}x + \frac{4}{3}$$

The equation of the line is $y = \frac{2}{3}x + \frac{4}{3}$.

4. $P_1(3, 1), \ P_2(5, 2)$

$$m = \frac{y_2 - y_1}{x_2 - x_1} = \frac{2 - 1}{5 - 3} = \frac{1}{2}$$

$$y - y_1 = m(x - x_1)$$
$$y - 1 = \frac{1}{2}(x - 3)$$
$$y - 1 = \frac{1}{2}x - \frac{3}{2}$$
$$y = \frac{1}{2}x - \frac{1}{2}$$

The equation of the line is $y = \frac{1}{2}x - \frac{1}{2}$.

5. $P_1(-2, 2), \ P_2(1, 3)$

$$m = \frac{y_2 - y_1}{x_2 - x_1} = \frac{3 - 2}{1 - (-2)} = \frac{1}{3}$$

$$y - y_1 = m(x - x_1)$$
$$y - 3 = \frac{1}{3}(x - 1)$$
$$y - 3 = \frac{1}{3}x - \frac{1}{3}$$
$$y = \frac{1}{3}x + \frac{8}{3}$$

The equation of the line is $y = \frac{1}{3}x + \frac{8}{3}$.

6. $P_1(-2, -2), \ P_2(1, 1)$

$$m = \frac{y_2 - y_1}{x_2 - x_1} = \frac{1 - (-2)}{1 - (-2)} = 1$$

$$y - y_1 = m(x - x_1)$$
$$y - 1 = 1(x - 1)$$
$$y - 1 = x - 1$$
$$y = x$$

The equation of the line is $y = x$.

7. $P_1(-2, -3), \ P_2(2, 3)$

$$m = \frac{y_2 - y_1}{x_2 - x_1} = \frac{3 - (-3)}{2 - (-2)} = \frac{6}{4} = \frac{3}{2}$$

$$y - y_1 = m(x - x_1)$$
$$y - 3 = \frac{3}{2}(x - 2)$$
$$y - 3 = \frac{3}{2}x - 3$$
$$y = \frac{3}{2}x$$

The equation of the line is $y = \frac{3}{2}x$.

8. $P_1(-2, 1)$, $P_2(1, 5)$

$m = \dfrac{y_2 - y_1}{x_2 - x_1} = \dfrac{5-1}{1-(-2)} = \dfrac{4}{3}$

$y - y_1 = m(x - x_1)$

$y - 5 = \dfrac{4}{3}(x - 1)$

$y - 5 = \dfrac{4}{3}x - \dfrac{4}{3}$

$y = \dfrac{4}{3}x + \dfrac{11}{3}$

The equation of the line is $y = \dfrac{4}{3}x + \dfrac{11}{3}$.

9. $P_1(1, 0)$, $P_2(0, -2)$

$m = \dfrac{y_2 - y_1}{x_2 - x_1} = \dfrac{-2-0}{0-1} = 2$

$y - y_1 = m(x - x_1)$

$y - 0 = 2(x - 1)$

$y - 0 = 2x - 2$

$y = 2x - 2$

The equation of the line is $y = 2x - 2$.

10. $P_1(0, 3)$, $P_2(-1, 0)$

$m = \dfrac{y_2 - y_1}{x_2 - x_1} = \dfrac{0-3}{-1-0} = 3$

$y - y_1 = m(x - x_1)$

$y - 3 = 3(x - 0)$

$y - 3 = 3x$

$y = 3x + 3$

The equation of the line is $y = 3x + 3$.

11. $P_1(0, 0)$, $P_2(2, 1)$

$m = \dfrac{y_2 - y_1}{x_2 - x_1} = \dfrac{1-0}{2-0} = \dfrac{1}{2}$

$y - y_1 = m(x - x_1)$

$y - 0 = \dfrac{1}{2}(x - 0)$

$y = \dfrac{1}{2}x$

The equation of the line is $y = \dfrac{1}{2}x$.

12. $P_1(1, -3)$, $P_2(0, 0)$

$m = \dfrac{y_2 - y_1}{x_2 - x_1} = \dfrac{0-(-3)}{0-1} = -3$

$y - y_1 = m(x - x_1)$

$y - 0 = -3(x - 0)$

$y = -3x$

The equation of the line is $y = -3x$.

13. $P_1(1, 0)$, $P_2(-2, 2)$

$m = \dfrac{y_2 - y_1}{x_2 - x_1} = \dfrac{2-0}{-2-1} = -\dfrac{2}{3}$

$y - y_1 = m(x - x_1)$

$y - 0 = -\dfrac{2}{3}(x - 1)$

$y = -\dfrac{2}{3}x + \dfrac{2}{3}$

The equation of the line is $y = -\dfrac{2}{3}x + \dfrac{2}{3}$.

14. $P_1(2, -4)$, $P_2(-1, 0)$

$m = \dfrac{y_2 - y_1}{x_2 - x_1} = \dfrac{0-(-4)}{-1-2} = -\dfrac{4}{3}$

$y - y_1 = m(x - x_1)$

$y - 0 = -\dfrac{4}{3}[x - (-1)]$

$y = -\dfrac{4}{3}x - \dfrac{4}{3}$

The equation of the line is $y = -\dfrac{4}{3}x - \dfrac{4}{3}$.

15. $P_1(3, -2)$, $P_2(-2, 4)$

$m = \dfrac{y_2 - y_1}{x_2 - x_1} = \dfrac{4-(-2)}{-2-3} = -\dfrac{6}{5}$

$y - y_1 = m(x - x_1)$

$y - 4 = -\dfrac{6}{5}[x - (-2)]$

$y - 4 = -\dfrac{6}{5}x - \dfrac{12}{5}$

$y = -\dfrac{6}{5}x + \dfrac{8}{5}$

The equation of the line is $y = -\dfrac{6}{5}x + \dfrac{8}{5}$.

16. $P_1(4, -5)$, $P_2(-3, 1)$

$m = \dfrac{y_2 - y_1}{x_2 - x_1} = \dfrac{1-(-5)}{-3-4} = -\dfrac{6}{7}$

$y - y_1 = m(x - x_1)$

$y - 1 = -\dfrac{6}{7}[x - (-3)]$

$y - 1 = -\dfrac{6}{7}x - \dfrac{18}{7}$

$y = -\dfrac{6}{7}x - \dfrac{11}{7}$

The equation of the line is $y = -\dfrac{6}{7}x - \dfrac{11}{7}$.

17. $P_1(-1, 3), \ P_2(2, 0)$

$$m = \frac{y_2 - y_1}{x_2 - x_1} = \frac{0 - 3}{2 - (-1)} = -1$$

$$y - y_1 = m(x - x_1)$$
$$y - 0 = -1(x - 2)$$
$$y = -x + 2$$

The equation of the line is $y = -x + 2$.

18. $P_1(2, -2), \ P_2(3, -2)$

$$m = \frac{y_2 - y_1}{x_2 - x_1} = \frac{-2 - (-2)}{3 - 2} = 0 \quad .$$

$$y - y_1 = m(x - x_1)$$
$$y - (-2) = 0(x - 2)$$
$$y + 2 = 0$$
$$y = -2$$

The equation of the line is $y = -2$.

Objective 5.4C – Solve application problems by using linear models

For each application, data are collected and the independent and dependent variables are selected. Then a linear function that models the data is determined.

Example

The value of a certain business machine decreases as time goes on. Suppose that the value depreciates such that the machine is worth $6750 two years after purchase and $4500 five years after purchase. Assuming that the depreciation is linear, describe this situation with a linear function. Use the function to predict the value of the machine 7 years after purchase.

Solution

Select independent and dependent variables.

  Because we are trying to determine the value of the machine, that quantity is the dependent variable, y. The number of years after purchase is the independent variable, x.

From the given data, two ordered pairs are
  (2, 6750) and (5, 4500).

After finding the linear function, evaluate that function when $x = 7$ to predict the value of the machine in 7 years.

Choose $P_1(2,\ 6750)$ and $P_2(5,\ 4500)$. Find the slope.

$$m = \frac{4500 - 6750}{5 - 2} = -\frac{2250}{3} = -750$$

Use the point-slope formula and substitute $m = -750$ and $(x_1, y_1) = (2,\ 6750)$ to find the linear function.

$$y - y_1 = m(x - x_1)$$
$$y - 6750 = -750(x - 2)$$
$$y - 6750 = -750x + 1500$$
$$y = -750x + 8250$$

The linear function is $f(x) = -750x + 8250$.

Evaluate the function at $x = 7$.
$$f(7) = -750(7) + 8250 = 3000$$

The value of the machine will be $3000 after 7 years.

Try it

1. A building contractor estimates that the cost to build a tiny home is $15,000 plus $95 for each square foot of floor space in the house. Determine a linear function that will give the cost of building a tiny home that contains a given number of square feet. Use this function to determine the cost to build a tiny home containing 250 sq. ft.

Reflect on it

• When determining a linear function,
   identify the dependent and independent variables,
   find the slope of the line, and then
   use the point-slope formula to find the equation of the line.

Quiz Yourself 5.4C

1. An office-supply store sells a roll of bubble wrap for $6.89.
   a. Write a linear function for the cost of bubble wrap in terms of the number of rolls purchased.
   b. Use your function to find the cost of purchasing 7 rolls of bubble wrap.

2. Under a certain cell-phone plan, the plan along with 17 minutes of roaming calls in Canada in a month costs $51.73, and the plan along with 29 minutes of roaming calls in Canada in a month costs $60.01.
   a. Write a linear function for the monthly cost of the plan in terms of the number of minutes of roaming calls in Canada in the month.
   b. Use your function to find the cost in a month of the plan when there are 38 minutes of roaming calls in Canada in the month.

3. The water rates in a certain community are such that using 5400 gallons in a month brings about a bill of $96.40, and using 9900 gallons in a month brings about a bill of $168.40.
   a. Write a linear function for the monthly cost of water in terms of the number of gallons of water used in the month.
   b. At these rates, what bill will result from using 12,000 gallons in a month?

Practice Sheet 5.4C

Solve.

1. A building contractor estimates that the cost to build a home is $35,000 plus $95 for each square foot of floor space in the house.
   a. Determine a linear function that will give the cost of building a house that contains a given number of square feet.
   b. Use this model to determine the cost to build a house containing 1500 sq. ft.

2. The gas tank of a certain car contains 13 gal when the driver of the car begins a trip. Each mile driven by the driver decreases the amount of gas in the tank by 0.025 gal.
   a. Write a linear function for the number of gallons of gas in the tank in terms of the number of miles driven.
   b. Use your equation to find the number of gallons in the tank after 180 miles are driven.

1. a. _____

   b. _____

2. a. _____

   b. _____

3. A manufacturer of cars determined that 75,000 cars per month can be sold at a price of $24,000. At a price of $23,500 the number of cars sold per month would increase to 80,000.
   a. Determine a linear function that will predict the number of cars that would be sold each month at a given price.
   b. Use this model to predict the number of cars that would be sold at a price of $23,250.

4. An account executive receives a base salary plus a commission. On $25,000 in monthly sales, the account executive receives $2450. On $40,000 in monthly sales, the account executive receives $3200.
   a. Determine a linear function that will yield the compensation of the account executive for a given amount of monthly sales.
   b. Use this model to determine the account executive's compensation for $70,000 in monthly sales.

3. a. _____

   b. _____

4. a. _____

   b. _____

5. There are approximately 90 Calories in a 4 oz serving of cottage cheese and approximately 135 Calories in a 6 oz serving.
   a. Determine a linear function for the number of Calories in a serving of cottage cheese in terms of the size of the serving.
   b. Use your equation to estimate the number of Calories in a 5 oz serving.

6. A cellular phone company offers a plan for people who plan to use the phone only in emergencies. The plan costs the user $7.95 per month plus $0.69 per minute used.
   a. Write a linear function for the monthly cost in terms of the number of minutes used.
   b. Use your equation to find the cost of using the cellular phone for 9 minutes in one month.

5. a. _____

   b. _____

6. a. _____

   b. _____

7. At sea level, the boiling point of water is 100°C. At an altitude of 1.5 km, the boiling point of water is 94.75°C.
   a. Write a linear function for the boiling point of water in terms of the altitude above sea level.
   b. Use your equation to predict the boiling point of water at an altitude of 5 km above sea level.

8. Let f be a linear function. If $f(-3) = -1$ and $f(2) = -11$, find $f(x)$.

7. a. _____

   b. _____

8. _____

Answers

Try it 5.4C

1. $f(x) = 95x + 15,000$

   $38,750

Quiz 5.4C

1. a. $f(x) = 6.89x$

   b. $48.23

2. a. $f(x) = 40 + 0.69x$

   b. $66.22

3. a. $f(x) = 10 + 0.016x$

   b. $202

Solutions to Practice Sheet 5.4C

1. **STRATEGY**

   Let x represent the number of square feet.
   Let y represent the cost.
   Use the slope-intercept form to find the equation of the line.

   SOLUTION

   a. y-intercept $= (0, 35,000)$; slope $= 95$

   $y = 95x + 35,000$

   The linear function is $f(x) = 95x + 35,000$.

   b. $f(1500) = 95(1500) + 35,000 = 177,500$

   The cost to build a house with 1500 sq. ft is $177,500.

2. **STRATEGY**

   Let x represent the number of miles driven.
   Let y represent the number of gallons of gasoline in the tank.
   Use the slope-intercept form to find the equation of the line.

   SOLUTION

   a. y-intercept $= (0, 13)$; slope $= -0.025$

   $y = -0.025x + 13$

   The linear function is $f(x) = -0.025x + 13$.

   b. $f(180) = -0.025(180) + 13 = 8.5$

   There will be 8.5 gallons of gasoline in the tank after driving 180 miles.

3. **STRATEGY**

   Let x represent the price of the car.
   Let y represent the number of cars sold.
   Use the point-slope formula to find the equation of the line.

   SOLUTION

   a. $P_1(24,000, \ 75,000)$, $P_2(23,500, \ 80,000)$

   $$m = \frac{y_2 - y_1}{x_2 - x_1} = \frac{80,000 - 75,000}{23,500 - 24,000} = -10$$

   $$y - y_1 = m(x - x_1)$$
   $$y - 75,000 = -10(x - 24,000)$$
   $$y - 75,000 = -10x + 240,000$$
   $$y = -10x + 315,000$$

   The linear function is
   $f(x) = -10x + 315,000$.

   b. $f(23,250) = -10(23,250) + 315,000$
   $\qquad\qquad\quad = 82,500$

   82,500 cars will be sold at a price of $23,250.

4. **STRATEGY**

   Let x represent monthly sales.
   Let y represent the commission earned.
   Use the point-slope formula to find the equation of the line.

   SOLUTION

   a. $P_1(25,000, \ 2450)$, $P_2(40,000, \ 3200)$

   $$m = \frac{y_2 - y_1}{x_2 - x_1} = \frac{3200 - 2450}{40,000 - 25,000} \approx 0.05$$

   $$y - y_1 = m(x - x_1)$$
   $$y - 2450 = 0.05(x - 25,000)$$
   $$y - 2450 = 0.05x - 1250$$
   $$y = 0.05x + 1200$$

   The linear function is $f(x) = 0.05x + 1200$.

   b. $f(70,000) = 0.05(70,000) + 1200$
   $\qquad\qquad\quad = 4700$

   The commission on $70,000 in sales will be $4700.

5. STRATEGY
    Let x represent the number of ounces.
    Let y represent the number of Calories.
    Use the point-slope formula to find the
    equation of the line.

    SOLUTION
    a. $P_1(4, 90)$, $P_2(6, 135)$

    $$m = \frac{y_2 - y_1}{x_2 - x_1} = \frac{135 - 90}{6 - 4} = 22.5$$

    $$y - y_1 = m(x - x_1)$$
    $$y - 90 = 22.5(x - 4)$$
    $$y - 90 = 22.5x - 90$$
    $$y = 22.5x$$

    The linear function is $f(x) = 22.5x$.

    b. $f(5) = 22.5(5) = 112.5$

    In a 6 oz serving, there are 112.5 Calories.

6. STRATEGY
    Let x represent the number minutes used.
    Let y represent the cost.
    Use the slope-intercept form to find the
    equation of the line.

    SOLUTION
    a. y-intercept $= (0, 7.95)$; slope $= 0.69$
    $$y = 0.69x + 7.95$$

    The linear function is $f(x) = 0.69x + 7.95$.

    b. $f(9) = 0.69(9) + 7.95 = 14.16$

    The cost for 9 minutes of use is $14.16.

7. STRATEGY
    Let x represent the altitude.
    Let y represent the boiling point of water.
    Find the slope
    Use the slope-intercept form to find the
    equation of the line.

    SOLUTION
    a. $P_1(0, 100)$, $P_2(1.5, 94.75)$

    $$m = \frac{y_2 - y_1}{x_2 - x_1} = \frac{94.75 - 100}{1.5 - 0} = -3.5$$

    $$y = -3.5x + 100$$

    The linear function is $f(x) = -3.5x + 100$.

    b. $f(5) = -3.5(5) + 100 = 82.5$

    At 5 km, the boiling point of water is 82.5°C.

8. STRATEGY
    Use the point-slope formula to find the
    equation of the line.

    SOLUTION
    $P_1(-3, -1)$, $P_2(2, -11)$

    $$m = \frac{y_2 - y_1}{x_2 - x_1} = \frac{-11 - (-1)}{2 - (-3)} = \frac{-10}{5} = -2$$

    $$y - y_1 = m(x - x_1)$$
    $$y - (-11) = -2(x - 2)$$
    $$y + 11 = -2x + 4$$
    $$y = -2x - 7$$

    The linear function is $f(x) = -2x - 7$.

Objective 5.5A – Graph a quadratic function

Key Words and Concepts

A **quadratic function** is a function that can be expressed by the equation $f(x) = ax^2 + bx + c, \ a \neq 0$.

The graph of a quadratic function is called a **parabola**.

The graph of a parabola opens _____ when $a > 0$ and opens _____ when $a < 0$.
                                     (up/down)   (up/down)

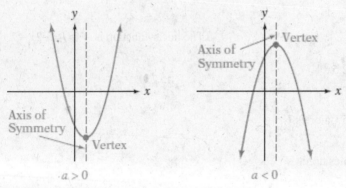

The **vertex** of a parabola is the point with the least y-coordinate when $a > 0$ and the point with the greatest y-coordinate when $a < 0$.

The line that passes through the vertex and is parallel to the y-axis is called the **axis of symmetry**.

Vertex and Axis of Symmetry of a Parabola

Let $f(x) = ax^2 + bx + c, \ a \neq 0,$ be the equation of a parabola.

The coordinates of the vertex are $\left(-\dfrac{b}{2a}, f\left(-\dfrac{b}{2a} \right) \right)$. The equation of the axis of symmetry is $x = -\dfrac{b}{2a}$.

Because $f(x) = ax^2 + bx + c,$ is a real number for all real numbers x, the domain of a quadratic function is all real numbers. The range of a quadratic function can be determined from the y-coordinate of the vertex.

Example

The quadratic function $f(x) = -2x^2 + 4x + 1$ has a graph that is a parabola. Find the coordinates of the vertex and the equation of the axis of symmetry for the parabola. Then sketch its graph. State the domain and range of the function.

Solution

Because a is negative ($a = -2$), the graph of f will open down.

For this function, $a = -2$, $b = 4$, and $c = 1$.
The x-coordinate of the vertex is

$$x = -\frac{b}{2a} = -\frac{4}{2(-2)} = 1.$$

The y-coordinate of the vertex is

$$f(1) = -2 \cdot 1^2 + 4 \cdot 1 + 1 = 3.$$

Thus, the coordinates of the vertex is (1, 3).

The axis of symmetry is

$$x = -\frac{b}{2a} = 1$$

Evaluate the function for various values of x.

x	$f(x) = -2x^2 + 4x + 1$	$f(x)$	(x, y)
-1	$f(-1) = -2(-1)^2 + 4(-1) + 1$	-5	$(-1, -5)$
0	$f(0) = -2(0)^2 + 4(0) + 1$	1	$(0, 1)$
2	$f(2) = -2(2)^2 + 4(2) + 1$	1	$(2, 1)$
3	$f(3) = -2(3)^2 + 4(3) + 1$	-5	$(3, -5)$

Graph the ordered-pair solutions, the vertex, and axis of symmetry. Draw a parabola through the points.

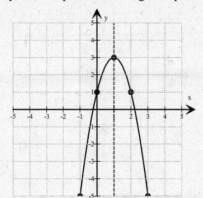

Because $f(x) = 2x^2 + 4x + 1$ is a real number for all values
of x, the domain of f is $\{x \mid x \in \text{ real numbers}\}$.

The vertex of the parabola is the highest point on the graph. Because the y-coordinate at the vertex is 3, the range
of f is $\{y \mid y \le 3\}$.

Try it

1. The quadratic function
$f(x) = -x^2 + 1$ has a graph
that is a parabola. Find the
coordinates of the vertex and the
equation of the axis of symmetry for
the parabola. Then sketch its graph.
State the domain and range of the
function.

Reflect on it

- The graph of a parabola opens up when $a > 0$ and opens down when $a < 0$.

Quiz Yourself 5.5A

In Exercises 1 through 3, find the coordinates of the vertex and the equation of the axis of symmetry for the parabola with the given equation. Then sketch its graph.

1. $y = x^2 - 1$

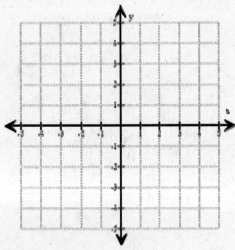

2. $y = -\frac{1}{2}x^2 + 2$

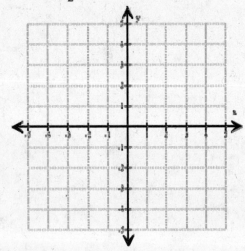

3. $y = -x^2 + 2x + 4$

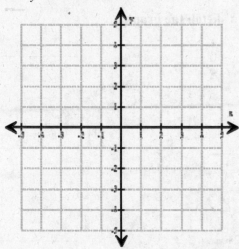

4. Graph $f(x) = -x^2 + 2x - 3$. State the domain and range of the function.

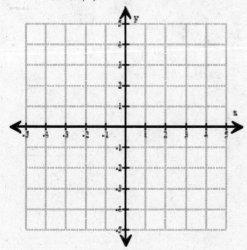

5. Graph $f(x) = 2x^2 + 4x - 3$. State the domain and range of the function.

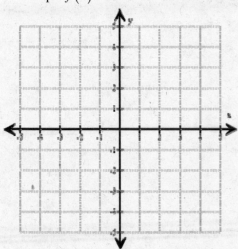

Practice Sheet 5.5A

Find the vertex and axis of symmetry of the parabola. Then sketch the graph.

1. $y = x^2 - 1$

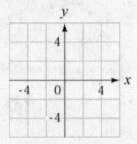

2. $y = \frac{1}{2}x^2 + 2$

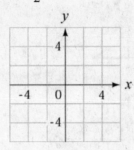

3. $y = x^2 + 3x$

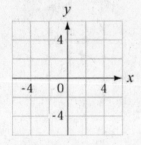

4. $y = -x^2 + 3x - 2$

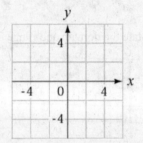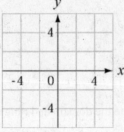

Graph the function. State the domain and range of the function.

5. $y = x^2 + x - 2$

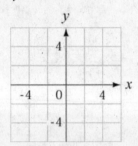

6. $y = -x^2 + 1$

7. $y = 2x^2 + x - 3$

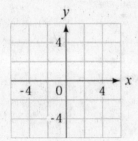

8. $y = x^2 - 3x$

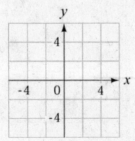

1. _____

2. _____

3. _____

4. _____

5. _____

6. _____

7. _____

8. _____

Answers

Try it 5.5A

1. Vertex: (0, 1); axis of symmetry $x = 0$

   Domain: $\{x \mid x \in \text{real numbers}\}$

   Range: $\{y \mid y \le 1\}$

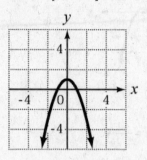

Quiz 5.5A

1. Vertex: $(0, -1)$; axis of symmetry: $x = 0$

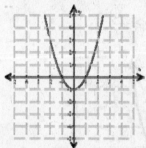

2. Vertex: $(0, 2)$; axis of symmetry: $x = 0$

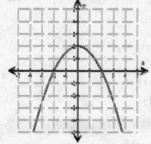

3. Vertex: $(1, 5)$; axis of symmetry: $x = 1$

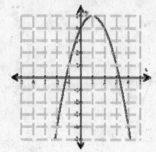

4. Domain: $\{x \mid x \in \text{real numbers}\}$

   Range: $\{y \mid y \le -2\}$

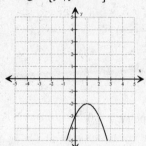

5. Domain: $\{x \mid x \in \text{real numbers}\}$

   Range: $\{y \mid y \ge -5\}$

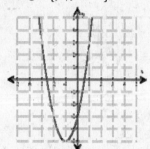

Solutions to Practice Sheet 5.5A

1. $y = x^2 - 1$

   $a = 1, \ b = 0$

   $-\dfrac{b}{2a} = -\dfrac{0}{2(1)} = 0$

   $y = 0^2 - 1 = -1$

   Vertex: $(0, -1)$

   Axis of symmetry: $x = 0$

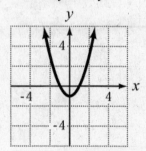

2. $y = \frac{1}{2}x^2 + 2$

$a = \frac{1}{2}, \; b = 0$

$-\frac{b}{2a} = -\frac{0}{2\left(\frac{1}{2}\right)} = 0$

$y = \frac{1}{2}(0)^2 + 2 = 2$

Vertex: $(0, 2)$

Axis of symmetry: $x = 0$

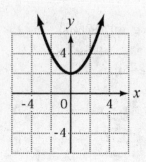

3. $y = x^2 + 3x$

$a = 1, \; b = 3$

$-\frac{b}{2a} = -\frac{3}{2(1)} = -\frac{3}{2}$

$y = \left(-\frac{3}{2}\right)^2 + 3\left(-\frac{3}{2}\right) = -\frac{9}{4}$

Vertex: $\left(-\frac{3}{2}, -\frac{9}{4}\right)$

Axis of symmetry: $x = -\frac{3}{2}$

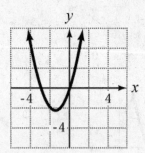

4. $y = -x^2 + 3x - 2$

$a = -1, \; b = 3$

$-\frac{b}{2a} = -\frac{3}{2(-1)} = \frac{3}{2}$

$y = -\left(\frac{3}{2}\right)^2 + 3\left(\frac{3}{2}\right) - 2 = \frac{1}{4}$

Vertex: $\left(\frac{3}{2}, \frac{1}{4}\right)$

Axis of symmetry: $x = \frac{3}{2}$

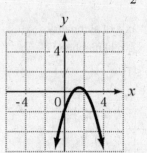

5. $y = x^2 + x - 2$

$a = 1, \; b = 1$

$-\frac{b}{2a} = -\frac{1}{2(1)} = -\frac{1}{2}$

$y = \left(-\frac{1}{2}\right)^2 + \left(-\frac{1}{2}\right) - 2 = -\frac{9}{4}$

Domain: $\{x \mid x \in \text{real numbers}\}$

Range: $\left\{y \mid y \geq -\frac{9}{4}\right\}$

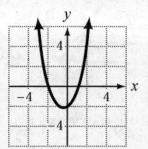

6. $y = -x^2 + 1$

$a = -1, \; b = 0$

$-\frac{b}{2a} = -\frac{0}{2(-1)} = 0$

$y = -(0)^2 + 1 = 1$

Domain: $\{x \mid x \in \text{real numbers}\}$

Range: $\{y \mid y \leq 1\}$

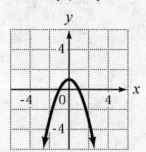

7. $y = 2x^2 + x - 3$

$a = 2, \ b = 1$

$-\dfrac{b}{2a} = -\dfrac{1}{2(2)} = -\dfrac{1}{4}$

$y = 2\left(-\dfrac{1}{4}\right)^2 + \left(-\dfrac{1}{4}\right) - 3 = -\dfrac{25}{8}$

Domain: $\{x \mid x \in \text{real numbers}\}$

Range: $\left\{y \mid y \geq -\dfrac{25}{8}\right\}$

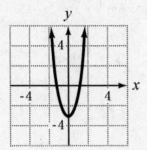

8. $y = x^2 - 3x$

$a = 1, \ b = -3$

$-\dfrac{b}{2a} = -\dfrac{-3}{2(1)} = \dfrac{3}{2}$

$y = \left(\dfrac{3}{2}\right)^2 - 3\left(\dfrac{3}{2}\right) = -\dfrac{9}{4}$

Domain: $\{x \mid x \in \text{real numbers}\}$

Range: $\left\{y \mid y \geq -\dfrac{9}{4}\right\}$

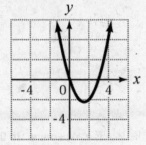

Objective 5.5B – Find the *x*-intercepts of a quadratic function

Key Words and Concepts

Recall that a point at which a graph crosses the *x*- or *y*-axis is called an *intercept* of the graph.
The *x*-intercepts of the graph of an equation occur when $y = 0$; the *y*-intercepts occur when $x = 0$.

Example

Find the coordinates of the *x*-intercepts of the parabola whose equation is $y = 2x^2 - 5x - 3$.

Solution
To find the *x*-intercepts, let $y = 0$ and solve for *x*.

$y = 2x^2 - 5x - 3$ Copy the equation

$0 = 2x^2 - 5x - 3$ Let $y = 0$

$0 = (2x + 1)(x - 3)$ Factor

$2x + 1 = 0$ or $x - 3 = 0$ Use the Principal of

$x = -\dfrac{1}{2}$ $x = 0$ Zero Products

The coordinates of the *x*-intercepts are $\left(-\dfrac{1}{2},\ 0\right)$ and $(3, 0)$.

See the graph below.

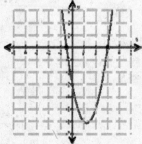

Try it

1. Find the coordinates of the *x*-intercepts of the parabola whose equation is $y = x^2 - 2x - 15$.

If $ax^2 + bx + c = 0$ has a double root, then the graph of $y = ax^2 + bx + c$ intersects the *x*-axis at one point. In that case, the graph is said to be **tangent** to the *x*-axis.

Example

Find the coordinates of the x-intercept of the parabola whose equation is $y = -9x^2 + 12x - 4$.

Solution

To find the x-intercepts, let $y = 0$ and solve for x.

$y = -9x^2 + 12x - 4$ Copy the equation

$0 = -9x^2 + 12x - 4$ Let $y = 0$

$0 = 9x^2 - 12x + 4$ Mulitply each side by -1

$0 = (3x - 2)(3x - 2)$ Factor

$3x - 2 = 0$ or $3x - 2 = 0$ Use the Principal of

$x = \dfrac{2}{3}$ $x = \dfrac{2}{3}$ Zero Products

The coordinates of the x-intercept is $\left(\dfrac{2}{3},\ 0\right)$.

See the graph below.

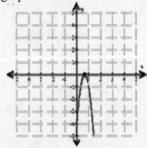

Try it

2. Find the coordinates of the x-intercept of the parabola whose equation is $y = x^2 - 10x + 25$.

Reflect on it

- When a quadratic equation has two solutions, then the graph intersects the x-axis at two different points.
- When a quadratic equation has one solution (a double root), then the graph intersects the x-axis at one point.
- How many times will a graph with no real solutions intersects the x-axis?

Quiz Yourself 5.5B

For Exercises 1 through 5, find the coordinates of the x-intercepts of the parabola given by the equation.

1. $y = x^2 - 3x - 4$

2. $y = 2x^2 - 10x$

3. $y = 3x^2 - 7x - 6$

4. $y = x^2 + x + 3$

5. $y = x^2 - 4x + 1$

Practice Sheet 5.5B

Find the coordinates of the *x*-intercepts of the parabola.

1. $y = x^2 - 1$

2. $y = x^2 - 16$

3. $y = x^2 - x$

4. $y = 4x^2 - 8x$

5. $y = 2x^2 - 3x - 2$

6. $y = x^2 + 5x - 6$

7. $y = 3x^2 - 5x - 2$

8. $y = x^2 + x + 2$

9. $y = 2x^2 - 9x - 5$

10. $y = 3x^2 + 2x + 2$

11. $y = 2x^2 + x - 1$

12. $y = 2x^2 + 5x - 3$

13. $y = 2x^2 - 3x - 2$

14. $y = x^2 - 4x + 4$

15. $y = x^2 - 6x + 9$

16. $y = 4x^2 + x - 3$

1. _____

2. _____

3. _____

4. _____

5. _____

6. _____

7. _____

8. _____

9. _____

10. _____

11. _____

12. _____

13. _____

14. _____

15. _____

16. _____

Answers

Try it 5.5B
1. $(-3, 0)$, $(5, 0)$
2. $(5, 0)$

Quiz 5.5B
1. $(-1, 0)$, $(4, 0)$
2. $(0, 0)$, $(5, 0)$
3. $\left(-\dfrac{2}{3}, 0\right)$, $(3, 0)$
4. No x-intercepts
5. $\left(2+\sqrt{3},\, 0\right)$, $\left(2-\sqrt{3},\, 0\right)$

Solutions to Practice Sheet 5.5B

1. $y = x^2 - 1$
   $0 = x^2 - 1$
   $0 = (x+1)(x-1)$
   $x+1 = 0$ or $x-1 = 0$
   $x = -1$ $\qquad$ $x = 1$
   The x-intercepts are $(-1, 0)$ and $(1, 0)$.

2. $y = x^2 - 16$
   $0 = x^2 - 16$
   $0 = (x+4)(x-4)$
   $x+4 = 0$ or $x-4 = 0$
   $x = -4$ $\qquad$ $x = 4$
   The x-intercepts are $(-4, 0)$ and $(4, 0)$.

3. $y = x^2 - x$
   $0 = x^2 - x$
   $0 = x(x-1)$
   $x = 0$ or $x-1 = 0$
   $\qquad\qquad$ $x = 1$
   The x-intercepts are $(0, 0)$ and $(1, 0)$.

4. $y = 4x^2 - 8x$
   $0 = 4x^2 - 8x$
   $0 = 4x(x-2)$
   $x = 0$ or $x-2 = 0$
   $\qquad\qquad$ $x = 2$
   The x-intercepts are $(0, 0)$ and $(2, 0)$.

5. $y = 2x^2 - 3x - 2$
   $0 = 2x^2 - 3x - 2$
   $0 = (2x+1)(x-2)$
   $2x+1 = 0$ or $x-2 = 0$
   $x = -\dfrac{1}{2}$ $\qquad$ $x = 2$
   The x-intercepts are $\left(-\dfrac{1}{2}, 0\right)$ and $(2, 0)$.

6. $y = x^2 + 5x - 6$
   $0 = x^2 + 5x - 6$
   $0 = (x+6)(x-1)$
   $x+6 = 0$ or $x-1 = 0$
   $x = -6$ $\qquad$ $x = 1$
   The x-intercepts are $(-6, 0)$ and $(1, 0)$.

7. $y = 3x^2 - 5x - 2$
   $0 = 3x^2 - 5x - 2$
   $0 = (3x+1)(x-2)$
   $3x+1 = 0$ or $x-2 = 0$
   $x = -\dfrac{1}{3}$ $\qquad$ $x = 2$
   The x-intercepts are $\left(-\dfrac{1}{3}, 0\right)$ and $(2, 0)$.

8. $y = x^2 + x - 2$
   $0 = x^2 + x - 2$
   $0 = (x+2)(x-1)$
   $x+2 = 0$ or $x-1 = 0$
   $x = -2$ $\qquad$ $x = 1$
   The x-intercepts are $(-2, 0)$ and $(1, 0)$.

9. $y = 2x^2 - 9x - 5$
   $0 = 2x^2 - 9x - 5$
   $0 = (2x+1)(x-5)$
   $2x+1 = 0$ or $x-5 = 0$
   $x = -\dfrac{1}{2}$ $\qquad$ $x = 5$
   The x-intercepts are $\left(-\dfrac{1}{2}, 0\right)$ and $(5, 0)$.

10. $y = 3x^2 + 2x + 2$
    $0 = 3x^2 + 2x + 2$
    The quadratic is not factorable over the real numbers.
    The parabola has no x-intercepts.

11. $y = 2x^2 + x - 1$

$0 = 2x^2 + x - 1$

$0 = (x+1)(2x-1)$

$x + 1 = 0 \quad$ or $\quad 2x - 1 = 0$

$x = -1 \qquad\qquad x = \dfrac{1}{2}$

The x-intercepts are $(-1, 0)$ and $\left(\dfrac{1}{2}, 0\right)$.

12. $y = 2x^2 + 5x - 3$

$0 = 2x^2 + 5x - 3$

$0 = (x+3)(2x-1)$

$x + 3 = 0 \quad$ or $\quad 2x - 1 = 0$

$x = -3 \qquad\qquad x = \dfrac{1}{2}$

The x-intercepts are $(-3, 0)$ and $\left(\dfrac{1}{2}, 0\right)$.

13. $y = 2x^2 - 3x - 2$

$0 = 2x^2 - 3x - 2$

$0 = (2x+1)(x-2)$

$2x + 1 = 0 \quad$ or $\quad x - 2 = 0$

$x = -\dfrac{1}{2} \qquad\qquad x = 2$

The x-intercepts are $\left(-\dfrac{1}{2}, 0\right)$ and $(2, 0)$.

14. $y = x^2 - 4x + 4$

$0 = x^2 - 4x + 4$

$0 = (x-2)(x-2)$

$x - 2 = 0 \quad$ or $\quad x - 2 = 0$

$x = 2 \qquad\qquad x = 2$

The x-intercept is $(2, 0)$.

15. $y = x^2 - 6x + 9$

$0 = x^2 - 6x + 9$

$0 = (x-3)(x-3)$

$x - 3 = 0 \quad$ or $\quad x - 3 = 0$

$x = 3 \qquad\qquad x = 3$

The x-intercept is $(3, 0)$.

16. $y = 4x^2 + x - 3$

$0 = 4x^2 + x - 3$

$0 = (x+1)(4x-3)$

$x + 1 = 0 \quad$ or $\quad 4x - 3 = 0$

$x = -1 \qquad\qquad x = \dfrac{3}{4}$

The x-intercepts are $(-1, 0)$ and $\left(\dfrac{3}{4}, 0\right)$.

Objective 5.5C – Find the minimum or maximum of a quadratic function

Key Words and Concepts

If $a > 0$ in $f(x) = ax^2 + bx + c$, then the graph of f is a parabola that opens _____.

(up/down)

The vertex of the parabola is the lowest point on the parabola.
It is the point that has the minimum y-coordinate.
Therefore, the value of the function at this point is a **minimum**.

If $a < 0$ in $f(x) = ax^2 + bx + c$, then the graph of f is a parabola that opens _____.

(up/down)

The vertex of the parabola is the highest point on the parabola.
It is the point that has the maximum y-coordinate.
Therefore, the value of the function at this point is a **maximum**.

To find the minimum or maximum value of a quadratic function,
First find the x-coordinate of the vertex.
Then evaluate the function at that value.

Example

Find the minimum or maximum value of $f(x) = 3x^2 + 6x - 1$.

Solution
Find the x-coordinate of the vertex.

$$x = -\frac{b}{2a} = -\frac{6}{2(3)} = -1.$$

Find the y-coordinate of the vertex.

$$f(x) = 3x^2 + 6x - 1$$

$$f(-1) = 3(-1)^2 + 6(-1) - 1$$

$$f(-1) = -4$$

Because $a = 3 > 0$, the graph of f opens up.
Therefore, the function has a minimum value.

The minimum value of the function is –4.

Try it

1. Find the minimum or maximum value of $f(x) = -2x^2 - 4x + 4$.

Reflect on it

- If $a > 0$ in $f(x) = ax^2 + bx + c$, then the graph of f is a parabola that opens up.
- If $a < 0$ in $f(x) = ax^2 + bx + c$, then the graph of f is a parabola that opens down.
- If the parabola opens up, it has a minimum.
- If the parabola opens down, it has a maximum.

Quiz Yourself 5.5C

In Exercises 1 through 4, find the maximum or minimum value of the function.

1. $f(x) = -x^2 - 2x + 12$

2. $f(x) = x^2 + 6x - 6$

3. $f(x) = 2x^2 + x + 14$

4. $f(x) = -2x^2 + 14x - 15$

Practice Sheet 5.5C

Find the minimum or maximum value of the quadratic function.

1. $f(x) = x^2 + 6x + 8$ **2.** $f(x) = x^2 + 2x - 3$ **1.** _____

     2. _____

3. $f(x) = -2x^2 + 8x + 5$ **4.** $f(x) = -x^2 - 2x + 1$ **3.** _____

     4. _____

5. $f(x) = x^2 + 6x - 2$ **6.** $f(x) = x^2 - 3x + 2$ **5.** _____

     6. _____

7. $f(x) = -3x^2 - 12x + 1$ **8.** $f(x) = 2x^2 + 4x - 5$ **7.** _____

     8. _____

9. $f(x) = -x^2 - x + 6$ **10.** $f(x) = x^2 - 4x - 21$ **9.** _____

     10. _____

11. $f(x) = 2x^2 + 6x - 1$ **12.** $f(x) = -2x^2 + 4x - 1$ **11.** _____

     12. _____

13. $f(x) = x^2 + 6x + 6$ **14.** $f(x) = x^2 - 3x + 6$ **13.** _____

     14. _____

15. $f(x) = -3x^2 + 4x + 6$ **16.** $f(x) = -2x^2 - 3x + 1$ **15.** _____

     16. _____

Answers

Try it 5.5C

1. Maximum: 6

Quiz 5.5C

1. Maximum: 13
2. Minimum: -15
3. Minimum: $\dfrac{111}{8}$
4. Maximum: $\dfrac{19}{2}$

Solutions to Practice Sheet 5.5C

1. $f(x) = x^2 + 6x + 8$

   $x = -\dfrac{b}{2a} = -\dfrac{6}{2(1)} = -3$

   $f(-3) = (-3)^2 + 6(-3) + 8$

   $f(-3) = -1$

   Because $a > 0$, the graph opens up.
   Therefore, the function has a minimum value.
   The minimum value of the function is -1.

2. $f(x) = x^2 + 2x - 3$

   $x = -\dfrac{b}{2a} = -\dfrac{2}{2(1)} = -1$

   $f(-1) = (-1)^2 + 2(-1) - 3$

   $f(-1) = -4$

   Because $a > 0$, the graph opens up.
   Therefore, the function has a minimum value.
   The minimum value of the function is -4.

3. $f(x) = -2x^2 + 8x + 5$

   $x = -\dfrac{b}{2a} = -\dfrac{8}{2(-2)} = 2$

   $f(2) = -2(2)^2 + 8(2) + 5$

   $f(2) = 13$

   Because $a < 0$, the graph opens down.
   Therefore, the function has a maximum value.
   The maximum value of the function is 13.

4. $f(x) = -x^2 - 2x + 1$

   $x = -\dfrac{b}{2a} = -\dfrac{-2}{2(-1)} = -1$

   $f(-1) = -(-1)^2 - 2(-1) + 1$

   $f(-1) = 2$

   Because $a < 0$, the graph opens down.
   Therefore, the function has a maximum value.
   The maximum value of the function is 2.

5. $f(x) = x^2 + 6x - 2$

   $x = -\dfrac{b}{2a} = -\dfrac{6}{2(1)} = -3$

   $f(-3) = (-3)^2 + 6(-3) - 2$

   $f(-3) = -11$

   Because $a > 0$, the graph opens up.
   Therefore, the function has a minimum value.
   The minimum value of the function is -11.

6. $f(x) = x^2 - 3x + 2$

   $x = -\dfrac{b}{2a} = -\dfrac{-3}{2(1)} = \dfrac{3}{2}$

   $f\left(\dfrac{3}{2}\right) = \left(\dfrac{3}{2}\right)^2 - 3\left(\dfrac{3}{2}\right) + 2$

   $f\left(\dfrac{3}{2}\right) = -\dfrac{1}{4}$

   Because $a > 0$, the graph opens up.
   Therefore, the function has a minimum value.
   The minimum value of the function is $-\dfrac{1}{4}$.

7. $f(x) = -3x^2 - 12x + 1$

   $x = -\dfrac{b}{2a} = -\dfrac{-12}{2(-3)} = -2$

   $f(-2) = -3(-2)^2 - 12(-2) + 1$

   $f(-2) = 13$

   Because $a < 0$, the graph opens down.
   Therefore, the function has a maximum value.
   The maximum value of the function is 13.

8. $f(x) = 2x^2 + 4x - 5$

   $x = -\dfrac{b}{2a} = -\dfrac{4}{2(2)} = -1$

   $f(-1) = 2(-1)^2 + 4(-1) - 5$

   $f(-1) = -7$

   Because $a > 0$, the graph opens up.
   Therefore, the function has a minimum value.
   The minimum value of the function is -7.

9. $f(x) = -x^2 - x + 6$

   $x = -\dfrac{b}{2a} = -\dfrac{-1}{2(-1)} = -\dfrac{1}{2}$

   $f\left(-\dfrac{1}{2}\right) = -\left(-\dfrac{1}{2}\right)^2 - \left(-\dfrac{1}{2}\right) + 6$

   $f\left(-\dfrac{1}{2}\right) = \dfrac{25}{4}$

   Because $a < 0$, the graph opens down.
   Therefore, the function has a maximum value.
   The maximum value of the function is $\dfrac{25}{4}$.

10. $f(x) = x^2 - 4x - 21$

$x = -\dfrac{b}{2a} = -\dfrac{-4}{2(1)} = 2$

$f(2) = 2^2 - 4(2) - 21$

$f(2) = -25$

Because $a > 0$, the graph opens up.

Therefore, the function has a minimum value.

The minimum value of the function is -25.

11. $f(x) = 2x^2 + 6x - 1$

$x = -\dfrac{b}{2a} = -\dfrac{6}{2(2)} = -\dfrac{3}{2}$

$f\left(-\dfrac{3}{2}\right) = 2\left(-\dfrac{3}{2}\right)^2 + 6\left(-\dfrac{3}{2}\right) - 1$

$f\left(-\dfrac{3}{2}\right) = -\dfrac{11}{2}$

Because $a > 0$, the graph opens up.

Therefore, the function has a minimum value.

The minimum value of the function is $-\dfrac{11}{2}$.

12. $f(x) = -2x^2 + 4x - 1$

$x = -\dfrac{b}{2a} = -\dfrac{4}{2(-2)} = 1$

$f(1) = -2(1)^2 + 4(1) - 1$

$f(1) = 1$

Because $a < 0$, the graph opens down.

Therefore, the function has a maximum value.

The maximum value of the function is 1.

13. $f(x) = x^2 + 6x + 6$

$x = -\dfrac{b}{2a} = -\dfrac{6}{2(1)} = -3$

$f(-3) = (-3)^2 + 6(-3) + 6$

$f(-3) = -3$

Because $a > 0$, the graph opens up.

Therefore, the function has a minimum value.

The minimum value of the function is -3.

14. $f(x) = x^2 - 3x + 6$

$x = -\dfrac{b}{2a} = -\dfrac{-3}{2(1)} = \dfrac{3}{2}$

$f\left(\dfrac{3}{2}\right) = \left(\dfrac{3}{2}\right)^2 - 3\left(\dfrac{3}{2}\right) + 6$

$f\left(\dfrac{3}{2}\right) = \dfrac{15}{4}$

Because $a > 0$, the graph opens up.

Therefore, the function has a minimum value.

The minimum value of the function is $\dfrac{15}{4}$.

15. $f(x) = -3x^2 + 4x + 6$

$x = -\dfrac{b}{2a} = -\dfrac{4}{2(-3)} = \dfrac{2}{3}$

$f\left(\dfrac{2}{3}\right) = -3\left(\dfrac{2}{3}\right)^2 + 4\left(\dfrac{2}{3}\right) + 6$

$f\left(\dfrac{2}{3}\right) = \dfrac{22}{3}$

Because $a < 0$, the graph opens down.

Therefore, the function has a maximum value.

The maximum value of the function is $\dfrac{22}{3}$.

16. $f(x) = -2x^2 - 3x + 1$

$x = -\dfrac{b}{2a} = -\dfrac{-3}{2(-2)} = -\dfrac{3}{4}$

$f\left(-\dfrac{3}{4}\right) = -2\left(-\dfrac{3}{4}\right)^2 - 3\left(-\dfrac{3}{4}\right) + 1$

$f\left(-\dfrac{3}{4}\right) = \dfrac{17}{8}$

Because $a < 0$, the graph opens down.

Therefore, the function has a maximum value.

The maximum value of the function is $\dfrac{17}{8}$.

Objective 5.5D – Solve applications of quadratic functions

Example

The perimeter of a rectangular area is to be 80 m. What dimensions yield maximum area? What is the maximum area?

Solution

Start with the rectangle perimeter formula, $P = 2L + 2W$, substitute 80 for P, and solve for L.

$$P = 2L + 2W \qquad \text{Perimeter formula}$$
$$80 = 2L + 2W \qquad \text{Perimeter is 80 m}$$
$$40 = L + W \qquad \text{Divide each side by 2}$$
$$40 - W = L \qquad \text{Solve for } L$$

Start with the rectangle area formula, $A = LW$, substitute $40 - W$ for L, and simplify.

$$A = LW \qquad \text{Area formula}$$
$$A = (40 - W)W \qquad \text{Substitute } 40 - W \text{ for } L$$
$$A = 40W - W^2 \qquad \text{Distribute}$$
$$A = -W^2 + 40W \qquad \text{Rewrite}$$

The graph of $-W^2 + 40W$ is a parabola that opens down and it has a maximum. Find the W-coordinate of the vertex.

$$W = -\frac{b}{2a} = -\frac{40}{2(-1)} = 20.$$

The width of the rectangle is 20 m.

Find L.

$$L = 40 - W \qquad \text{From above}$$
$$= 40 - 20 \qquad \text{Substitute 20 for } W$$
$$= 20$$

The length of the rectangle is 20 m.

Thus the dimensions that yield maximum area are 20 m by 20 m.

The maximum area is $(20 \text{ m})(20 \text{ m}) = 400 \text{ m}^2$.

Try it

1. Find two numbers whose difference is 60 and whose product is a minimum.

Example

Suppose that the height (in meters) above ground of a hanging cable is modeled by $h(x) = 0.5x^2 - 0.7x + 30$, where x is the horizontal distance from one of the points at which the cable is attached. What is the minimum height of the cable? Round to the nearest tenth.

Solution

First, to find the horizontal distance that will minimize the height, find the x-coordinate of the vertex.

$$h(x) = 0.5x^2 - 0.7x + 30$$

$$x = -\frac{b}{2a} = -\frac{-0.7}{2(0.5)} = 0.7$$

Then, to find the minimum height, evaluate the function at the x-coordinate of the vertex.

$$f(0.7) = 0.5(0.7)^2 - 0.7(0.7) + 30 \quad \text{Substitute 0.7}$$

$$= 29.755 \qquad \text{Exact value}$$

$$\approx 29.8 \qquad \text{Round to the nearest tenth.}$$

The minimum height of the cable is about 29.8 m.

Try it

2. A tour operator believes that the profit P, in dollars, from selling x tickets per trip is given by $P(x) = 200x - 5x^2$. Using this model, what is the maximum profit the tour operator can expect?

Quiz Yourself 5.5D

1. The height s, in feet, of a rock thrown upward at an initial speed of 64 ft/s from a cliff 50 ft above an ocean beach is given by the function $s(t) = -16t^2 + 64t + 50$, where t is the time in seconds. Find the maximum height above the beach that the rock will attain.

2. A tour operator believes that the profit P, in dollars, from selling x tickets is given by $P(x) = 40x - 0.25x^2$. Using this model, what is the maximum profit the tour operator can expect?

3. Suppose the height h, in meters, of an airplane that produces weightlessness is modeled by the equation $h(t) = -1.42t^2 + 119t + 6000,$ where t is the number of seconds elapsed since the plane entered its parabolic path. Find to the nearest hundred meters the maximum height of this airplane.

4. A company has determined that the cost, c, in dollars per dozen widgets of producing a widget is given by $c(x) = 0.25x^2 - 3x + 10.$ Find the minimum cost of producing a dozen widgets.

Practice Sheet 5.5D

Solve.

1. The height in feet (s) of a rock thrown straight up is given by the function $s(t) = -16t^2 + 96t$, where t is the time in seconds. Find the maximum height above the ground that the rock will attain.

2. The height in feet (s) of a ball thrown upward at an initial speed of 72 ft/s from a platform 50 ft high is given by the function $s(t) = -16t^2 + 72t + 50$, where t is the time in seconds. Find the maximum height above the ground that the ball will attain.

1. _____

2. _____

3. A pool is treated with a chemical to reduce the amount of algae. The amount of algae in the pool t days after the treatment can be approximated by the function $A(t) = 25t^2 - 200t + 425$. How many days after treatment will the pool have the least amount of algae?

4. A manufacturer of video cassette recorders believes that the revenue the company receives is related to the price (P) of a recorder by the function $R(P) = 75P - \frac{1}{5}P^2$. What price will give the maximum revenue?

3. _____

4. _____

5. The height in feet (s) of a rocket after t seconds is given by the formula $s(t) = 224t - 16t^2$. After how many seconds does the rocket reach its maximum height?

6. Find the two numbers whose sum is 50 and whose product is a maximum.

5. _____

6. _____

7. Find two numbers whose sum is 30 and whose product is a maximum.

8. Find two numbers whose difference is 20 and whose product is a minimum.

7. _____

8. _____

9. A rectangle has a perimeter of 56 ft. Find the dimensions of the rectangle with this perimeter that will have the maximum area. What is the maximum area?

10. The perimeter of a rectangular window is 32 ft. Find the dimensions of the window with this perimeter that will enclose the largest area. What is the maximum area?

9. _____

10. _____

Answers

Try it 5.5D

1. −30 and 30
2. $2000

Quiz 5.5D

1. 114 ft
2. $1600
3. 8500 m
4. $1

Solutions to Practice Sheet 5.5D

1. **STRATEGY**
   To find the maximum height, find the
   t-coordinate of the vertex and evaluate the
   function at that value.

   SOLUTION
   $$s(t) = -16t^2 + 96t$$
   $$t = -\frac{b}{2a} = -\frac{96}{2(-16)} = 3$$
   $$s(3) = -16(3)^2 + 96(3) = 144$$
   The maximum height of the rock is 144 ft.

2. **STRATEGY**
   To find the maximum height, find the
   t-coordinate of the vertex and evaluate the
   function at that value.

   SOLUTION
   $$s(t) = -16t^2 + 72t + 50$$
   $$t = -\frac{b}{2a} = -\frac{72}{2(-16)} = 2.25$$
   $$s(2.25) = -16(2.25)^2 + 72(2.25) + 50 = 131$$
   The maximum height of the ball is 131 ft.

3. **STRATEGY**
   To find the number of days to minimize the
   algae, find the t-coordinate of the vertex.

   SOLUTION
   $$A(t) = 25t^2 - 200t + 425$$
   $$t = -\frac{b}{2a} = -\frac{-200}{2(25)} = 4$$
   The number of days that will minimize the
   algae is 4 days.

4. **STRATEGY**
   To find the price to maximize the revenue,
   find the P-coordinate of the vertex.

SOLUTION
$$R(P) = 75P - \frac{1}{5}P^2 = -\frac{1}{5}P^2 + 75P$$
$$P = -\frac{b}{2a} = -\frac{75}{2\left(-\frac{1}{5}\right)} = 187.5$$

The price that will maximize revenue
is $187.50.

5. **STRATEGY**
   To find the time when the rocket reaches the
   maximum height, find the t-coordinate of the
   vertex.

   SOLUTION
   $$s(t) = 224t - 16t^2 = -16t^2 + 224t$$
   $$t = -\frac{b}{2a} = -\frac{224}{2(-16)} = 7$$
   After 7 seconds, the rocket will reach its
   maximum height.

6. **STRATEGY**
   Let x and y represent the two numbers.
   Express y in terms of x.
   $$x + y = 50$$
   $$y = 50 - x$$
   Express the product of the numbers in
   terms of x.
   $$xy = x(50 - x)$$
   $$f(x) = -x^2 + 50x$$
   To find one of the two numbers, find the
   x-coordinate of the vertex of
   $$f(x) = -x^2 + 50x.$$
   To find the other number, replace x in $50 - x$
   by the x-coordinate of the vertex and evaluate.

   SOLUTION
   $$f(x) = -x^2 + 50x$$
   $$x = -\frac{b}{2a} = -\frac{50}{2(-1)} = 25$$
   $$50 - x = 50 - 25 = 25$$
   The numbers are 25 and 25.

7. **STRATEGY**
   Let x and y represent the two numbers.
   Express y in terms of x.
   $$x + y = 30$$
   $$y = 30 - x$$
   Express the product of the numbers in
   terms of x.
   $$xy = x(30 - x)$$
   $$f(x) = -x^2 + 30x$$

To find one of the two numbers, find the x-coordinate of the vertex of

$$f(x) = -x^2 + 30x.$$

To find the other number, replace x in $30 - x$ by the x-coordinate of the vertex and evaluate.

SOLUTION

$$f(x) = -x^2 + 30x$$
$$x = -\frac{b}{2a} = -\frac{30}{2(-1)} = 15$$
$$30 - x = 30 - 15 = 15$$

The numbers are 15 and 15.

8. **STRATEGY**

Let x and y represent the two numbers. Express y in terms of x.

$$y - x = 20$$
$$y = x + 20$$

Express the product of the numbers in terms of x.

$$xy = x(x + 20)$$
$$f(x) = x^2 + 20x$$

To find one of the two numbers, find the x-coordinate of the vertex of

$$f(x) = x^2 + 20x.$$

To find the other number, replace x in $x + 20$ by the x-coordinate of the vertex and evaluate.

SOLUTION

$$f(x) = x^2 + 20x$$
$$x = -\frac{b}{2a} = -\frac{20}{2(1)} = -10$$
$$x + 20 = -10 + 20 = 10$$

The numbers are –10 and 10.

9. **STRATEGY**

Let L and W represent the length and width. Express L in terms of W.

$$2L + 2W = 56$$
$$2L = 56 - 2W$$
$$L = 28 - W$$

Express the (area) product in terms of W.

$$LW = (28 - W)W$$
$$A(W) = -W^2 + 28W$$

To find the width, find the

W-coordinate of the vertex of

$$A(W) = -W^2 + 28W.$$

To find the length, replace W in $28 - W$ by the W-coordinate of the vertex and evaluate.

To find the maximum area, evaluate the function at that value.

SOLUTION

$$A(W) = -W^2 + 28W$$
$$W = -\frac{b}{2a} = -\frac{28}{2(-1)} = 14$$
$$28 - W = 28 - 14 = 14$$
$$A(14) = -(14)^2 + 28(14) = 196$$

The dimensions are 14 ft by 14 ft.

The maximum area is 196 ft^2.

10. **STRATEGY**

Let L and W represent the length and width. Express L in terms of W.

$$2L + 2W = 32$$
$$2L = 32 - 2W$$
$$L = 16 - W$$

Express the (area) product in terms of W.

$$LW = (16 - W)W$$
$$A(W) = -W^2 + 16W$$

To find the width, find the W-coordinate of the vertex of

$$A(W) = -W^2 + 16W.$$

To find the length, replace W in $16 - W$ by the W-coordinate of the vertex and evaluate.

To find the maximum area, evaluate the function at that value.

SOLUTION

$$A(W) = -W^2 + 16W$$
$$W = -\frac{b}{2a} = -\frac{16}{2(-1)} = 8$$
$$16 - W = 16 - 8 = 8$$
$$A(8) = -(8)^2 + 16(8) = 64$$

The dimensions are 8 ft by 8 ft.

The maximum area is 64 ft^2.

Objective 5.6A – Graph an exponential function

Key Words and Concepts

The **exponential function** with base b is defined by $f(x) = b^x$ where $b > 0$, $b \neq 1$, and x is any real number.

If $b > 1$, the graph of f depicts **exponential** _____.
(growth/decay)

If $0 < b < 1$, the graph of f depicts **exponential** _____.
(growth/decay)

Because $f(x) = b^x$ ($b > 0$, $b \neq 1$) can be evaluated at both rational and irrational numbers,

the domain of f is all real numbers. And because $b^x > 0$ for all values of x,
the range of f is the positive real numbers.

Example

Evaluate $f(x) = 5^x$ at (a) $x = 2$ and (b) $x = -2$.

Solution
Substitute each value for x and simplify.

(a) $\quad f(2) = 5^2 = 25$

(b) $\quad f(-2) = 5^{-2} = \dfrac{1}{5^2} = \dfrac{1}{25}$

Try it

1. Evaluate $f(x) = 6^x$ at
(a) $x = 2$ and (b) $x = -2$.

A frequently used base in applications of exponential functions is an irrational number designated by e. The number e is approximately 2.71828183. It is an irrational number, so it has a nonterminating, nonrepeating decimal representation.

Natural Exponential Function

The function defined by $f(x) = e^x$ is called the **natural exponential function.**

Example

Evaluate $f(x) = e^x$ at (a) $x = 2$ and (b) $x = \sqrt{2}$. Round to the nearest ten-thousandth.

Solution
Substitute each value for x using the calculator and round.

(a) $\quad f(2) = e^2 \approx 7.3891$

(b) $\quad f(\sqrt{2}) = e^{\sqrt{2}} \approx 4.1133$

Try it

2. Evaluate $f(x) = e^x$ at (a) $x = 4$
and (b) $x = \sqrt{5}$. Round to the nearest ten-thousandth.

Some of the properties of an exponential function can be seen by considering its graph.

Example

Graph: $f(x) = 3^x$

Try it

3. Graph: $f(x) = 2^x + 1$

Solution

Think of the function as $y = 3^x$.

Choose values of x. Find the corresponding values of y.

x	$f(x) = 3^x$	y
-2	$f(-2) = 3^{-2} = \dfrac{1}{3^2} = \dfrac{1}{9}$	$\dfrac{1}{9}$
-1	$f(-1) = 3^{-1} = \dfrac{1}{3^1} = \dfrac{1}{3}$	$\dfrac{1}{3}$
0	$f(0) = 3^0 = 1$	1
1	$f(1) = 3^1 = 3$	3
2	$f(2) = 3^2 = 9$	9

Graph the ordered pairs on a rectangular coordinate system.

Connect the points with a smooth curve.

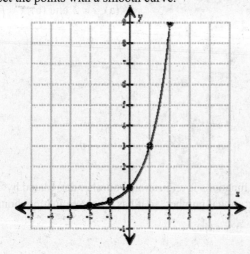

Example

Graph: $g(x) = \left(\frac{1}{3}\right)^{x-1}$

Solution

Proceed as in the previous example,

thinking of the function as $y = \left(\frac{1}{3}\right)^{x-1}$.

x	$f(x) = \left(\frac{1}{3}\right)^{x-1}$	y
-1	$f(-1) = \left(\frac{1}{3}\right)^{-1-1} = \left(\frac{1}{3}\right)^{-2} = 3^2 = 9$	9
0	$f(0) = \left(\frac{1}{3}\right)^{0-1} = \left(\frac{1}{3}\right)^{-1} = 3$	3
1	$f(1) = \left(\frac{1}{3}\right)^{1-1} = \left(\frac{1}{3}\right)^{0} = 1$	1
2	$f(2) = \left(\frac{1}{3}\right)^{2-1} = \left(\frac{1}{3}\right)^{1} = \frac{1}{3}$	$\frac{1}{3}$
3	$f(3) = \left(\frac{1}{3}\right)^{3-1} = \left(\frac{1}{3}\right)^{2} = \frac{1}{9}$	$\frac{1}{9}$

Graph the points and connect with a smooth curve.

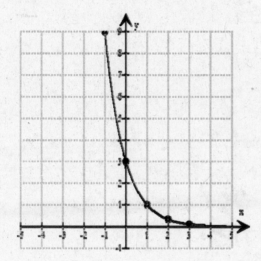

Try it

4. Graph: $f(x) = \left(\frac{1}{3}\right)^{-x-1}$

Reflect on it

- If $b > 1$, the graph of f depicts exponential growth.
- If $0 < b < 1$, the graph of f depicts exponential decay.
- The irrational number e is approximately 2.71828183.

Quiz Yourself 5.6A

1. Given $g(x) = 2^{3x-1}$, evaluate:

a. $g(-1)$ b. $g(0)$ c. $g(1)$

2. Given $h(x) = e^{2-2x}$, evaluate the following. Round to the nearest ten-thousandth.

a. $h(-1)$ b. $h\left(\dfrac{1}{2}\right)$ d. $h\left(\dfrac{3}{2}\right)$

3. Graph: $f(x) = 3^{1-x}$

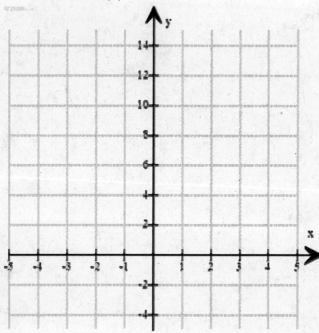

4. Graph: $f(x) = 3^{\frac{1}{2}x}$

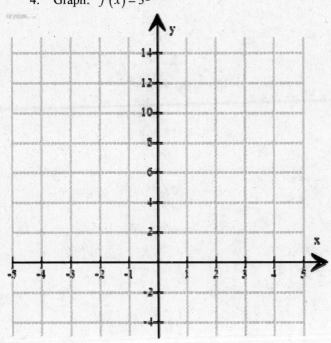

5. Graph: $f(x) = 3^x - 2$

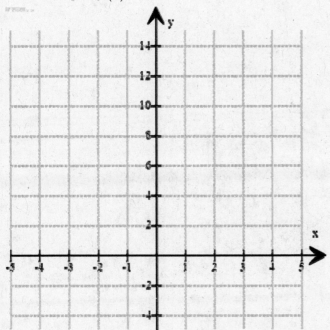

6. Graph: $f(x) = e^{2-x}$

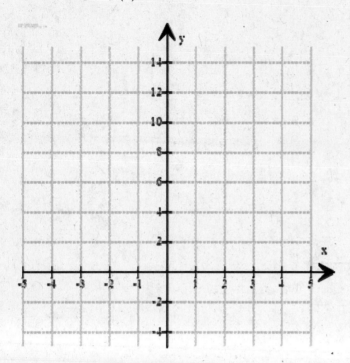

Practice Sheet 5.6A

Given the function $f(x) = 4^x$, **find:**

1. $f(1)$	**2.** $f(-1)$	**3.** $f(0)$

1. _____

2. _____

3. _____

4. $f(2)$	**5.** $f(-2)$	**6.** $f(3)$

4. _____

5. _____

6. _____

Given the function $f(x) = \left(\frac{1}{3}\right)^{2x}$, **find:**

7. $f(0)$	**8.** $f(1)$	**9.** $f(-1)$

7. _____

8. _____

9. _____

10. $f\left(\frac{1}{2}\right)$	**11.** $f\left(-\frac{1}{2}\right)$	**12.** $f(2)$

10. _____

11. _____

12. _____

Graph.

13. $f(x) = 2^{-x}$

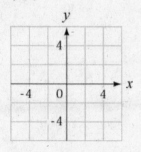

14. $f(x) = 3^{x+2}$

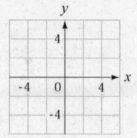

15. $f(x) = \left(\frac{1}{2}\right)^x$

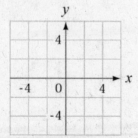

16. $f(x) = 2^x - 3$

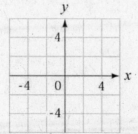

Answers

Try it 5.6A

1. a. 36   b. $\dfrac{1}{36}$

2. a. 54.5982   b. 9.3565

3.

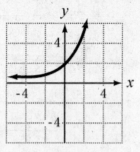

4.

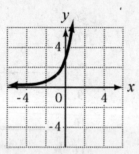

Quiz 5.6A

1. a. $\dfrac{1}{16}$ b. $\dfrac{1}{2}$ c. 4

2. a. 54.5982   b. 2.7183   c. 0.3679

3.

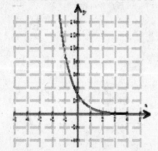

4.

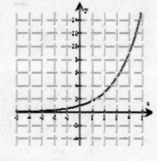

5.

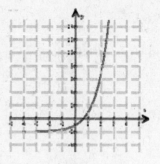

6.

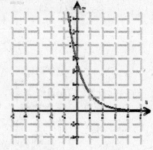

Solutions to Practice Sheet 5.6A

1. $f(x) = 4^x$
   $f(1) = 4^1 = 4$

2. $f(x) = 4^x$
   $f(-1) = 4^{-1} = \dfrac{1}{4}$

3. $f(x) = 4^x$
   $f(0) = 4^0 = 1$

4. $f(x) = 4^x$
   $f(2) = 4^2 = 16$

5. $f(x) = 4^x$
   $f(-2) = 4^{-2} = \dfrac{1}{16}$

6. $f(x) = 4^x$
   $f(3) = 4^3 = 64$

7. $f(x) = \left(\dfrac{1}{3}\right)^{2x}$
   $f(0) = \left(\dfrac{1}{3}\right)^{2(0)} = 1$

8. $f(x) = \left(\dfrac{1}{3}\right)^{2x}$
   $f(1) = \left(\dfrac{1}{3}\right)^{2(1)} = \dfrac{1}{9}$

9. $f(x) = \left(\dfrac{1}{3}\right)^{2x}$

$f(-1) = \left(\dfrac{1}{3}\right)^{2(-1)} = \left(\dfrac{1}{3}\right)^{-2} = 3^2 = 9$

10. $f(x) = \left(\dfrac{1}{3}\right)^{2x}$

$f\left(\dfrac{1}{2}\right) = \left(\dfrac{1}{3}\right)^{2\left(\frac{1}{2}\right)} = \left(\dfrac{1}{3}\right)^{1} = \dfrac{1}{3}$

11. $f(x) = \left(\dfrac{1}{3}\right)^{2x}$

$f\left(-\dfrac{1}{2}\right) = \left(\dfrac{1}{3}\right)^{2\left(-\frac{1}{2}\right)} = \left(\dfrac{1}{3}\right)^{-1} = 3$

12. $f(x) = \left(\dfrac{1}{3}\right)^{2x}$

$f(2) = \left(\dfrac{1}{3}\right)^{2(2)} = \left(\dfrac{1}{3}\right)^{4} = \dfrac{1}{81}$

13. $f(x) = 2^{-x}$

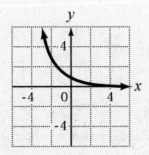

14. $f(x) = 3^{x+2}$

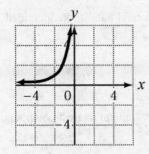

15. $f(x) = \left(\dfrac{1}{2}\right)^{x}$

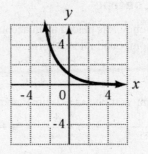

16. $f(x) = 2^x - 3$

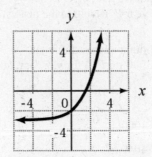

Objective 5.6B – Solve applications of exponential functions

Key Words and Concepts

When an amount of money P is placed in an account that earns compound interest, the value A of the money after t years is given by the compound interest formula

$$A = P\left(1+\frac{r}{n}\right)^{nt}$$

where r is the annual interest rate as a decimal and n is the number of compounding periods per year.

Example

An investor deposits $3000 into an account that earns 6% annual interest compounded quarterly. What will be the value of the investment after 5 years? Round to the nearest cent.

> Solution
> Use the compound interest formula.
> Substitute 3000 for P, 4 for n, 5 for t, and 0.06 for r.
>
> $$A = P\left(1+\frac{r}{n}\right)^{nt}$$
> $$A = 3000\left(1+\frac{0.06}{4}\right)^{4\cdot5}$$
> $$A \approx 4040.57$$
>
> The value after 5 years is approximately $4040.57.

Try it

1. A $7000 certificate of deposit (CD) earns 4% annual interest, compounded daily. What is the value of the investment after 7 years?

Quiz Yourself 5.6B

In Exercises 1 and 2, use the compound interest formula $A = P\left(1+\frac{r}{n}\right)^{nt}$, where P is the amount deposited, A is the value of the money after t years, r is the annual interest rate as a decimal and n is the number of compounding periods per year.

1. A hospital administrator deposits $10,000 into an account that earns 9% annual interest compounded monthly. What is the value of the account after 15 years?

2. Eighteen years ago, your parents put $6000 into an account that earns 2.5% annual interest compounded daily. What is the current value of the account?

3. The current I, in amperes, in an electric circuit is given by $I = 6\left(1-e^{-2t}\right)$, where t is the time in seconds. Using this model, what is the current after 0.6 seconds? Round to the nearest tenth.

Practice Sheet 5.6B

Use the compound interest formula $A = P\left(1 + \dfrac{r}{n}\right)^{nt}$, where P is the amount deposited, A is the value of the money after t years, r is the annual interest rate as a decimal, and n is the number of compounding periods.

1. An investor deposits $8000 into an account that earns 12% interest compounded semi-annually. What is the value of the investment after 4 years?

2. A $5000 certificate of deposit (CD) earns 6% annual interest, compounded daily. What is the value of the investment after 10 years?

1. _____

2. _____

3. An investor deposits $10,000 into an account that earns 4.5% interest compounded monthly. What is the value of the investment after 6 years?

4. A $3000 certificate of deposit (CD) earns 5% annual interest, compounded daily. What is the value of the investment after 15 years?

3. _____

4. _____

The percent of correct welds a student can make will increase with practice and can be approximated by the equation $P = 100[1 - (0.75)^t]$, where P is the percent of correct welds and t is the number of weeks of practice.

5. Find the percent of correct welds a student will make after four weeks of practice.

6. Find the percent of correct welds a student will make after six weeks of practice.

5. _____

6. _____

7. Find the percent of correct welds a student will make after ten weeks of practice.

8. Find the percent of correct welds a student will make after twelve weeks of practice.

7. _____

8. _____

Answers

Try it 5.6B

1. $9261.77

Quiz 5.6B

1. $38,380.43
2. $9409.73
3. 4.2 amps

Solutions to Practice Sheet 5.6B

1. **STRATEGY**

   To find the value of the investment, solve the compound interest formula for A.

   Use $P = 8000$, $r = 12\% = 0.12$, $n = 2$, $t = 4$.

   SOLUTION

   $$A = P\left(1 + \frac{r}{n}\right)^{nt}$$

   $$A = 8000\left(1 + \frac{0.12}{2}\right)^{2 \cdot 4}$$

   $$A \approx 12,750.78$$

   The value of the investment after 4 years is $12,750.78.

2. **STRATEGY**

   To find the value of the CD, solve the compound interest formula for A.

   Use $P = 5000$, $r = 6\% = 0.06$, $n = 365$, $t = 10$.

   SOLUTION

   $$A = P\left(1 + \frac{r}{n}\right)^{nt}$$

   $$A = 5000\left(1 + \frac{0.06}{365}\right)^{365 \cdot 10}$$

   $$A \approx 9110.14$$

   The value of the CD after 10 years is $9110.14.

3. **STRATEGY**

   To find the value of the investment, solve the compound interest formula for A.

   Use $P = 10,000$, $r = 4.5\% = 0.045$, $n = 12$, and $t = 6$.

   SOLUTION

   $$A = P\left(1 + \frac{r}{n}\right)^{nt}$$

   $$A = 10,000\left(1 + \frac{0.045}{12}\right)^{12 \cdot 6}$$

   $$A \approx 13,093.03$$

   The value of the investment after 6 years is $13,093.03.

4. **STRATEGY**

   To find the value of the CD, solve the compound interest formula for A.

   Use $P = 3000$, $r = 5\% = 0.05$, $n = 365$, and $t = 15$.

   SOLUTION

   $$A = P\left(1 + \frac{r}{n}\right)^{nt}$$

   $$A = 3000\left(1 + \frac{0.05}{365}\right)^{365 \cdot 15}$$

   $$A \approx 6350.67$$

   The value of the CD after 15 years is $6350.67.

5. **STRATEGY**

   Solve for P, where $t = 4$.

   SOLUTION

   $$P = 100\left[1 - (0.75)^t\right]$$

   $$P = 100\left[1 - (0.75)^4\right] \approx 68$$

   The student will make about 68% correct welds after 4 weeks of practice.

6. **STRATEGY**

   Solve for P, where $t = 6$.

   SOLUTION

   $$P = 100\left[1 - (0.75)^t\right]$$

   $$P = 100\left[1 - (0.75)^6\right] \approx 82$$

   The student will make about 82% correct welds after 6 weeks of practice.

7. **STRATEGY**

   Solve for P, where $t = 10$.

   SOLUTION

   $$P = 100\left[1 - (0.75)^t\right]$$

   $$P = 100\left[1 - (0.75)^{10}\right] \approx 94$$

   The student will make about 94% correct welds after 10 weeks of practice.

8. **STRATEGY**

   Solve for P, where $t = 12$.

   SOLUTION

   $$P = 100\left[1 - (0.75)^t\right]$$

   $$P = 100\left[1 - (0.75)^{12}\right] \approx 97$$

   The student will make about 97% correct welds after 12 weeks of practice.

Application and Activities

Discussion and reflection questions

1. Is revenue the same as profit?

2. If you owned a business how would you go about predicting your future revenue?

3. Given a scatter diagram, how would you decide which shape best fit the points?

4. Discuss the differences in shape between a linear function, a quadratic function and an exponential function.

Group Activity (2-3 people)

Objectives for the lesson:

In this lesson you will use and enhance your understanding of the following:

- Graphing points on a rectangular coordinate system
- Graphing a scatter diagram
- Graphing an equation in two variables
- Graphing a function
- Finding the slope of a line given two points
- Finding the equation of a line given two points
- Graphing a linear equation
- Graphing a quadratic function
- Graphing an exponential function
- Solve applications of exponential functions

Sam started a new business in January 2014. He measured the revenue at the end of each month in thousands of dollars, and came up with the following table of data:

Month in the year 2014 (x)	Profits in thousands (y)
1	3
2	3.6
3	3.2
4	3.6
5	3.5
6	4.2
7	5
8	5.4
9	6.1

1. Create a scatter diagram of the data Sam collected.

2. Use the points (2,3.6) and (7,5) to find a linear equation that models the data. Label the model $L(x)$, where x is the month in 2014 and $L(x)$ is the revenue in thousands.

3. Graph the line $L(x)$ in *blue*, on the top of the existing graph of the scatter diagram.

4. Sam fit a quadratic equation to his data and arrived at the function $Q(x) = 0.1x^2 - x + 6$, where x is the month in 2014 and $Q(x)$ is the revenue in thousands. Graph $Q(x)$ in *red* on the same rectangular coordinate system as the scatter diagram and $L(x)$.

5. Sam then created an exponential equation for his data. $E(x) = 2.5(1.1)^x$, where x is the month in 2014 and $E(x)$ is the revenue in thousands. Graph $E(x)$ in *green* on the same rectangular coordinate system as the other functions and the scatter diagram.

6. Looking at the scatter diagram, along with the 3 functions graphed with it, decide which function visually fits the data best.

7. Use the function you chose in #6 to predict Sam's revenue for December of 2014.

8. If Sam's modeled his costs over the year 2014 using the function $C(x) = 1.5(1.05)^x$, where x is the month in 2014 and $C(x)$ is the cost in thousands, find his expected profit for December 2014.

Objective 6.1A – Organizing Data

Key Words and Concepts

Define statistics. _____

_____

Data are collected from a **population**, which is the set of all observations of interest.

A **frequency distribution** is one method of organizing the data collected from a population. A frequency distribution is constructed by dividing the data gathered from the population into **classes**.

Example

The corporate room rates per night for 50 hotels are shown here.

Corporate Room Rates for 50 Hotels

160	187	177	217	214	182	191	165	169	163
206	171	174	186	206	178	201	200	207	209
157	206	203	200	195	168	199	212	207	177
164	168	199	212	207	176	216	200	182	186
181	198	192	178	195	189	191	202	215	227

Make a frequency distribution for the data. Use 7 classes. What percent of hotels charge a corporate room rate that is between $201 and $211 per night?

Solution

Find the smallest number (157) and the largest number (227) in the table. The difference between these two numbers,
$$227 - 157 = 70,$$
is the **range** of the data.

Divide this range by the number of classes, 7.
(If necessary, round this quotient to a whole number.)

This number, $\frac{70}{7} = 10$, is called the **class width**.

Form the classes by repeatedly adding 7, starting with the smallest number, until a class contains the largest number in the set of data. Tabulate the data for each class.

For this data we get:

Class	Tally	Frequency
157 – 167	/////	5
168 – 178	//////////	10
179 – 189	///////	7
190 – 200	///////////	11
201 – 211	//////////	10
212 – 222	•//////	6
223 – 233	/	1

$\frac{10}{50} = 0.2 = 20\%$, so 20% of hotels charge a corporate

rate that is between $201 and $211 per night.

Try it

1. From the table of corporate room rates given to the left, make a frequency distribution that has 5 classes.

Reflect on it
- Statistics is the study of collecting, organizing, and interpreting data.

Quiz Yourself 6.1A
The annual sales at 40 restaurants are given in the table. The amounts listed are in thousands of dollars. Use the data to answer the questions.

Annual Sales for 40 Restaurants (in thousands of dollars)

731	512	428	732	680
982	1312	584	1412	1402
812	943	201	893	634
912	1021	1040	923	398
550	487	983	882	503
612	900	1121	862	1312
1377	857	893	1490	903
198	784	504	875	429

1. What is the range of the data?

2. Construct a frequency distribution for the data. Use 6 classes.

3. What percent of the restaurants had annual sales between $750,000 and $1,000,000?

Practice Sheet 6.1A

Use the table below.

Number of hours worked at a part-time job per week for 40 students

10	21	16	8	12	10	24	2
26	6	11	18	31	3	14	22
11	7	5	32	27	30	4	10
14	16	23	3	8	18	21	13
27	10	1	22	18	25	30	12

1. What is the range of the data in the table?

2. Make a frequency distribution table for the number of hour worked. Use 8 classes.

1. _____

2. _____

3. Which class has the greatest frequency?

4. How many students worked between 22 and 32 hours?

3. _____

4. _____

5. How many students worked between 9 and 16 hours?

6. What percent of students worked between 5 and 12 hours?

5. _____

6. _____

7. What percent of students worked between 17 and 21 hours?

8. How many students worked 21 hours or less?

7. _____

8. _____

Answers

Try it 6.1A

1.

Class	Frequency
157 – 170	8
171 – 184	10
185 – 198	10
199 – 212	17
213 – 227	5

Quiz 6.1A

1. 1292

2.

Class	Frequency
0 – 250,000	2
250,000 – 500,000	4
500,000 – 750,000	10
750,000 – 1,000,000	15
1,000,000 – 1,250,000	3
1,250,000 – 1,500,000	6

3. 37.5%

Solutions to Practice Sheet 6.1A

1. Range = 32 – 1 = 31

2. STRATEGY
   To prepare the frequency distribution table, first find the range. Then divide the range by 8, the number of classes. If necessary, round the quotient to a whole number. This is the class width.

   SOLUTION
   Range = 32 – 1 = 31

   Class width $= \dfrac{31}{8} = 3.875$

Classes	Tally	Frequency
1 – 4	/////	5
5 – 8	/////	5
9 – 12	////////	8
13 – 16	/////	5
17 – 21	/////	5
22 – 24	////	4
25 – 28	////	4
29 – 32	////	4

3. The class with the greatest frequency is 9–12, with 8.

4. 12 students worked between 22 and 32 hours.

5. 13 students worked between 9 and 16 hours.

6. 13 students worked between 5 and 12 hours.
   $\dfrac{13}{40} = 0.325 = 32.5\%$
   32.5% of students worked between 5 and 12 hours.

7. 5 students worked between 17 and 21 hours.
   $\dfrac{5}{40} = 0.125 = 12.5\%$
   12.5% of students worked between 7 and 21 hours.

8. 28 students worked 21 hours or less.

Objective 6.1B – Read histograms

Key Words and Concepts

A **histogram** is a bar graph that represents the data in a frequency distribution. The width of each bar represents a class, and the height of the bar corresponds to the frequency of the class.

For example, a research group measured the fuel efficiency of 92 cars. The results are recorded in the histogram. The height of the fourth bar shows that 17 cars get between 24 and 26 miles per gallon.

Class Interval (miles per gallon)	Class Frequency (number of cars)
18–20	12
20–22	19
22–24	24
24–26	17
26–28	15
28–30	5

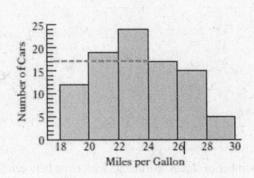

Example

The times, in minutes, for 100 runners in a marathon were recorded. A histogram of the data is shown below.

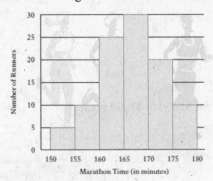

Use the histogram above to find the number of runners with times between 160 and 170 minutes.

Solution
To find the number of runners:
Read the histogram to find the number of runners who had times between 160 and 165 min and the number of runners who had times between 165 and 170 min.
Add the two numbers.

Number between 160 and 165 min: 25
Number between 165 and 170 min: 30
    25 + 30 = 55
Fifty-five runners had times between 160 and 170 minutes.

Try it

1. Use the histogram above to find the number of cars that had fuel efficiency between 26 and 30 mpg.

Reflect on it
• A histogram is a visual representation of a frequency distribution.

Quiz Yourself 6.1B

A total of 40 apartment complexes were surveyed to find the monthly rent for a studio apartment. A histogram of these data is shown below. Use the histogram to answer Exercises 1 through 4.

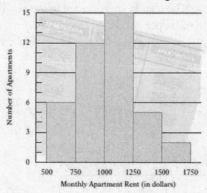

1.   Find the number of studio apartments with rents between $750 and $1250.

2.   How many studio apartments have rents less than $1000?

3.   What percent of the complexes surveyed rent studio apartments for $1250 to $1500 per month?

4.   What percent of the complexes surveyed rent studio apartments for $1500 to $1750 per month?

Practice Sheet 6.1B

The fuel usage of 100 cars was measured by a research group. The results are recorded in the histogram below.

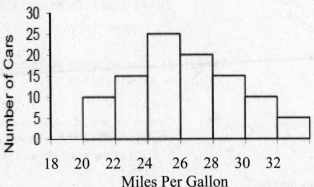

1. Find the number of cars which get between 24 and 26 miles per gallon.

2. Find the number of cars which get between 30 and 32 miles per gallon.

3. Find the number of cars which get 24 or more miles per gallon.

4. What is the percent of the number of cars which get between 18 and 22 miles per gallon?

1. _____

2. _____

3. _____

4. _____

The hourly wages of the 100 employees of a company are recorded in the histogram below.

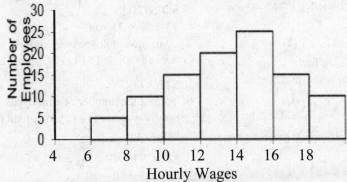

5. Find the number of employees whose hourly wage is between $6 and $10.

6. Find the ratio of the number of employees whose hourly wage is between $12 and $14 to the total number of employees.

7. Find the number of employees whose hourly wages is between $8 and $14.

8. How many employees earn $12 or more per hour?

5. _____

6. _____

7. _____

8. _____

Answers

Try it 6.1B
1. 20 cars

Quiz 6.1B
1. 27 apartments
2. 18 apartments
3. 12.5%
4. 5%

Solutions to Practice Sheet 6.1B

1. STRATEGY
    To find the number of cars with fuel efficiency between 24 and 26 mpg, read the histogram.

    SOLUTION
    There are 20 cars with fuel efficiency between 24 and 26 mpg.

2. STRATEGY
    To find the number of cars with fuel efficiency between 30 and 32 mpg, read the histogram.

    SOLUTION
    There are 5 cars with fuel efficiency between 30 and 32 mpg.

3. STRATEGY
    To find the number of cars with fuel efficiency above 24, read the histogram.

    SOLUTION
    $20 + 15 + 10 + 5 = 50$
    There are 50 cars with fuel efficiency greater than 24 mpg.

4. STRATEGY
    To find the percent, find the efficiency for 18 to 20 and 20 to 22 mpg from the histogram, and divide the sum by 100.

    SOLUTION
    $10 + 15 = 25$

    $\dfrac{25}{100} = 0.25 = 25\%$

    25% of the cars have fuel efficiency between 18 and 22 mpg.

5. STRATEGY
    To find the number of employees whose hourly wage is between $6 and $10, read the histogram and add $6 to $8 and $8 to $10.

    SOLUTION
    $10 + 15 = 25$
    25 of the employees have hourly wages between $6 and $10.

6. STRATEGY
    To find the ratio, divide the number of employees who earn between $12 and $14 (25) by the total number of employees (100).

    SOLUTION
    $\text{Ratio} = \dfrac{25}{100} = \dfrac{1}{4}$

    The ratio is $\dfrac{1}{4}$.

7. STRATEGY
    To find the number of employees whose hourly wage is between $8 and $14, read the histogram and add $8 to $10, $10 to $12, and $12 to $14.

    SOLUTION
    $15 + 20 + 25 = 60$
    60 of the employees have hourly wages between $8 and $14.

8. STRATEGY
    To find the number of employees whose hourly wage is $12 or more, read the histogram.

    SOLUTION
    $25 + 15 + 10 = 50$
    There are 50 employees who earn $12 or more per hour.

Objective 6.1C – Read frequency polygons

Key Words and Concepts

A **frequency polygon** is a graph that displays information in a manner similar to a histogram. A dot is placed above the center of each class interval at a height corresponding to that class's frequency. The dots are then connected to form a broken-line graph. The center of a class interval is called the **class midpoint**.

Example

The frequency polygon below shows the approximate numbers of participants to finish in each of the given time slots in the 2011 Boston Marathon. (Times are given in hours and minutes.) Use the figure to answer the questions.

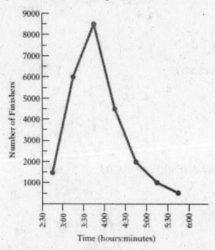

Determine the approximate number of participants who finished with times between 2:30 and 6 hours.

Solution

The seven marked points correspond to, in thousands,
$$1.5 + 6 + 8.5 + 4.5 + 2 + 1 + 0.5 = 24$$
so there were approximately 24,000 participants.

Try it

1. Find the approximate number of marathoners who finished with times of more than 4 hours.

Reflect on it

• Which is easier to read, a histogram or a frequency polygon?

Quiz Yourself 6.1C

The frequency polygon shows the distribution of scores of the approximately 1,080,000 students who took an SAT exam. Use the figure for Exercises 1 through 3.

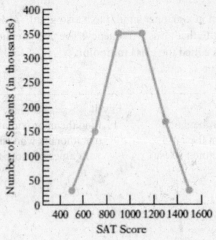

1. How many students scored between 1200 and 1400 on the exam?

2. What percent of the students who took the exam scored between 800 and 1000?
    Round to the nearest tenth of a percent.

3. How many students scored below 1000?

Practice Sheet 6.1C

A radio rating service surveyed 105 families to find the number of hours they listened to the radio. The results are recorded in the figure below.

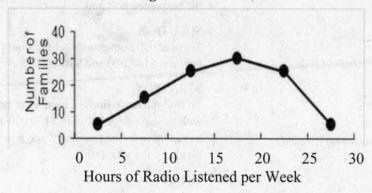

Hours of Radio Listened per Week

1. How many families listened between 10 and 15 hours a week?

2. What is the ratio of the number of families who listened between 20 and 25 hours a week to the total number in the survey?

1. _____

2. _____

3. How many families listened between 20 and 30 hours a week?

4. What is the ratio of the number of families who listened 15 or more hours a week to the total number in the survey?

3. _____

4. _____

A real estate company sold 100 homes during the last three months. The selling prices are recorded in the figure below.

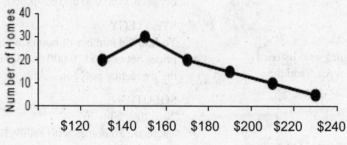

Sales Price (in thousands of dollars)

5. How many homes sold for between $140,000 and $160,000?

6. Find the ratio of the number of homes which sold for between $120,000 and $140,000 to the total number of homes sold during the three months.

5. _____

6. _____

7. How many homes sold for between $160,000 and $240,000?

8. How many homes sold for between $120,000 and $180,000?

7. _____

8. _____

Answers

Try it 6.1C
1. 8000 participants

Quiz 6.1C
1. 170,000
2. 32.4%
3. 530,000

Solutions to Practice Sheet 6.1C

1. **STRATEGY**
   To find the number of families who listened between 10 and 15 hours a week, read the frequency polygon.

   SOLUTION
   There are 25 families who listened between 10 and 15 hours a week.

2. **STRATEGY**
   To find the ratio, find the number of families who listened between 20 and 25 hours a week from the frequency polygon and divide by the total (105).

   SOLUTION
   $$\text{Ratio} = \frac{25}{105} = \frac{5}{21}$$
   The ratio is $\frac{5}{21}$.

3. **STRATEGY**
   To find the number of families who listened between 20 and 30 hours a week, read the frequency polygon.

   SOLUTION
   $25 + 5 = 30$
   There are 30 families who listened between 20 and 30 hours a week.

4. **STRATEGY**
   To find the ratio, find the number of families who listened 15 or more hours a week from the frequency polygon and divide by the total (105).

   SOLUTION
   $30 + 25 + 5 = 60$
   $$\text{Ratio} = \frac{60}{105} = \frac{4}{7}$$
   The ratio is $\frac{4}{7}$.

5. **STRATEGY**
   To find the number of homes with selling prices between $140,000 and $160,000, read the frequency polygon.

   SOLUTION
   There are 30 homes with selling prices between $140,000 and $160,000.

6. **STRATEGY**
   To find the ratio, find the number of homes with selling prices between $120,000 and $140,000 from the frequency polygon and divide by the total (100).

   SOLUTION
   $$\text{Ratio} = \frac{20}{100} = \frac{1}{5}$$
   The ratio is $\frac{1}{5}$.

7. **STRATEGY**
   To find the number of homes with selling prices between $160,000 and $240,000, read the frequency polygon.

   SOLUTION
   $20 + 15 + 10 + 5 = 50$
   There are 50 homes with selling prices between $160,000 and $240,000.

8. **STRATEGY**
   To find the number of homes with selling prices between $120,000 and $180,000, read the frequency polygon.

   SOLUTION
   $20 + 30 + 20 = 70$
   There are 70 homes with selling prices between $120,000 and $180,000.

Objective 6.2A – Find the mean, median, and mode of a distribution

Key Words and Concepts

The **mean** of a data set is the sum of the measurements divided by the number of measurements.

Note   The symbol for the mean is $\bar{x}$.

Formula for the Mean

$$\text{Mean} = \bar{x} = \frac{\text{sum of all data values}}{\text{number of data values}}$$

Example

The exam scores for 12 students are shown here. Find the mean score.

82	46	92	77	91	70
88	79	77	50	68	62

Solution

To find the mean score, divide the total of the scores by 12.

$$\text{Mean} = \frac{82+46+92+77+91+70+88+79+77+50+68+62}{12}$$

$$= \frac{882}{12}$$

$$= 73.5$$

The mean score is 73.5.

Try it

1. The bowling scores for 10 games are shown here. Find the mean score.

126	150	206	98	173
230	165	190	95	217

The **median** of a data set is the number that separates the data into two equal parts when the numbers are arranged from smallest to largest (or largest to smallest). There are always an equal number of values above the median and below the median. If the data contain an even number of values, the median is the sum of the two middle numbers, divided by 2.

Example

Find the median of the 12 scores used in the previous example.

Solution

In order, the 12 scores are

46 50 62 68 70 77 77 79 82 88 91 92

Since the data contain an even number of values, the median is the sum of two middle numbers, 77 and 77, divided by 2

$$\text{median} = \frac{77+77}{2} = 77$$

The median score is 77.

Try it

2. Find the median of the 10 scores.

126	150	206	98	173
230	165	190	95	217

The **mode** of a set of numbers is the value that occurs most frequently.
If a set of numbers has no number that occurs more than once, then the data have *no mode*.

481

Example

Find the mode of the data set used in the previous two examples.

Solution

The list of scores in order is

46 50 62 68 70 77 77 79 82 88 91 92

The value 77 occurs twice, and no other value occurs twice.

The mode is 77.

Try it

3. Find the mode of the 10 scores.

126   150   206   98   173
230   165   190   95   217

Although any of the averages can be used when the data collected consist of numbers, the mean and median are not appropriate for qualitative (non-numeric) data. For qualitative data, the **modal response** is the category that receives the greatest number of responses.

Reflect on it

- The mean is the average of a list of numbers.
- The median is the number in the middle of an ordered list.
- The mode is the most frequent number of a list of numbers.
- The modal response is the category with greatest number of responses.

Quiz Yourself 6.2A

1. A consumer research group purchased identical items at eight grocery stores. The costs for the purchased items were $85.89, $92.12, $81.43, $80.67, $88.73, $82.45, $87.81, and $85.82. Calculate the mean and the median costs of the purchased items.

2. The times, in seconds, for a 100-meter dash at a college track meet were 10.45, 10.23, 10.57, 11.01, 10.26, 10.90, 10.74, 10.64, 10.52, and 10.78. Calculate the mean and the median time for the 100-meter dash.

3.   One measure of a computer's hard-drive speed is called access time; this is measured in milliseconds (thousandths of a second). Find the mean and median access times for 11 hard drives whose access times, in milliseconds, were 5, 4.5, 4, 4.5, 5, 5.5, 6, 5.5, 3, 4.5, and 4.5. Round to the nearest tenth.

4.   The numbers of unforced errors a tennis player made in four sets of tennis were recorded. The numbers were 15, 22, 24, and 18. How many unforced errors did this player make in the fifth set if the mean number of unforced errors for the five sets was 20?

5.   The patrons of a restaurant were asked to rate the quality of the food. The responses were bad, 8; good, 21; very good, 43; excellent, 21. What was the modal response for this survey?

Practice Sheet 6.2A

Solve.

1. The prices of a scientific calculator at five stores were $16.25, $15.75, $16.15, $16.50, and $16.95. Find the mean price of the calculator.

2. A student received grades of 83, 89, 85, 91, and 92 on five mathematics exams. Find the mean grade of the student's mathematics exams.

1. _____

2. _____

3. The six sales representatives for an advertising agency received weekly bonuses of $345, $275, $190, $221, $335, and $260. Find the mean bonus.

4. The number of pizzas sold in five different stores over a three-day period was 212, 246, 205, 252, and 245. Find the mean number of pizzas sold during this period of time.

3. _____

4. _____

5. The prices of identical vintage clock radios at each of five stores were $33.25, $39.00, $36.75, $36.00, and $37.50. Find the median price of the vintage clock radio.

6. The number of miles driven during each of five days of a business trip was 107, 96, 151, 103, and 99. Find the median number of miles driven.

5. _____

6. _____

7. The hourly wages for seven job classifications at a company are $7.63, $10.43, $10.09, $7.59, $8.45, $7.47, and $8.38. Find the median hourly wage.

8. The ages of the seven most recently hired employees at a fast food store are 22, 41, 19, 21, 20, 26, and 29. Find the median age.

7. _____

8. _____

9. The number of responses to a discount coupon for a carpet cleaner during a six-day period was 35, 29, 33, 37, 28 and 26. What is the mode of the data?

10. The scores on eight math exams at a job placement service were 79, 93, 95, 69, 93, 96, 88, and 75. What is the mode of the data?

9. _____

10. _____

Answers

Try it 6.2A
1. 165
2. 169
3. no mode

Quiz 6.2A
1. Mean: $85.615; median: $85.855
2. Mean: 10.61 seconds; median: 10.605 seconds
3. Mean: 4.7 msec; median: 4.5 msec
4. 21
5. very good

Solutions to Practice Sheet 6.2A

1. STRATEGY
   To find the mean price:
   - Determine the sum of the prices.
   - Divide the sum by 5.

   SOLUTION
   $$\bar{x} = \frac{16.25 + 15.75 + 16.15 + 16.50 + 16.95}{5} = 16.32$$
   The mean price is $16.32.

2. STRATEGY
   To find the mean grade:
   - Determine the sum of the grades.
   - Divide the sum by 5.

   SOLUTION
   $$\bar{x} = \frac{83 + 89 + 85 + 91 + 92}{5} = 88$$
   The mean grade is 88.

3. STRATEGY
   To find the mean bonus:
   - Determine the sum of the bonuses.
   - Divide the sum by 6.

   SOLUTION
   $$\bar{x} = \frac{345 + 275 + 190 + 221 + 335 + 260}{6} = 271$$
   The mean bonus is $271.

4. STRATEGY
   To find the mean number of pizzas sold:
   - Determine the sum of the pizzas sold.
   - Divide the sum by 5.

   SOLUTION
   $$\bar{x} = \frac{212 + 246 + 205 + 252 + 245}{5} = 232$$
   The mean number of pizzas sold is 232.

5. STRATEGY
   To find the median price:
   - Arrange the numbers from smallest to largest.
   The median is the middle number.

   SOLUTION
   33.25 36.00 36.75 37.50 39.00
   The median price is $36.75.

6. STRATEGY
   To find the median number of miles:
   - Arrange the numbers from smallest to largest.
   The median is the middle number.

   SOLUTION
   96 99 103 107 151
   The median number of miles is 103.

7. STRATEGY
   To find the median hourly wage:
   - Arrange the numbers from smallest to largest.
   The median is the middle number.

   SOLUTION
   7.47 7.59 7.63 8.38 8.45 10.09 10.43
   The median hourly wage is $8.38.

8. STRATEGY
   To find the median age:
   - Arrange the numbers from smallest to largest.
   The median is the middle number.

   SOLUTION
   19 20 21 22 26 29 41
   The median age is 22.

9. STRATEGY
   To find the mode, write down the category that received the most responses.

   SOLUTION
   Since none of the categories were listed more than once, there is no mode.

10. STRATEGY
    To find the mode, write down the grade that was recorded the most.

    SOLUTION
    Since 93 was recorded most frequently, the mode is 93.

Objective 6.2B – Find the standard deviation of a distribution

Key Words and Concepts

One measure of the consistency, or "clustering," of data near the mean is the **standard deviation**.

To calculate the standard deviation:
1. Sum the squares of the differences between each data value and the mean.
2. Divide the result in Step 1 by the number of data values.
3. Take the square root of the result in Step 2.

Note   The symbol for the standard deviation is the Greek letter *sigma*, denoted by σ.

For this text, standard deviations are rounded to the nearest thousandth.

Example

The scores for five students on a test were 73, 78, 81, 80, and 70. Find the standard deviation of the scores.

   Solution

      Step 1:

         The mean $\bar{x}$ of the scores is $\dfrac{73+78+81+80+70}{5} = 76.4$.

x	$(x-\bar{x})$	$(x-\bar{x})^2$
73	$(73-76.4)$	$(-3.4)^2 = 11.56$
78	$(78-76.4)$	$1.6^2 = 2.56$
81	$(81-76.4)$	$4.6^2 = 21.16$
80	$(80-76.4)$	$3.6^2 = 12.96$
70	$(70-76.4)$	$(-6.4)^2 = 40.96$

Total = 89.2

      Step 2: $\dfrac{89.2}{5} = 17.84$

      Step 3: $\sigma = \sqrt{17.84} \approx 4.224$

      The standard deviation is $\sigma \approx 4.224$.

Try it

1. The number of miles logged by a runner for the last five days of running were 8, 5, 7, 3, and 10. Find the standard deviation of the of the number of miles run. Round to the nearest thousandth.

Reflect on it

* For large amounts of data, computers are used to make standard deviation calculations.

Quiz Yourself 6.2B

   1.   The weights in ounces of newborn infants were recorded by a hospital. The weights were 96, 105, 84, 90, 102, and 99. Find the standard deviation of the weights. Round to the nearest thousandth.

   2.   The scores for five college basketball games were 56, 68, 60, 72, and 64. The scores for five professional basketball games were 106, 118, 110, 122, and 114. Which scores have the greater standard deviation?

   3.   The high temperatures for five consecutive days at a desert resort were 95°, 98°, 98°, 104°, and 97°. For the same days, the high temperatures in Antarctica were 27°, 28°, 28°, 30°, and 28°. Which location has the greater standard deviation of high temperatures?

Practice Sheet 6.2B

Solve.

1. A student received grades of 84, 90, 83, 76 and 92 on five exams.   1. _____
    Find the standard deviation of these exams.

2. The ages of six students are 18, 21, 22, 19, 19, and 21.   2. _____
    Find the standard deviation of these ages.

3. Seven coins were tossed 20 times. The number of heads recorded were 8,   3. _____
    12, 11, 9, 12, 9, and 9. Find the standard deviation of the number of heads.

4. The scores for five professional golfers were 68, 70, 75, 69 and 73. The scores   4. _____
    for five amateur golfers were 84, 81, 89, 86, and 85. Which scores had the
    greater standard deviation?

Answers

Try it 6.2B
1. 2.417 mi

Quiz 6.2B
1. 7.141 oz
2. The standard deviations are the same.
3. desert resort

Solutions to Practice Sheet 6.2B

1. STRATEGY
   To calculate the standard deviation:
   - Find the mean of the grades.
   - Use the procedure for calculating standard deviation.

   SOLUTION
   $$\bar{x} = \frac{84+90+83+76+92}{5} = 85$$

   Step 1

x	$(x-\bar{x})^2$
84	$(84-85)^2 = 1$
90	$(90-85)^2 = 25$
83	$(83-85)^2 = 4$
76	$(76-85)^2 = 81$
92	$(92-85)^2 = \underline{49}$
	Total $= 160$

   Step 2
   $$\frac{160}{5} = 32$$

   Step 3
   $$\sigma = \sqrt{32} \approx 5.657$$
   The standard deviation of the exam grades is approximately 5.657.

2. STRATEGY
   To calculate the standard deviation:
   - Find the mean of the ages.
   - Use the procedure for calculating standard deviation.

   SOLUTION
   $$\bar{x} = \frac{18+21+22+19+19+21}{6} = 20$$

   Step 1

x	$(x-\bar{x})^2$
18	$(18-20)^2 = 4$
21	$(21-20)^2 = 1$
22	$(22-20)^2 = 4$
19	$(19-20)^2 = 1$
19	$(19-20)^2 = 1$
21	$(21-20)^2 = \underline{1}$
	Total $= 12$

   Step 2
   $$\frac{12}{6} = 2$$

   Step 3
   $$\sigma = \sqrt{2} \approx 1.414$$
   The standard deviation of the ages is approximately 1.414.

3. STRATEGY
   To calculate the standard deviation:
   - Find the mean of the heads.
   - Use the procedure for calculating standard deviation.

   SOLUTION
   $$\bar{x} = \frac{8+12+11+9+12+9+9}{7} = 10$$

   Step 1

x	$(x-\bar{x})^2$
8	$(8-10)^2 = 4$
12	$(12-10)^2 = 4$
11	$(11-10)^2 = 1$
9	$(9-10)^2 = 1$
12	$(12-10)^2 = 4$
9	$(9-10)^2 = 1$
9	$(9-10)^2 = \underline{1}$
	Total $= 16$

   Step 2
   $$\frac{16}{7}$$

   Step 3
   $$\sigma = \sqrt{\frac{16}{7}} \approx 1.512$$

   The standard deviation of the ages is approximately 1.512.

4. STRATEGY
Find and compare the standard deviation for the professional golfers' scores to the standard deviation of the amateur golfers' scores.

SOLUTION
Professional
$$\bar{x} = \frac{68+70+75+69+73}{5} = 71$$

Step 1

x	$(x-\bar{x})^2$
68	$(68-71)^2 = 9$
70	$(70-71)^2 = 1$
75	$(75-71)^2 = 16$
69	$(69-71)^2 = 4$
73	$(73-71)^2 = 4$
	Total $= 34$

Step 2

$$\frac{34}{5} = 6.8$$

Step 3

$$\sigma = \sqrt{6.8} \approx 2.608$$

Amateur
$$\bar{x} = \frac{84+81+89+86+85}{5} = 85$$

Step 1

x	$(x-\bar{x})^2$
84	$(84-85)^2 = 1$
81	$(81-85)^2 = 16$
89	$(89-85)^2 = 16$
86	$(86-85)^2 = 1$
85	$(85-85)^2 = 0$
	Total $= 34$

Step 2

$$\frac{34}{5} = 6.8$$

Step 3

$$\sigma = \sqrt{6.8} \approx 2.608$$

The standard deviations are the same.

Objective 6.3A – Calculate percentiles and quartiles

Key Words and Concepts

Most standardized examinations provide scores in terms of *percentiles*, which are defined as follows:

*p*th Percentile
A value x is called the **pth percentile** of a data set provided $p\%$ of the data values are less than x.

The following formula can be used to find the percentile that corresponds to a particular data value in a set of data.

Percentile for a Given Data Value
Given a set of data and a data value x,

$$\text{Percentile of score } x = \frac{\text{number of data values less than } x}{\text{total number of data values}} \cdot 100$$

Example
On a placement examination, Sarah's score of 480 was higher than the scores of 4513 of the 7326 students who took the test. Find the percentile, rounded to the nearest percent for Sarah's score.

Solution

To find the percentile, divide the number of scores less than Sarah's score by the total number of scores, and multiply by 100 to convert to a percentile.

$$\text{Percentile} = \frac{\text{number of scores less than } 480}{\text{total number of scores}}$$

$$= \frac{4513}{7326} \cdot 100$$

$$\approx 62$$

Sarah's score was at the 62nd percentile.

Try it
1. On a placement examination, Jacob's score was higher than the scores of 3752 of the 6708 students who took the test. Find the percentile, rounded to the nearest percent for Jacob's score.

The three numbers Q_1, Q_2, and Q_3 that partition a data set into four (approximately) equal groups are called the **quartiles** of the data.

The quartile Q_1 is called the *first quartile*.

The quartile Q_2 is called the *second quartile*. It is the median of the data.

The quartile Q_3 is called the *third quartile*.

The Median Procedure for Finding Quartiles
1. Arrange the data from smallest to largest or largest to smallest. This is called *ranking* the data.
2. Find the median of the data. This is the second quartile, Q_2.
3. The first quartile, Q_1, is the median of the data values less than Q_2. The third quartile, Q_3, is the median of the data values greater than Q_2.

Example

The number of calories per 10 oz of fifteen popular snack foods are listed below. Find the quartiles for the data.

20   100   50   82   63   200   120   75

93   165   80   55   98   140   105

Solution

Step 1: Put the data in order from smallest to largest.

20   50   55   63   75   80   82   93

98   100   105   120   140   165   200

Step 2: The median of these 15 data values is the eighth value of the list. The median is 93.

The second quartile Q_2 is the median of the data so,

$$Q_2 = 93.$$

Step 3: There are 7 data values less than the median and 7 data values greater than the median.

The first quartile is the middle score of the lower 7 data values, or the fourth value starting from the *bottom* of the list. The first quartile is 63.

$$Q_1 = 63.$$

The third quartile is the middle score of the upper 7 data values, or the fourth value starting from the *top* of the list. The third quartile is 120.

$$Q_3 = 120.$$

Try it

2. The weights, in ounces, of fifteen apples are listed below. Find the quartiles for the data.

6.8   5.2   7.1   6.5   4.9

8.2   5.9   6.7   5.6   7.6

6.6   7.5   7.7   8.1   4.3

Reflect on it

- The first quartile, Q_1, is the median of the lower half of the data.
- The third quartile, Q_3, is the median of the upper half of the data.

Quiz Yourself 6.3A

1. On a placement examination, Hope's score of 422 was higher than the scores of 5075 of the 8904 students who took the test. Find the percentile, rounded to the nearest percent for Hope's score.

2. On a science test, Dylan scored lower than 1642 of the 13,489 students who took the test. Find the percentile, rounded to the nearest percent for Dylan's score.

3. Li scored at the 70th percentile on a test given to 8970 students. How many students scored lower than Li?

4. The scores of nineteen exams are listed below. Find the quartiles for the data.

    81   92   78   60   85   72   77   95   86   83
    90   57   84   79   93   88   91   82   75

Practice Sheet 6.3A

Solve.

1. On a placement examination, Phoebe's score of 512 was higher than the scores of 4680 of the 7252 students who took the test. Find the percentile, rounded to the nearest percent for Phoebe's score.

2. On a science test, Hal scored lower than 1349 of the 11,620 students who took the test. Find the percentile, rounded to the nearest percent for Hal's score.

1. _____

2. _____

3. Bill scored at the 55th percentile on a test given to 9280 students. How many students scored lower than Bill?

4. Brielle scored at the 80th percentile on a test given to 8645 students. How many students scored higher than Brielle?

3. _____

4. _____

5. The scores of nineteen exams are listed below. Find the quartiles for the data.

    56   79   82   68   59   93   88   91   76   90

    78   83   74   89   73   84   96   85   65

5. _____

6. The weights, in pounds, of eleven babies are listed below. Find the quartiles for the data.

    7.8   9.9   8.5   10.2   8.3   9.4   9.7   7.4   8.1   9.0   9.1

6. _____

Answers

Try it 6.3A
1. 56th percentile
2. $Q_1 = 5.6$, $Q_2 = 6.7$, $Q_3 = 7.6$

Quiz 6.3A
1. 57th percentile
2. 88th percentile
3. 6279 students
4. $Q_1 = 77$, $Q_2 = 83$, $Q_3 = 90$

Solutions to Practice Sheet 6.3A

1. $\text{Percentile} = \dfrac{\text{number of scores less than 512}}{\text{total number of scores}}$

   $= \dfrac{4680}{7252} \cdot 100$

   ≈ 65

   Phoebe's score was at the 65th percentile.

2. Number of scores less than Hal's score
   $= 11{,}620 - 1349 = 10{,}271$

   $\text{Percentile} = \dfrac{\text{number of scores less than Hal's score}}{\text{total number of scores}}$

   $= \dfrac{10{,}271}{11{,}620} \cdot 100$

   ≈ 88

   Hal's score was at the 88th percentile.

3. Bill's percentile $= \dfrac{x}{9280} \cdot 100 = 55$

   $x = \dfrac{55 \cdot 9280}{100} = 5104$

   5104 students scored lower than Bill.

4. Brielle's percentile $= \dfrac{x}{8645} \cdot 100 = 80$

   $x = \dfrac{80 \cdot 8645}{100} = 6916$

   Number of scores greater than Brielle's score
   $= 8645 - 6916 = 1729$

   1729 students scored higher than Brielle.

5. Step 1: Put the data in order from smallest to largest.
   56 59 65 68 73 74 76 78 79 82
   83 84 85 88 89 90 91 93 96

   Step 2: The median of the nineteen scores is the tenth score. So $Q_2 = 82$.

   Step 3: The first quartile is the median of the data values less than the median. There are nine data values less than the median, so the first quartile is the value that is the fifth from the bottom of the list. So $Q_1 = 73$.

   The third quartile is the median of the data values greater than the median. There are nine data values greater than the median, so the third quartile is the value that is the fifth from the top of the list. So $Q_3 = 89$.

6. Step 1: Put the data in order from smallest to largest.
   7.4 7.8 8.1 8.3 8.5 9.0
   9.1 9.4 9.7 9.9 10.2

   Step 2: The median of the eleven scores is the sixth score. So $Q_2 = 9.0$.

   Step 3: The first quartile is the median of the data values less than the median. There are five data values less than the median, so the first quartile is the value that is the third from the bottom of the list. So $Q_1 = 8.1$.

   The third quartile is the median of the data values greater than the median. There are five data values greater than the median, so the third quartile is the value that is the third from the top of the list. So $Q_3 = 9.7$.

Objective 6.3B – Create a box-and-whiskers plot

Key Words and Concepts

A box-and-whiskers plot is a graph that gives a more comprehensive picture of the data. A box-and-whiskers plot shows five numbers: the smallest value, the first quartile, the median, the third quartile, and the largest value. The first quartile, symbolized by Q_1, is the number below which one-quarter of the data lie. The third quartile, symbolized by Q_3, is the number above which one-quarter of the data lie.

What is the interquartile range? _____

_____

A box-and-whiskers plot shows the data in the interquartile range as a box. To draw the box-and-whiskers plot for data, think of a number line that includes the five values noted above. With this in mind, mark off the five values. Draw a box that spans the distance from Q_1 to Q_3. Draw a vertical line the height of the box at the median, Q_2.

Example

The ages of the accountants who passed the certified public accountant (CPA) exam at one test center are recorded in the table below. Find the first quartile and the third quartile, and draw a box-and-whiskers plot of the data.

Ages of Accountants Passing the CPA Exam

24	42	35	26	24	37	27	26	28
34	43	46	29	34	25	30	28	

Solution

There are 17 data values. In order they are:
24 24 25 26 26 27 28 28 29 30 34 34 35 37 42 43 46

The median is the 9th data value, or 29.

The first quartile, Q_1, is the median of the lower half of the data. Here, the average of the 4th and 5th data values in the lower half of the data $Q_1 = \frac{26+26}{2} = 26$.

The third quartile, Q_3, is the median of the upper half of the data. Here, the average of the 4th and 5th data values in the upper half of the data $Q_3 = \frac{35+37}{2} = 36$.

Draw the box-and-whiskers plot using the five values: the smallest value, Q_1, the median, Q_3, and the largest value.

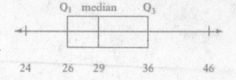

Try it

1. The ages of 14 people who attended a convention are listed below.

38	27	24	36	25	46	24
28	30	33	26	27	40	32

Draw a box-and-whiskers plot of the data.

Reflect on it
- The interquartile range is the difference $Q_3 - Q_1$.
- To draw a box-and-whiskers plot, you need five values: the smallest value, Q_1, the median, Q_3, and the largest value.

Quiz Yourself 6.3B

1. The times for new employees to learn how to assemble a toy are recorded in the table below. Find the first quartile and the third quartile, and draw a box-and-whiskers plot of the data.

Times to Train Employees (in hours)

4.3	3.1	5.3	8.0	2.6	3.5	4.9	4.3
6.2	6.8	5.4	6.0	5.1	4.8	5.3	6.7

2. A manufacturer of light bulbs tested the lives of 20 light bulbs. The results are recorded in the table below. Find the first quartile and the third quartile, and draw a box-and-whiskers plot of the data.

Lives of 20 Light Bulbs (in hours)

1010	1235	1200	998	1400	789	986	905	1050	1100
1180	1020	1381	992	1106	1298	1268	1309	1390	890

Practice Sheet 6.3B

The ages of the 300 accountants who passed the certified public accountant (CPA) exam at one test center were recorded. The box-and-whiskers plot below shows the distribution of their scores.

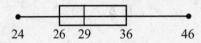

24   26 29   36   46

1. What is the youngest age?   2. What is the greatest oldest age?   1. _____

                                                                                2. _____

3. What is the first quartile?   4. What is the third quartile?   3. _____

                                                                                4. _____

5. What is the median?   6. What is the range?   5. _____

                                                                                6. _____

7. What is the interquartile range? 8. How many of the accountants were   7. _____
                                          older than 36?

                                                                                8. _____

9. How many of the accountants were 10. How many accountants are represented 9. _____
    younger than 29?   in each quartile?

                                                                                10. _____

11. What percent of the accountants were 12. What percent of the accountants were 11. _____
    younger than 36?   older than 46?

                                                                                12. _____

Answers

Try it 6.3B

1.
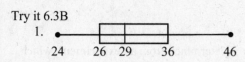
24   26 29   36   46

Quiz 6.3B

1.
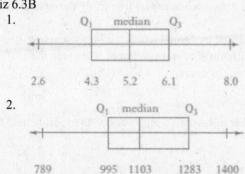
Q_1 median Q_3

2.6   4.3   5.2   6.1   8.0

2.
Q_1 median Q_3

789   995 1103   1283   1400

Solutions to Practice Sheet 6.3B

1. From the box-and-whisker plot, the youngest age is 24 years.

2. From the box-and-whisker plot, the youngest age is 46 years.

3. From the box-and-whisker plot, the first quartile is 26 years.

4. From the box-and-whisker plot, the third quartile is 36 years.

5. From the box-and-whisker plot, the median is 29 years.

6. The range is the difference between the largest and the smallest.
   $46 - 24 = 22$
   The range is 22.

7. The interquartile range is the difference between the first and third quartiles.
   $36 - 26 = 10$
   The interquartile range is 10.

8. Since 36 years is the third quartile, $\frac{1}{4}$ or 25% of the accountants were older than 36.
   $\frac{1}{4}(300) = 75$
   75 of the accountants were older than 36.

9. Since 29 years is the median, $\frac{1}{2}$ or 50% of the accountants were younger than 29.
   $\frac{1}{2}(300) = 150$
   150 of the accountants were younger than 29.

10. Since each quartile is one quarter of the data, and $\frac{1}{4}(300) = 75$,
    there are 75 accountants represented in each quartile.

11. Since 36 years is the third quartile, $\frac{3}{4}$ or 75% of the accountants were younger than 36.

12. Since 46 is the largest number of the data, there are no accountants, or 0% of the accountants older than 46.

Objective 6.4A – Calculate the probability of simple events

Key Words and Concepts

The likelihood of something happening is its **probability**. A probability is determined from an **experiment**, which is any activity that has an observable outcome. All of the possible outcomes of an experiment are called the **sample space** of the experiment. The outcomes of an experiment are listed between braces and frequently designated by *S*.

An **event** is one or more outcomes of an experiment. Events are denoted by capital letters. When discussing experiments and events, it is convenient to refer to the **favorable outcomes** of an experiment. These are the outcomes of the experiment that satisfy the requirements of the particular event.

Probability Formula

The probability of an event *E*, written $P(E)$, is the ratio of the number of favorable outcomes of an experiment to the total number of possible outcomes of the experiment.

$$P(E) = \frac{\text{number of favorable outcomes}}{\text{number of possible outcomes}}$$

This probability formula applies to experiments for which outcomes are **equally likely**—outcomes that are just as likely to occur as each other.

Example

In a box are 14 yellow balls, 13 blue balls, and 9 green balls. If a ball is selected at random, find the probability that it is yellow.

Solution

Let *E* be the event that a yellow ball is selected.
Then there are 14 favorable outcomes and $14+13+9 = 36$ possible outcomes.

The probability that the ball is yellow is

$$P(E) = \frac{14}{36} = \frac{7}{18}.$$

Try it

1. In a box are 8 red balls, 14 purple balls, and 7 orange balls. If a ball is selected at random, find the probability that it is red.

Example

A fair coin is tossed twice. Find the probability of getting tails each time.

Solution

The sample space is $S = \{HH, HT, TH, TT\}$.

We want to find the probability of event *E*, which is *TT*.
There is only one favorable outcome of four total outcomes, so

$$P(E) = \frac{1}{4}.$$

Try it

2. A fair coin is tossed twice. Find the probability of getting heads each time.

The probabilities that we have calculated so far are referred to as *mathematical* or *theoretical probabilities*. The calculations are based on theory. Empirical *probabilities* are based on observations of certain events.

Empirical Probability Formula

The empirical probability of an event E is the ratio of the number of observations of E to the total number of observations.

$$P(E) = \frac{\text{number of observations of } E}{\text{total number of observations}}$$

Example

Observations of students entering a school by a certain time on Valentine's Day show that 72 of 130 students are wearing red. What is the empirical probability that the next student entering the school that morning will be wearing red?

Solution

The empirical probability of E, where E is the event that the next student entering the school that morning will be wearing red, is the ratio of the number of students wearing red to the total number of students who have entered.

$$P(E) = \frac{72}{130} \approx 0.55.$$

Try it

3. A baseball player has had a hit in 167 of his last 482 at-bats. Find the empirical probability that the player will earn a hit in his next at-bat. Round to the nearest thousandth.

Reflect on it

- The probability of an event will never be more than 1.

Quiz Yourself 6.4A

1. In a box are 20 green pens, 32 red pens, and 12 blue pens. If a pen is drawn at random from the box, find the probability that it is green.

2.   A fair coin is tossed three times. Find the probability that the coin lands on tails exactly twice.

3.   A fair coin is tossed three times. Find the probability that the coin lands on heads more often than it lands on tails.

4.   Two fair six-sided dice are rolled. Find the probability that the number of dots are greater than or equal to four.

5.   A baseball player has had a hit in 243 of his last 751 at-bats. Find the empirical probability that the player will earn a hit in his next at-bat. Round to the nearest thousandth.

Practice Sheet 6.4A

Solve.

1. Two dice are rolled. What is the probability that the sum of the dots on the upwards faces is 4?

2. Two dice are rolled. What is the probability that the sum of the dots on the upwards faces is 9?

3. Two dice are rolled. What is the probability that the sum of the dots on the upwards faces is greater than 1?

4. A coin is tossed 3 times. What is the probability that the outcomes of the tosses consist of two tails and one head?

5. Each of the letters of the word *MISSISSIPPI* is written on a card, and the cards are placed in a hat. One card is drawn at random from the hat. What is the probability that the card has the letter *S* on it?

6. Each of the letters of the word *MISSISSIPPI* is written on a card, and the cards are placed in a hat. One card is drawn at random from the hat. What is the probability that the card has the letter *P* on it?

7. Which has a greater probability, drawing a 5, 6, or 10 from a deck of cards or drawing a diamond?

8. In a psychology class, a set of exams earned the following grades: 6 A's, 9 B's, 16 C's, 5 D's, and 3 F's. If a single student's exam is chosen from this class, what is the probability that it received an A?

9. Six purple marbles, four orange marbles, and eight blue marbles are placed in a bag. One marble is chosen at random. What is the probability that the marble chosen is purple?

10. Seven purple marbles, five orange marbles, and three blue marbles are placed in a bag. One marble is chosen at random. What is the probability that the marble chosen is orange?

1. _____

2. _____

3. _____

4. _____

5. _____

6. _____

7. _____

8. _____

9. _____

10. _____

Answers

Try it 6.4A

1. $\dfrac{8}{29}$

2. $\dfrac{1}{4}$

3. 0.346

Quiz 6.4A

1. $\dfrac{5}{16}$

2. $\dfrac{3}{8}$

3. $\dfrac{1}{2}$

4. $\dfrac{1}{4}$

5. 0.324

Solutions to Practice Sheet 6.4A

1. STRATEGY
   To calculate the probability:
   • Count the number of possible outcomes of the experiment.
   • Count the number of outcomes that are favorable.
   • Use the probability formula.

   SOLUTION
   There are 36 possible outcomes.
   There are 3 favorable outcomes: (1, 3), (3, 1), and (2, 2).
   $$P(E) = \frac{3}{36} = \frac{1}{12}$$
   The probability that the sum of the dots on the upwards faces is 4 is $\dfrac{1}{12}$.

2. STRATEGY
   To calculate the probability:
   • Count the number of possible outcomes of the experiment.
   • Count the number of outcomes that are favorable.
   • Use the probability formula.

   SOLUTION
   There are 36 possible outcomes.
   There are 4 favorable outcomes: (4, 5), (5, 4), (3, 6), and (6, 3).

$$P(E) = \frac{4}{36} = \frac{1}{9}$$
The probability that the sum of the dots on the upwards faces is 9 is $\dfrac{1}{9}$.

3. STRATEGY
   To calculate the probability: .
   • Count the number of possible outcomes of the experiment.
   • Count the number of outcomes that are favorable.
   • Use the probability formula.

   SOLUTION
   There are 36 possible outcomes.
   All the outcomes are favorable.
   $$P(E) = \frac{36}{36} = 1$$
   The probability that the sum of the dots on the upwards faces is greater than 1 is 1.

4. STRATEGY
   To calculate the probability:
   • Count the number of possible outcomes of the experiment.
   • Count the number of outcomes that are favorable.
   • Use the probability formula.

   SOLUTION
   There are 8 possible outcomes.
   There are 3 favorable outcomes: (T, T, H), (T, H, T), and (H, H, T).
   $$P(E) = \frac{3}{8}$$
   The probability that a coin tossed 3 times results in two tails and one head is $\dfrac{3}{8}$.

5. STRATEGY
   To calculate the probability:
   • Count the number of possible outcomes of the experiment.
   • Count the number of outcomes that are favorable.
   • Use the probability formula.

   SOLUTION
   There are 11 possible outcomes.
   There are 4 favorable outcomes: S
   $$P(E) = \frac{4}{11}$$

The probability that the card drawn is the letter S is $\dfrac{4}{11}$.

6. STRATEGY
To calculate the probability:
 • Count the number of possible outcomes of the experiment.
 • Count the number of outcomes that are favorable.
 • Use the probability formula.

SOLUTION
There are 11 possible outcomes.
There are 2 favorable outcomes: P

$$P(E) = \dfrac{2}{11}$$

The probability that the card drawn is the letter P is $\dfrac{2}{11}$.

7. STRATEGY
To calculate the probability:
 • Count the number of possible outcomes of the experiment.
 • Count the number of outcomes that are favorable.
 • Use the probability formula.
 • Compare the probabilities.

SOLUTION
There are 52 possible outcomes.
There are 12 favorable outcomes: 5, 6, 10
There are 13 favorable outcomes: diamond.

$$P(E) = \dfrac{12}{52}$$

$$P(E) = \dfrac{13}{52}$$

$$\dfrac{12}{52} < \dfrac{13}{52}$$

Drawing a diamond has a greater probability than drawing a 5, 6, or 10.

8. STRATEGY
To calculate the probability:
 • Count the number of possible outcomes of the experiment.
 • Count the number of outcomes that are favorable.
 • Use the probability formula.

SOLUTION
There are $6 + 9 + 16 + 5 + 3 = 39$ possible outcomes.

There are 6 favorable outcomes: A

$$P(E) = \dfrac{6}{39} = \dfrac{2}{13}$$

The probability that the exam chosen was an A is $\dfrac{2}{13}$.

9. STRATEGY
To calculate the probability:
 • Count the number of possible outcomes of the experiment.
 • Count the number of outcomes that are favorable.
 • Use the probability formula.

SOLUTION
There are $6 + 4 + 8 = 18$ possible outcomes.
There are 6 favorable outcomes: A

$$P(E) = \dfrac{6}{18} = \dfrac{1}{3}$$

The probability that the marble chosen was purple is $\dfrac{1}{3}$.

10. STRATEGY
To calculate the probability:
 • Count the number of possible outcomes of the experiment.
 • Count the number of outcomes that are favorable.
 • Use the probability formula.

SOLUTION
There are $7 + 5 + 3 = 15$ possible outcomes.
There are 5 favorable outcomes: A

$$P(E) = \dfrac{5}{15} = \dfrac{1}{3}$$

The probability that the marble chosen was orange is $\dfrac{1}{3}$.

Objective 6.4B – Calculate the probability of compound events

Key Words and Concepts

Two evens that cannot occur at the same time are called **mutually exclusive events**.

Mutually Exclusive Events

Two events A and B are mutually exclusive if they cannot occur at the same time. That is, A and B are mutually exclusive when $A \cap B = \varnothing$.

The probability that either of two mutually exclusive events will occur can be determined by adding the probabilities of the individual events.

Probability of Mutually Exclusive Events

If A and B are two mutually exclusive events, then the probability that either A or B will occur is

$$P(A \text{ or } B) = P(A) + P(B)$$

Example

Suppose a single card is drawn from a standard deck of playing cards. Find the probability of drawing a 3 or a jack.

Solution

Let A = {3 of spades, 3 of hearts, 3 of diamonds, 3 of clubs} and B = {jack of spades, jack of hearts, jack of diamonds, jack of clubs}

There are 52 cards in a standard deck of playing cards so

$$n(S) = 52$$

The probability of drawing a 3 is

$$P(A) = \frac{4}{52} = \frac{1}{13}$$

The probability of drawing a jack is

$$P(B) = \frac{4}{52} = \frac{1}{13}$$

Because the events are mutually exclusive, use the formula for the probability of mutually exclusive events.

$$P(A \text{ or } B) = P(A) + P(B)$$
$$= \frac{1}{13} + \frac{1}{13}$$
$$= \frac{2}{13}$$

The probability of drawing a 3 or a jack is $\frac{2}{13}$.

Try it

1. Suppose a single card is drawn from a standard deck of playing cards. Find the probability of drawing a 7 or a queen.

Now consider events that are not mutually exclusive.

Addition Rule for Probabilities

If A and B are two events in a sample space S, then

$$P(A \text{ or } B) = P(A) + P(B) - P(A \text{ and } B)$$

Example

The table below shows the unemployment status of individuals in a particular town by age group.

Age	Full-time	Part-time	Unemployed
0–17	22	152	363
18–25	164	214	153
26–34	361	85	31
35–49	603	182	107
50+	419	151	181

If the person is randomly chosen from the town's population, what is the probability that the person is over 50 years old or employed full time?

Solution

Let A = {people over 50 years old} and
F = {people employed full time}.
These events are not mutually exclusive because there are 419 people who are over 50 years old and work full time.
The sample space S consists of
$22 + 152 + 363 + 164 + 214 + 153 + 361 + 85 + 31 + 603 + 182 + 107 + 419 + 151 + 181 = 3188$ people.
From the table,

$$n(A) = 419 + 151 + 181 = 751$$

$$n(F) = 22 + 164 + 361 + 603 + 419 = 1569$$

$$n(A \text{ and } F) = 419$$

Use the addition rule for probabilities.

$$P(A \text{ or } F) = P(A) + P(F) - P(A \text{ and } F)$$

$$= \frac{751}{3188} + \frac{1569}{3188} - \frac{419}{3188}$$

$$= \frac{1901}{3188} \approx 0.596$$

The probability that the person is over 50 years old or employed full time is $\frac{1901}{3188}$ or approximately 59.6%.

Try it

2. Use the table at the left. If the person is randomly chosen from the town's population, what is the probability that the person is under 18 years old or employed part time?

What is the complement of an event? _____

_____

Probability of the Complement of an Event

If E is an event and E^c is the complement of the event, then

$$P(E^c) = 1 - P(E)$$

Example

The probability of drawing a diamond from a standard deck of playing cards is $\frac{1}{13}$. What is the probability of not drawing a diamond?

Solution

$E = \{$draw a card that is a diamond$\}$. Then,

$E^c = \{$draw a card that is not a diamond$\}$.

Use the formula for the probability of the complement of an event.

$$P(E^c) = 1 - P(E)$$
$$= 1 - \frac{1}{13}$$
$$= \frac{12}{13}$$

The probability of not drawing a diamond is $\frac{12}{13}$.

Try it

3. The probability of drawing a face card from a standard deck of playing cards is $\frac{3}{13}$. What is the probability of not drawing a red card?

Reflect on it

- The complement of an event is the opposite of the event.
- The Addition Rule for Probability is used when events are not mutually exclusive.

Quiz Yourself 6.4B

1. Two dice are rolled. Find the probability of rolling a 3 or an 11.

2. If $P(A) = 0.4$, $P(B) = 0.5$, and $P(A \text{ and } B) = 0.1$, find $P(A \text{ or } B)$.

3. Suppose you ask a friend to randomly choose an integer between 1 and 10, inclusive. What is the probability that the number your friend chooses will be less than 3 or even?

4. Suppose you roll two fair dice. Find the probability of rolling an 11 or a number greater than 9.

5. The table below shows the unemployment status of individuals in a particular town by age group.

Age	Full-time	Part-time	Unemployed
0–17	22	152	363
18–25	164	214	153
26–34	361	85	31
35–49	603	182	107
50+	419	151	181

If a person is randomly chosen from the town's population, what is the probability that the person is between 18 and 25 years old or unemployed?

6. Suppose the probability that it will rain tomorrow is 0.46. What is the probability that it will not rain tomorrow?

Practice Sheet 6.4B

Solve.

1. Two fair dice are rolled. Find the probability of rolling the sum of 5 or 6.

2. Two fair dice are rolled. Find the probability of rolling the sum of 2 or 12.

1. _____

2. _____

3. If $P(A) = 0.2$, $P(B) = 0.6$, and $P(A \text{ and } B) = 0.2$, find $P(A \text{ or } B)$.

4. Suppose you roll two fair dice. Find the probability of rolling the sum of 4 or an even number.

3. _____

4. _____

The table below shows the unemployment status of individuals in a particular town by age group.

Age	Full-time	Part-time	Unemployed
0 – 17	22	152	363
18 – 25	164	214	153
26 – 34	361	85	31
35 – 49	603	182	107
50 +	419	151	181

5. If a person is randomly chosen from the town's population, what is the probability that the person is between 35 and 49 years old or employed part time?

6. If a person is randomly chosen from the town's population, what is the probability that the person is between 26 and 34 years old or employed full time?

5. _____

6. _____

7. Suppose the probability that it will snow tomorrow is 0.16. What is the probability that it will not snow tomorrow?

8. The probability of drawing a red card from a standard deck of playing cards is $\frac{1}{2}$. What is the probability of not drawing a red card?

7. _____

8. _____

Answers

Try it 6.4B

1. $\dfrac{2}{13}$

2. $\dfrac{1169}{3188}$

3. $\dfrac{10}{13}$

Quiz 6.4B

1. $\dfrac{1}{9}$

2. 0.8

3. $\dfrac{6}{10}$

4. $\dfrac{1}{6}$

5. $\dfrac{1213}{3188}$

6. 0.54

Solutions to Practice Sheet 6.4B

1. The probability of rolling a 5: $P(A) = \dfrac{4}{36}$

   The probability of rolling a 6: $P(B) = \dfrac{5}{36}$

   The events are mutually exclusive.
   $$P(A \text{ or } B) = P(A) + P(B)$$
   $$= \dfrac{4}{36} + \dfrac{5}{36}$$
   $$= \dfrac{9}{36} = \dfrac{1}{4}$$

   The probability of rolling a 5 or a 6 is $\dfrac{1}{4}$.

2. The probability of rolling a 2: $P(A) = \dfrac{1}{36}$

   The probability of rolling a 12: $P(B) = \dfrac{1}{36}$

   The events are mutually exclusive.
   $$P(A \text{ or } B) = P(A) + P(B)$$
   $$= \dfrac{1}{36} + \dfrac{1}{36}$$
   $$= \dfrac{2}{36} = \dfrac{1}{18}$$

   The probability of rolling a 2 or a 12 is $\dfrac{1}{18}$.

3. Given $P(A) = 0.2$, $P(B) = 0.6$, and
   $P(A \text{ and } B) = 0.2$,
   $$P(A \text{ or } B) = P(A) + P(B) - P(A \text{ and } B)$$
   $$= 0.2 + 0.6 - 0.2$$
   $$= 0.6$$

4. The probability of rolling a 4: $P(A) = \dfrac{3}{36}$

   The probability of rolling an even: $P(B) = \dfrac{18}{36}$

   The events are not mutually exclusive since 4 is an even number.
   $$P(A \text{ or } B) = P(A) + P(B) - P(A \text{ and } B)$$
   $$= \dfrac{3}{36} + \dfrac{18}{36} - \dfrac{3}{36}$$
   $$= \dfrac{1}{2}$$

5. Let A = {people 35–49 years old} and
   P = {people employed part time}.
   Sample space is 3188 people.
   From the table, $n(A) = 892$, $n(P) = 784$,
   and $n(A \text{ and } P) = 182$
   $$P(A \text{ or } P) = P(A) + P(P) - P(A \text{ and } P)$$
   $$= \dfrac{892}{3188} + \dfrac{784}{3188} - \dfrac{182}{3188}$$
   $$= \dfrac{1494}{3188} \approx 0.469$$

   The probability that the person is 35–49 years old or employed part time is $\dfrac{1494}{3188}$ or approximately 46.9%.

6. Let A = {people 26–34 years old} and
   F = {people employed full time}.
   Sample space is 3188 people.
   From the table, $n(A) = 477$, $n(F) = 1569$,
   and $n(A \text{ and } F) = 361$
   $$P(A \text{ or } F) = P(A) + P(F) - P(A \text{ and } F)$$
   $$= \dfrac{477}{3188} + \dfrac{1569}{3188} - \dfrac{361}{3188}$$
   $$= \dfrac{1685}{3188} \approx 0.529$$

   The probability that the person is 26–34 years old or employed full time is $\dfrac{1685}{3188}$ or approximately 52.9%.

7. Let $E = \{\text{snow tomorrow}\}$, then

    $E^c = \{\text{not snow tomorrow}\}$

    $P(E^c) = 1 - P(E) = 1 - 0.16 = 0.84$

    The probability that it will not snow tomorrow is 0.84, or 84%.

8. Let $E = \{\text{red card}\}$, then

    $E^c = \{\text{not a red card}\}$

    $P(E^c) = 1 - P(E) = 1 - \dfrac{1}{2} = \dfrac{1}{2}$

    The probability of not drawing a red card is $\dfrac{1}{2}$.

Objective 6.4C – Calculate the odds of an event

Key Words and Concepts

Sometimes the chance of an event occurring is given in terms of *odds*.

Odds in Favor of an Event

The **odds in favor** of an event is the ratio of the number of favorable outcomes of an experiment to the number of unfavorable outcomes.

$$\text{Odds in favor} = \frac{\text{number of favorable outcomes}}{\text{number of unfavorable outcomes}}$$

Odds Against an Event

The **odds against** an event is the ratio of the number of unfavorable outcomes of an experiment to the number of favorable outcomes.

$$\text{Odds against} = \frac{\text{number of unfavorable outcomes}}{\text{number of favorable outcomes}}$$

Example

In a box are 14 yellow balls, 13 blue balls, and 9 green balls. If a ball is selected at random, find
(a) the odds that it is yellow; and
(b) the odds against it being green.
Solution
(a) There are 14 favorable outcomes and
$13 + 9 = 22$ unfavorable outcomes.

The odds in favor of the ball being yellow are $\frac{14}{22}$, or $\frac{7}{11}$.

These odds may be expressed using the word *to*: the odds in favor of the ball being yellow are 7 to 11.

(b) There are $14 + 13 = 27$ unfavorable outcomes and 9 favorable outcomes.

The odds against the ball being green are $\frac{27}{9} = \frac{3}{1}$.

This may be expressed by saying that the odds against the ball being green are 3 to 1.

Try it

1. In a box are 8 red balls, 14 purple balls, and 7 orange balls. If a ball is selected at random, find
   (a) the odds that it is red; and
   (b) the odds against it being purple.

It is possible to compute the probability of an event from the odds-in-favor fraction. The probability of an event is the ratio of the numerator to the sum of the numerator and denominator.

Example

The odds of winning a prize in a certain contest are 1 to 8. Find the probability of winning a prize.

Solution

The probability of winning a prize is the ratio of the numerator to the sum of the numerator and denominator.

$$\frac{1}{1+8} = \frac{1}{9}$$

The probability of winning is $\frac{1}{9}$.

Try it

2. The odds of winning a prize in a certain contest are 1 to 12. Find the probability of winning a prize.

Reflect on it

- What is the sum of the odds in favor of an event and the odds against an the same event?

Quiz Yourself 6.4C

1. A fair coin is tossed three times. What are the odds in favor of having exactly two tails?

2. A fair coin is tossed three times. What are the odds in favor of having more tails than heads?

3. A fair six-sided die is rolled twice. What are the odds against rolling a sum of six?

4. The odds of a certain event are 7 to 2. What is the probability of the event occurring?

5. The odds *against* a certain event are 8 to 3. What is the probability that the event *does* occur?

Practice Sheet 6.4C

Solve.

1. A coin is tossed twice. What are the odds of its showing heads both times?

2. Two dice are rolled. What are the odds in favor of rolling the sum of 2?

3. Two dice are rolled. What are the odds in favor of rolling the sum of 6?

4. A single card is selected from a standard deck of playing cards. What are the odds against its being a jack?

5. A single card is selected from a standard deck of playing cards. What are the odds against its being a spade?

6. The odds in favor of a candidate winning an election is 4 to 3. What is the probability of the candidate winning the election?

7. Two dice are rolled. What are the odds against rolling the sum of 3?

8. At the beginning of the professional football season, one team was given 20 to 1 odds against its winning the Super Bowl. What is the probability of this team winning the Super Bowl?

1. _____

2. _____

3. _____

4. _____

5. _____

6. _____

7. _____

8. _____

Answers

Try it 6.4C
1. a. 8 to 21; b. 15 to 14
2. $\dfrac{1}{13}$

Quiz 6.4C
1. 3 to 5
2. 1 to 1
3. 31 to 5
4. $\dfrac{7}{9}$
5. $\dfrac{3}{11}$

Solutions to Practice Sheet 6.4C

1. STRATEGY
   To find the odds in favor:
   • Count the number of favorable outcomes.
   • Count the number of unfavorable outcomes.
   • Use the formula for the odds in favor of an event.

   SOLUTION
   Number of favorable outcomes: 1
   (H, H)
   Number of unfavorable outcomes: 3

   Odds in favor $= \dfrac{1}{3}$

   The odds in favor of showing a heads both times are 1 to 3.

2. STRATEGY
   To find the odds in favor:
   • Count the number of favorable outcomes.
   • Count the number of unfavorable outcomes.
   • Use the formula for the odds in favor of an event.

   SOLUTION
   Number of favorable outcomes: 1
   (1, 1)
   Number of unfavorable outcomes: 35

   Odds in favor $= \dfrac{1}{35}$

   The odds in favor of rolling a 2 are 1 to 35.

3. STRATEGY
   To find the odds in favor:
   • Count the number of favorable outcomes.
   • Count the number of unfavorable outcomes.
   • Use the formula for the odds in favor of an event.

   SOLUTION
   Number of favorable outcomes: 5
   (1, 5), (5, 1), (2, 4), (4, 2), (3, 3)
   Number of unfavorable outcomes: 31

   Odds in favor $= \dfrac{5}{31}$

   The odds in favor of rolling a 6 are 5 to 31.

4. STRATEGY
   To find the odds against:
   • Count the number of unfavorable outcomes.
   • Count the number of favorable outcomes.
   • Use the formula for the odds against an event.

   SOLUTION
   Number of unfavorable outcomes: 48
   Number of favorable outcomes: 4
   Jack (hearts, spades, clubs, diamonds)

   Odds against $= \dfrac{48}{4} = \dfrac{12}{1}$

   The odds against a card being a jack are 12 to 1.

5. STRATEGY
   To find the odds against:
   • Count the number of unfavorable outcomes.
   • Count the number of favorable outcomes.
   • Use the formula for the odds against an event.

   SOLUTION
   Number of unfavorable outcomes: 39
   Number of favorable outcomes: 13
   Spade

   Odds against $= \dfrac{39}{13} = \dfrac{3}{1}$

   The odds against a card being a spade are 3 to 1.

6. STRATEGY
   To calculate the probability of winning, use the odds-in-favor fraction. The probability of winning is the ratio of the numerator to the sum of the numerator and denominator.

   SOLUTION
   Probability of winning $= \dfrac{4}{4+3} = \dfrac{4}{7}$

   The probability of winning the election is $\dfrac{4}{7}$.

7. **STRATEGY**
    To find the odds against:
    - Count the number of unfavorable outcomes.
    - Count the number of favorable outcomes.
    - Use the formula for the odds against
    an event.

    SOLUTION
    Number of unfavorable outcomes: 34
    Number of favorable outcomes: 2
    $(1, 2), (2, 1)$

    Odds against $= \dfrac{34}{2} = \dfrac{17}{1}$

    The odds against a rolling a 3
    are 17 to 1.

8. **STRATEGY**
    To calculate the probability of winning, use
    the odds-in-favor fraction. The probability of
    winning is the ratio of the numerator to the
    sum of the numerator and denominator.

    SOLUTION

    Probability of winning $= \dfrac{1}{1+20} = \dfrac{1}{21}$

    The probability of winning the Super Bowl

    is $\dfrac{1}{21}$.

Application and Activities

Discussion and reflection questions

1. Define what quality control might mean for a company.

2. Turbine engines for helicopters need to be able to produce a pre-determined horsepower without exceeding an exhaust temperature limit. Why do you think that might be important?

3. Once a batch of 20 turbine engines is tested, how do you think statistics would be used to aid the engineer before the engines that pass are released?

Group Activity (2-3 people)

Objectives for the lesson:

In this lesson you will use and enhance your understanding of the following:

- Creating a frequency distribution
- Finding the mean, median and mode of a distribution
- Finding the standard deviation of a distribution
- Calculating quartiles
- Creating a box and whiskers plot
- Calculating the probability of a compound event
- Calculating the odds of an event

New production turbine engines go through an acceptance test procedure after they are built to make sure they are healthy. It is desirable for the engine to be able to produce a pre-determined horsepower without exceeding an exhaust temperature limit. The data below represents 20 helicopter turbine engines Jon was testing that must deliver 2200 hp without exceeding $1690\,^{\circ}F$ exhaust temperature.

Engine Serial Number (x)	Exhaust Temperature °F (y)	Engine Serial Number (x)	Exhaust Temperature °F (y)
001	1669	011	1668
002	1664	012	1661
003	1697	013	1656
004	1687	014	1688
005	1693	015	1664
006	1697	016	1661
007	1686	017	1659
008	1684	018	1669
009	1664	019	1688
010	1665	020	1688

1. Find the range for the data in the table.

2. Create a frequency distribution for the data collected. Create 10 classes for the data.

Complete the table:

Classes	Tally	Frequency

3. Using the data Jon collected for his 20 engines, find the mean, median and the mode of the data and explain their meaning in the data.

4. Find the standard deviation of the engine data Jon collected, and explain its meaning to the data.

5. Calculate the Quartiles for the data set.

6. Use the quartiles you found in #5 to create a box and whiskers plot of the data.

7. Using the data Jon collected on the first 20 engines, determine the probability that one of those engine will have a temperature greater than $1690°F$ when delivering 2200 hp.

8. Calculate the odds of an engine in Jon's batch having a temperature that does not exceed $1690°F$ at 2200 hp.

Objective 7.1A – Calculate simple interest

Key Words and Concepts

The amount deposited in a bank or borrowed from a bank is called the **principal**. The amount of interest paid is usually given as a percent of a principal. The percent used to determine the amount of interest is called the **interest rate**.

Interest paid on the original principal is called **simple interest**.

Simple Interest Formula
The simple interest formula is

$$I = Prt$$

where I is the interest, P is the principal, r is the interest rate, and t is the time period.

Example
Calculate the simple interest due on a 3-month loan of $3000 if the annual simple interest rate is 7.5%.

Solution
Because t is measured in years,

$$t = \frac{3 \text{ months}}{1 \text{ years}} = \frac{3 \text{ months}}{12 \text{ months}} = \frac{3}{12}.$$

Use the simple interest formula, and substitute the values $P = \$3000$ and $r = 7.5\% = 0.075$ into the formula.

$$I = Prt$$
$$I = (3000)(0.075)\left(\frac{3}{12}\right)$$
$$I = 56.25$$

The simple interest due is $56.25.

Try it
1. Calculate the simple interest due on a 3-month loan of $4000 if the annual simple interest rate is 9.5%.

There are two methods for converting time from days to years: the exact method and the ordinary method.

Using the exact method, the number of days of the loan is divided by 365, the number of days in a year.

Exact method: $t = \dfrac{\text{number of days}}{365}$

The ordinary method is based on there being an average of 30 days in a month and 12 months in a year $(30 \cdot 12) = 360$. Using this method, the number of days of the loan is divided by 360.

Ordinary method: $t = \dfrac{\text{number of days}}{360}$

Example

Calculate the simple interest due on a 45-day loan of $4000 if the annual simple interest rate is 9%.

Solution

Use the simple interest formula. Substitute the following values into the formula. $P = \$4000$, $r = 9\% = 0.09$, and

$$t = \frac{\text{number of days}}{360} = \frac{45}{360} .$$

$$I = Prt$$

$$I = (4000)(0.09)\left(\frac{45}{360}\right)$$

$$I = 45$$

The simple interest due is $45.

Try it

2. Calculate the simple interest due on a 45-day loan of $2000 if the annual simple interest rate is 7%.

The simple interest formula can be used to find the interest rate on a loan when the interest, principal, and time period of the loan are known.

Example

The simple interest charged on a 6-month loan of $4000 is $160. Find the annual simple interest rate.

Solution

Use the simple interest formula. Solve the equation for r.

Substitute $P = 4000$, $I = 160$, and $t = \frac{6}{12}$.

$$I = Prt$$

$$160 = (4000)(r)\left(\frac{6}{12}\right)$$

$$160 = 2000r$$

$$0.08 = r$$

$$r = 8\%$$

The annual simple interest rate on the loan is 8%.

Try it

3. The simple interest charged on a 6-month loan of $5000 is $225. Find the annual simple interest rate.

Reflect on it

- Before performing calculations involving interest rates, write the interest rate as a decimal.

Quiz Yourself 7.1A

For Exercises 1 to 3, calculate the simple interest earned on the investment.

1. $P = \$8000$, $r = 7.5\%$, $t = 6$ months

2. $P = \$6000$, $r = 8\%$, $t = 4$ months

3. $P = \$1000$, $r = 2.5\%$, $t = 9$ months

4. Calculate the annual simple interest rate if $P = \$3000$, $I = \$75$, $t = 6$ months.

5. You deposit $1600 in an account earning 2.8% annual simple interest. Calculate simple interest in 6 months.

6. Calculate the simple interest due on a 45-day loan of $2200 if the annual simple interest rate is 4.2%.

Practice Sheet 7.1A

Calculate the simple interest earned on the investment.

1. $P = \$6000$, $r = 10\%$, $t = 4$ months   2. $P = \$5000$, $r = 7.2\%$, $t = 4$ months   1. _____

   2. _____

3. $P = \$4000$, $r = 5.1\%$, $t = 2$ months   4. $P = \$5500$, $r = 6.6\%$, $t = 2$ months   3. _____

   4. _____

5. $P = \$3500$, $r = 8.3\%$, $t = 6$ months   6. $P = \$1500$, $r = 6.7\%$, $t = 6$ months   5. _____

   6. _____

7. $P = \$2500$, $r = 3.2\%$, $t = 45$ days   8. $P = \$1800$, $r = 2.6\%$, $t = 45$ days   7. _____

   8. _____

Calculate the simple interest rate.

9. $P = \$2000$, $I = \$48$, $t = 6$ months   10. $P = \$4000$, $I = \$72$, $t = 6$ months   9. _____

   10. _____

Answers

Try it 7.1A
1. $95
2. $17.50
3. 9%

Quiz 7.1A
1. $300
2. $160
3. $18.75
4. 5%
5. $22.40
6. $11.55

Solutions to Practice Sheet 7.1A

1. $I = Prt$

   $I = (6000)(0.10)\left(\dfrac{4}{12}\right)$

   $I = \$200.00$

2. $I = Prt$

   $I = (5000)(0.072)\left(\dfrac{4}{12}\right)$

   $I = \$120.00$

3. $I = Prt$

   $I = (4000)(0.051)\left(\dfrac{2}{12}\right)$

   $I = \$34.00$

4. $I = Prt$

   $I = (5500)(0.066)\left(\dfrac{2}{12}\right)$

   $I = \$60.50$

5. $I = Prt$

   $I = (3500)(0.083)\left(\dfrac{6}{12}\right)$

   $I = \$145.25$

6. $I = Prt$

   $I = (1500)(0.067)\left(\dfrac{6}{12}\right)$

   $I = \$50.25$

7. $I = Prt$

   $I = (2500)(0.032)\left(\dfrac{45}{360}\right)$

   $I = \$10.00$

8. $I = Prt$

   $I = (1800)(0.026)\left(\dfrac{45}{360}\right)$

   $I = \$5.85$

9. $I = Prt$

   $48 = (2000)(r)\left(\dfrac{6}{12}\right)$

   $48 = 1000r$

   $0.048 = r$

   $r = 4.8\%$

10. $I = Prt$

    $72 = (4000)(r)\left(\dfrac{6}{12}\right)$

    $72 = 2000r$

    $0.036 = r$

    $r = 3.6\%$

Objective 7.1B – Calculate future value and maturity value

Key Words and Concepts

When you borrow money, the total amount to be repaid to the lender is the sum of the principal and interest.

Future Value or Maturity Value Formula for Simple Interest
The future value or maturity value formula for simple interest is
$$A = P + I$$
where A is the amount after the interest I has been added to the principal P.

The formula can be used for loans or investments. When used for a loan, A is the total amount to be repaid to the lender; this sum is called the **maturity value** of the loan.

For an investment, such as a deposit in a bank savings account, A is the total amount on deposit after the interest earned has been added to the principal. This sum is called the **future value** of the investment.

Example
Calculate the maturity value of a simple interest, 3-month loan of $3000 if the annual simple interest rate is 7.5%.

Solution
Step 1: Find the interest. Use the simple interest formula, and substitute the values $P = \$3000$ and

$r = 7.5\% = 0.075$, and $t = \dfrac{3}{12}$ into the formula.

$$I = Prt$$
$$I = (3000)(0.075)\left(\frac{3}{12}\right)$$
$$I = 56.25$$

Step 2: Find the maturity value. Use the maturity value formula for simple interest. Substitute the values $P = 3000$ and $I = 56.25$ into the formula.
$$A = P + I$$
$$A = 3000 + 56.25$$
$$A = 3056.25$$

The maturity value is $3056.25.

Try it
1. Calculate the maturity value of a simple interest 3-month loan of $4000 if the annual simple interest rate is 9.5%.

Recall that the simple interest formula states that $I = Prt$. We can substitute Prt for I in the future value or maturity value formula as follows.
$$A = P + I$$
$$A = P + Prt$$
$$A = P(1 + rt)$$

In the final equation, A is the future value of an investment or the maturity value of a loan, P is the principal, r is the interest rate, and t is the time period.

Example

Find the future value after 1 year of $4000 in an account earning 2% annual simple interest.

Solution

Substitute the following values into the formula.
$P = \$4000$, $r = 2\% = 0.02$, and, $t = 1$.
$$A = P(1 + rt)$$
$$A = 4000[1 + 0.02(1)]$$
$$A = 4080$$
The future value of the account after 1 year is $4080.

Try it

2. Find the future value after 1 year of $2000 in an account earning 3% annual simple interest.

Recall that the formula $A = P + I$ states that A is the amount after the interest has been added to the principal. Subtracting P from each side of this equation yields the following formula.
$$I = A - P$$
This formula states that the amount of interest paid is equal to the total amount minus the principal.

Example

The maturity value of a 6-month loan of $4000 is $4160. What is the annual simple interest rate?

Solution

First find the amount of interest paid.
Subtract the principal from the maturity value.
$$I = A - P$$
$$I = 4160 - 4000$$
$$I = 160$$
Find the annual simple interest rate by solving the simple interest formula for r.

Substitute $P = 4000$, $I = 160$, and $t = \dfrac{6}{12}$.
$$I = Prt$$
$$160 = (4000)(r)\left(\frac{6}{12}\right)$$
$$160 = 2000r$$
$$0.08 = r$$
$$r = 8\%$$
The annual simple interest rate on the loan is 8%.

Try it

3. The maturity value of a 6-month loan of $5000 is $5225. What is the annual simple interest rate?

Reflect on it

• Maturity value is associated with loans. Future value is associated with deposits.

Quiz Yourself 7.1B

For Exercises 1 and 2, use the formula $A = P(1 + rt)$ to calculate the maturity value of the simple interest loan.

1. $P = \$8000$, $r = 7.5\%$, $t = 6$ months

2. $P = \$1000$, $r = 2.5\%$, $t = 9$ months

3. Calculate the maturity value of a simple-interest 8-month loan of $2200 if the annual simple interest rate is 4.2%.

4. You deposit $1600 in an account earning 2.8% annual simple interest.
    Find the future value of the investment after 1 year.

5. Find the annual simple interest rate on a 6-month loan of $3000 if the maturity value of the loan is $3075.

Practice Sheet 7.1B

Use the formula $A = P(1 + rt)$ to calculate the maturity value of the simple interest loan.

1. $P = \$6000$, $r = 10\%$, $t = 4$ months

2. $P = \$5000$, $r = 7.2\%$, $t = 4$ months

1. _____

2. _____

3. $P = \$4000$, $r = 5.1\%$, $t = 1$ year

4. $P = \$5500$, $r = 6.6\%$, $t = 1$ year

3. _____

4. _____

5. $P = \$3500$, $r = 8.3\%$, $t = 6$ months

6. $P = \$1500$, $r = 6.7\%$, $t = 6$ months

5. _____

6. _____

7. $P = \$2500$, $r = 3.2\%$, $t = 1$ year

8. $P = \$1800$, $r = 2.6\%$, $t = 1$ year

7. _____

8. _____

Calculate the simple interest rate.

9. $P = \$2000$, $A = \$2048$, $t = 6$ months

10. $P = \$4000$, $A = \$4072$, $t = 6$ months

9. _____

10. _____

Answers

Try it 7.1B

1. $4095
2. $2060
3. 9%

Quiz 7.1B

1. $8300
2. $1018.75
3. $2261.60
4. $1644.80
5. 5%

Solutions to Practice Sheet 7.1B

1. $A = P(1+rt)$

   $A = 6000\left[1+0.10\left(\dfrac{4}{12}\right)\right]$

   $A = \$6200.00$

2. $A = P(1+rt)$

   $A = 5000\left[1+0.072\left(\dfrac{4}{12}\right)\right]$

   $A = \$5120.00$

3. $A = P(1+rt)$

   $A = 4000[1+0.051(1)]$

   $A = \$4204.00$

4. $A = P(1+rt)$

   $A = 5500[1+0.066(1)]$

   $A = \$5863.00$

5. $A = P(1+rt)$

   $A = 3500\left[1+0.083\left(\dfrac{6}{12}\right)\right]$

   $A = \$3645.25$

6. $A = P(1+rt)$

   $A = 1500\left[1+0.067\left(\dfrac{6}{12}\right)\right]$

   $A = \$1550.25$

7. $A = P(1+rt)$

   $A = 2500[1+0.032(1)]$

   $A = \$2580.00$

8. $A = P(1+rt)$

   $A = 1800[1+0.026(1)]$

   $A = \$1846.80$

9. $I = A - P$

   $I = 2048 - 2000$

   $I = 48$

   $I = Prt$

   $48 = (2000)(r)\left(\dfrac{6}{12}\right)$

   $48 = 1000r$

   $0.048 = r$

   $r = 4.8\%$

10. $I = A - P$

    $I = 4072 - 4000$

    $I = 72$

    $I = Prt$

    $72 = (4000)(r)\left(\dfrac{6}{12}\right)$

    $72 = 2000r$

    $0.036 = r$

    $r = 3.6\%$

Objective 7.2A – Calculate the future value of an investment using compound interest

Key Words and Concepts

Simple interest is generally used for loans of 1 year or less. For loans of more than 1 year, the interest paid on the money borrowed is called *compound interest*. **Compound interest** is interest calculated not only on the original principal but also on any interest that has already been earned.

What is a compounding period? _____

_____

Future Value Formula

The future value A of an investment is given by

$$A = P\left(1+\frac{r}{n}\right)^{nt}$$

where P is the amount of money invested or deposited (also called the *present value*),
  r is the annual interest rate as a decimal,
  n is the number of compounding periods per year, and
  t is the number of years.

Example

Calculate the future value of $8000 deposited in an account earning 7% annual interest compounded monthly, for 3 years.

  Solution
    Use the future value formula.
    $P = 8000, \ r = 7\% = 0.07, \ n = 12, \ t = 3$

$$A = P\left(1+\frac{r}{n}\right)^{nt}$$

$$A = 8000\left(1+\frac{0.07}{12}\right)^{12\cdot 3}$$

$$A \approx 8000(1.005833333)^{36}$$

$$A \approx 9863.40$$

  The future value after 3 years is approximately $9863.40.

Try it

1. Calculate the future value of $5000 deposited in an account earning 5% annual interest compounded semiannually, for 5 years.

What is n when the compounding period is

    annually? _____   semiannually? _____

    quarterly? _____   monthly? _____

    weekly? _____   daily? _____

The formula $I = A - P$ is also used to calculate compound interest.

Example
How much interest is earned in 3 years on $6000 deposited in an account paying 4% interest, compounded quarterly?

Solution

Use the future value formula.

$P = 6000$, $r = 4\% = 0.04$, $n = 4$, $t = 3$

$$A = P\left(1+\frac{r}{n}\right)^{nt}$$

$$A = 6000\left(1+\frac{0.04}{4}\right)^{4\cdot3}$$

$$A = 6000(1.01)^{12}$$

$$A \approx 6000(1.126825)$$

$$A \approx 6760.95$$

Calculate the interest earned. Use the formula $I = A - P$.

$$I = A - P$$

$$I = 6760.95 - 6000$$

$$I = 760.95$$

The amount of interest earned is approximately $760.95.

Try it
2. How much interest is earned in 4 years on $9000 deposited in an account paying 3% interest, compounded daily?

Reflect on it
• A compounding period is the frequency with which the interest is compounded.

Quiz Yourself 7.2A
For Exercises 1 to 3, use the future value formula to calculate the future value of the investment.

1. $P = \$600$, $r = 5\%$, compounded daily, $t = 4$ years

2. $P = \$7600$, $r = 4\%$, compounded monthly, $t = 5$ years

3. $P = \$6500$, $r = 3\%$, compounded quarterly, $t = 2$ years

4. How much interest is earned in 10 years on $7000 deposited in an account paying 6% annual interest, compounded semiannually?

5. How much interest is earned in 3 years on $10,000 deposited in an account paying 2% annual interest, compounded weekly?

Practice Sheet 7.2A

Calculate the future value the investment.

1. $P = \$700$, $r = 8\%$, compounded quarterly, $t = 4$ years

2. $P = \$1550$, $r = 7\%$, compounded monthly, $t = 5$ years

1. _____

2. _____

3. $P = \$3440$, $r = 7\%$, compounded semiannually, $t = 8$ years

4. $P = \$5400$, $r = 9\%$, compounded daily, $t = 6$ years

3. _____

4. _____

5. $P = \$9875$, $r = 9\%$, compounded quarterly, $t = 10$ years

6. $P = \$9500$, $r = 7\%$, compounded monthly, $t = 4$ years

5. _____

6. _____

Calculate the interest of the investment.

7. How much interest is earned on an investment of $7500 with an interest rate of 8% compounded semiannually for 5 years?

8. How much interest is earned on an investment of $7500 with an interest rate of 8% compounded quarterly for 5 years?

7. _____

8. _____

9. How much interest is earned on an investment of $7500 with an interest rate of 8% compounded monthly for 5 years?

10. How much interest is earned on an investment of $7500 with an interest rate of 8% compounded daily for 5 years?

9. _____

10. _____

Answers

Try it 7.2A
1. $6400.42
2. $1147.42

Quiz 7.2A
1. $732.83
2. $9279.57
3. $6900.39
4. $5642.78
5. $618.24

Solutions to Practice Sheet 7.2A

1. $A = P\left(1+\dfrac{r}{n}\right)^{nt}$

   $A = 700\left(1+\dfrac{0.08}{4}\right)^{4\cdot4}$

   $A \approx 960.95$
   The future value is about $960.95.

2. $A = P\left(1+\dfrac{r}{n}\right)^{nt}$

   $A = 1550\left(1+\dfrac{0.07}{12}\right)^{12\cdot5}$

   $A \approx 2197.32$
   The future value is about $2197.32.

3. $A = P\left(1+\dfrac{r}{n}\right)^{nt}$

   $A = 3440\left(1+\dfrac{0.07}{2}\right)^{2\cdot8}$

   $A \approx 5964.91$
   The future value is about $5964.91.

4. $A = P\left(1+\dfrac{r}{n}\right)^{nt}$

   $A = 5400\left(1+\dfrac{0.09}{365}\right)^{365\cdot6}$

   $A \approx 9265.82$
   The future value is about $9265.82.

5. $A = P\left(1+\dfrac{r}{n}\right)^{nt}$

   $A = 9875\left(1+\dfrac{0.09}{4}\right)^{4\cdot10}$

   $A \approx 24,047.49$
   The future value is about $24,047.49.

6. $A = P\left(1+\dfrac{r}{n}\right)^{nt}$

   $A = 9500\left(1+\dfrac{0.07}{12}\right)^{12\cdot4}$

   $A \approx 12,559.51$
   The future value is about $12,559.51.

7. $A = P\left(1+\dfrac{r}{n}\right)^{nt}$

   $A = 7500\left(1+\dfrac{0.08}{2}\right)^{2\cdot5}$

   $A \approx 11,101.83$
   $I = A - P$
   $I = 11,101.83 - 7500$
   $I = 3601.83$
   The interest earned is $3601.83.

8. $A = P\left(1+\dfrac{r}{n}\right)^{nt}$

   $A = 7500\left(1+\dfrac{0.08}{4}\right)^{4\cdot5}$

   $A \approx 11,144.61$
   $I = A - P$
   $I = 11,144.61 - 7500$
   $I = 3644.61$
   The interest earned is $3644.61.

9. $A = P\left(1+\dfrac{r}{n}\right)^{nt}$

   $A = 7500\left(1+\dfrac{0.08}{12}\right)^{12\cdot5}$

   $A \approx 11,173.84$
   $I = A - P$
   $I = 11,173.84 - 7500$
   $I = 3673.84$
   The interest earned is $3673.84.

10. $A = P\left(1+\dfrac{r}{n}\right)^{nt}$

    $A = 7500\left(1+\dfrac{0.08}{365}\right)^{365\cdot5}$

    $A \approx 11,188.19$
    $I = A - P$
    $I = 11,188.19 - 7500$
    $I = 3688.19$
    The interest earned is $3688.19.

Objective 7.2B – Calculate the present value of an investment using compound interest

Key Words and Concepts

The **present value** of an investment is the original principal invested, or the value of the investment before it earns any interest.

Present Value Formula
The present value formula is

$$P = \frac{A}{\left(1 + \frac{r}{n}\right)^{nt}}$$

where P is the present value of an investment or deposit,
  A is the future value of the investment or deposit,
  r is the annual interest rate as a decimal,
  n is the number of compounding periods per year, and
  t is the number of years.

Example

How much money should be invested in an account that earns 7% annual interest, compounded monthly, in order to have $20,000 in 8 years?

  Solution
    Use the present value formula.
    $A = 20,000, \ r = 8\% = 0.08, \ n = 12, \ t = 8$

$$P = \frac{A}{\left(1 + \frac{r}{n}\right)^{nt}}$$

$$P = \frac{20,000}{\left(1 + \frac{0.08}{12}\right)^{12 \cdot 8}} \approx \frac{20,000}{(1.00666667)^{96}} \approx \frac{20,000}{1.89245722}$$

$$P \approx 10,568.27$$

    $10,568.27 should be invested in the account in order to have $20,000 in 8 years.

Try it

1. How much money should be invested in an account that earns 6% annual interest, compounded semiannually, in order to have $50,000 in 10 years?

Reflect on it

• Present value is used to determine how much money must be invested today in order for an investment to have a specific value at a future date.

Quiz Yourself 7.2B

1. How much money should be invested in an account that earns 4% annual interest, compounded quarterly, in order to have $20,000 in 8 years?

2. How much money should be invested in an account that earns 3% annual interest, compounded monthly, in order to have $12,000 in 6 years?

3. You want to buy a new car in 6 years. What principal must be deposited now if you wish to have $15,000 when you're ready to buy the new car? Assume the money earns 2.5% compounded daily.

4. How much money would you have to invest today at 8% annual interest, compounded daily, in order to have $500,000 in 30 years?

5. What principal must be deposited by the parents of a newborn if they wish to have $60,000 in 18 years? Assume the money earns 3.5% annual interest, compounded daily.

Practice Sheet 7.2B

Solve.

1. How much money should be invested in an account that earns 5% annual interest compounded quarterly in order to have $5000 in 10 years?

2. How much money should be invested in an account that earns 7% annual interest compounded monthly in order to have $4000 in 6 years?

3. How much money should be invested in an account that earns 8% annual interest compounded semiannually in order to have $25,000 in 20 years?

4. How much money should be invested in an account that earns 9% annual interest compounded daily in order to have $20,000 in 8 years?

5. At age 20, a student opens a Roth IRA with $5000. If the money earns 9% annual interest compounded daily, what is the value at age 60?

6. At age 20, a student opens a Roth IRA with $5000. If the money earns 9% annual interest compounded daily, what is the interest that has accrued by age 60?

7. How much money must you invest today at 6% annual interest compounded monthly to have $100,000 in 25 years?

8. What principal must be deposited to have $65,000 in 18 years? Assume the money earns 5.5% compounded monthly.

1. _____

2. _____

3. _____

4. _____

5. _____

6. _____

7. _____

8. _____

Answers

Try it 7.2B
1. $27,683.79

Quiz 7.2B
1. $14,546.08
2. $10,025.49
3. $12,910.69
4. $45,370.91
5. $31,956.47

Solutions to Practice Sheet 7.2B

1. $P = \dfrac{A}{\left(1+\dfrac{r}{n}\right)^{nt}}$

   $P = \dfrac{5000}{\left(1+\dfrac{0.05}{4}\right)^{4\cdot 10}}$

   $P \approx 3042.07$

   The present value is $3042.07.

2. $P = \dfrac{A}{\left(1+\dfrac{r}{n}\right)^{nt}}$

   $P = \dfrac{4000}{\left(1+\dfrac{0.07}{12}\right)^{12\cdot 6}}$

   $P \approx 2631.40$

   The present value is $2631.40.

3. $P = \dfrac{A}{\left(1+\dfrac{r}{n}\right)^{nt}}$

   $P = \dfrac{25,000}{\left(1+\dfrac{0.08}{2}\right)^{2\cdot 20}}$

   $P \approx 5207.23$

   The present value is $5207.23.

4. $P = \dfrac{A}{\left(1+\dfrac{r}{n}\right)^{nt}}$

   $P = \dfrac{20,000}{\left(1+\dfrac{0.09}{365}\right)^{365\cdot 8}}$

   $P \approx 9735.91$

   The present value is $9735.91.

5. $A = P\left(1+\dfrac{r}{n}\right)^{nt}$

   $A = 5000\left(1+\dfrac{0.09}{365}\right)^{365\cdot 40}$

   $A \approx 182,909.98$

   The future value is $182,909.98.

6. From Exercise 5, $A = 182,909.98$.

   $I = A - P$

   $I = 182,909.98 - 5000$

   $I = 177,909.98$

   The interest earned is $177,909.98.

7. $P = \dfrac{A}{\left(1+\dfrac{r}{n}\right)^{nt}}$

   $P = \dfrac{100,000}{\left(1+\dfrac{0.06}{12}\right)^{12\cdot 25}}$

   $P \approx 22,396.57$

   You must invest $22,396.57.

8. $P = \dfrac{A}{\left(1+\dfrac{r}{n}\right)^{nt}}$

   $P = \dfrac{65,000}{\left(1+\dfrac{0.055}{12}\right)^{12\cdot 18}}$

   $P \approx 24,207.18$

   $24,207.18 should be deposited.

Objective 7.2C – Calculate the effects of inflation

Key Words and Concepts

Inflation is an economic condition in which there are increases in the costs of goods and services.

    Note   Inflation is expressed as a percent.

To calculate the effects of inflation, we use the same procedure we used to calculate future value.

In this objective, we will assume *constant annual inflation rates*, and we will use annual compounding in solving inflation problems, i.e. $n = 1$ for these exercises.

Example

Suppose your annual salary today is $40,000. You want to know what an equivalent salary will be in 15 years—that is, a salary that will have the same purchasing power. Assume a 5% inflation rate.

    Solution

        Use the future value formula.

        $P = 40,000$, $r = 5\% = 0.05$, and $t = 15$.

        The inflation rate is an annual rate, so $n = 1$.

$$A = P\left(1 + \frac{r}{n}\right)^{nt}$$

$$A = 40,000\left(1 + \frac{0.05}{1}\right)^{1 \cdot 15}$$

$$A = 40,000(1.05)^{15}$$

$$A \approx 83,157.13$$

Fifteen years from now, you need to earn an annual salary of approximately $83,157.13 in order to have the same purchasing power.

Try it

1. Suppose your annual salary today is $45,000. You want to know what an equivalent salary will be in 25 years—that is, a salary that will have the same purchasing power. Assume a 3% inflation rate.

The present value formula can be used to determine the effect of inflation on the future purchasing power of a given amount of money. Substitute the inflation rate for the interest rate in the present value formula. The compounding period is 1 year. Again we will assume a constant rate of inflation.

Example

Suppose you purchase an insurance policy in 2015 that will provide you with $300,000 when you retire in 45 years. Assuming an annual inflation rate of 6%, what will be the purchasing power of the $300,000 in 2060?

Solution

Use the present value formula.
$A = 300,000$, $r = 6\% = 0.06$, $t = 45$
The inflation rate is an annual rate, so $n = 1$.

$$P = \frac{A}{\left(1 + \dfrac{r}{n}\right)^{nt}}$$

$$P = \frac{300,000}{\left(1 + \dfrac{0.06}{1}\right)^{1 \cdot 45}}$$

$$P = \frac{300,000}{(1.06)^{45}}$$

$$P \approx \frac{300,000}{13.76461083}$$

$$P \approx 21,795.02$$

Assuming an annual inflation rate of 6%, the purchasing power of $300,000 will be about $21,795.02 in 2060.

Try it

2. Suppose you purchase an insurance policy in 2015 that will provide you with $400,000 when you retire in 40 years. Assuming an annual inflation rate of 5%, what will be the purchasing power of the $400,000 in 2055?

Reflect on it

• For how long is a constant inflation rate realistic?

Quiz Yourself 7.2C

1. Suppose your current salary is $45,000. Assuming an annual inflation rate of 6%, what salary will you need 5 years from now in order to have the same purchasing power? Round to the nearest dollar.

2. The median price of a house in a certain city is $210,000. If the annual inflation rate for the price of a house is 8%, find the median price of a house in the same city 5 years from now. Round to the nearest dollar.

3. The annual health care premium for an employee at a corporation is $12,000. If the annual inflation rate for health care premiums at this corporation is 4.5%, what will be the annual health care premium for the same employee in 4 years? Round to the nearest dollar.

4. A retired couple has a fixed income of $48,000 per years. Assuming an annual inflation rate of 6.5%, what will be the purchasing power of the couple's annual income in 8 years? Round to the nearest dollar.

Practice Sheet 7.2C

Solve.

1. A professor's current salary is $45,000. Assuming an annual inflation rate of 5%, what salary will he need in 6 years from now in order to have the same purchasing power? Round to the nearest dollar.

2. Suppose your current salary is $65,000. Assuming an annual inflation rate of 4%, what salary will you need 10 years from now in order to have the same purchasing power? Round to the nearest dollar.

1. _____

2. _____

3. The median price of a condo in a certain city is $250,000. If the annual inflation rate for the price of a condo is 6%, find the median price of a condo in the same city 5 years from now. Round to the nearest dollar.

4. Assume the new truck sticker price is $34,000. Use an annual inflation rate of 4% to estimate the average new truck sticker price in 8 years.

3. _____

4. _____

5. An insurance policy will provide $750,000 when you retire in 45 years. Assuming an annual inflation rate of 6%, what will be the purchasing power of the $750,000 at retirement? Round to the nearest dollar.

6. An insurance policy will provide $2,000,000 when you retire in 48 years. Assuming an annual inflation rate of 5%, what will be the purchasing power of the two million dollars at retirement? Round to the nearest dollar.

5. _____

6. _____

7. A veteran has a fixed income of $34,000 per year. Assuming an annual interest rate of 4%, what purchasing power will he have in 10 years?

8. A widow has a fixed income of $39,000 per year. Assuming an annual interest rate of 7%, what purchasing power will she have in 8 years?

7. _____

8. _____

Answers

Try it 7.2C
1. $94,220.01
2. $56,818.27

Quiz 7.2C
1. $60,220
2. $308,559
3. $14,310
4. $29,003

Solutions to Practice Sheet 7.2C

1. $A = P\left(1 + \dfrac{r}{n}\right)^{nt}$

   $A = 45,000\left(1 + \dfrac{0.05}{1}\right)^{1 \cdot 6}$

   $A \approx 60,304$

   He will need a salary of about $60,304.

2. $A = P\left(1 + \dfrac{r}{n}\right)^{nt}$

$A = 65,000\left(1 + \dfrac{0.04}{1}\right)^{1\cdot10}$

$A \approx 96,216$

You will need a salary of about \$96,216.

3. $A = P\left(1 + \dfrac{r}{n}\right)^{nt}$

$A = 250,000\left(1 + \dfrac{0.06}{1}\right)^{1\cdot5}$

$A \approx 334,556$

The median price will be about \$334,556.

4. $A = P\left(1 + \dfrac{r}{n}\right)^{nt}$

$A = 34,000\left(1 + \dfrac{0.04}{1}\right)^{1\cdot8}$

$A \approx 46,531$

The new sticker price will be about \$46,531.

5. $P = \dfrac{A}{\left(1 + \dfrac{r}{n}\right)^{nt}}$

$P = \dfrac{750,000}{\left(1 + \dfrac{0.06}{1}\right)^{1\cdot45}}$

$P \approx 54,488$

The purchasing power of \$750,000 will be about \$54,488 after 45 years.

6. $P = \dfrac{A}{\left(1 + \dfrac{r}{n}\right)^{nt}}$

$P = \dfrac{2,000,000}{\left(1 + \dfrac{0.05}{1}\right)^{1\cdot48}}$

$P \approx 192,284$

The purchasing power of \$2,000,000 will be about \$192,284 after 48 years.

7. $P = \dfrac{A}{\left(1 + \dfrac{r}{n}\right)^{nt}}$

$P = \dfrac{34,000}{\left(1 + \dfrac{0.04}{1}\right)^{1\cdot10}}$

$P \approx 22,969$

The purchasing power of \$34,000 will be about \$22,969 after 10 years.

8. $P = \dfrac{A}{\left(1 + \dfrac{r}{n}\right)^{nt}}$

$P = \dfrac{39,000}{\left(1 + \dfrac{0.07}{1}\right)^{1\cdot8}}$

$P \approx 22,698$

The purchasing power of \$39,000 will be about \$22,698 after 8 years.

Objective 7.3A – Calculate interest on a credit card bill

Key Words and Concepts

When a customer uses a credit card to make a purchase, the customer is actually receiving a loan. A **finance charge** is an amount paid in excess of the cash price; it is the cost to the customer for the use of credit.

Most credit card companies issue monthly bills. If the bill is paid in full by the due date, the customer pays no finance charge. If the bill is not paid in full by the due date, a finance charge is added to the next bill.

The most common method of determining finance charges is the **average daily balance method.**

Average Daily Balance

$$\text{Average daily balance} = \frac{\text{sum of the total amounts owed each day of the month}}{\text{number of days in the billing period}}$$

Example

An unpaid bill for $580 had a due date of June 10. A purchase of $184 was made on June 16, and $120 was charged on June 29. A payment of $200 was made on June 21. The interest on the average daily balance is 2.5% per month. Find the finance charge on the July 10 bill.

Solution

First calculate the sum of the total amounts owed each day of the month.

Date	Payments or Purchases	Balance Each Day	Number of Days until Balance Changes	Unpaid Balance Times Number of Days
June 10-15		$580	6	$580 \cdot 6 = \$3480$
June 16-20	$184	$580 + 184 = 764$	5	$764 \cdot 5 = \$3820$
June 21-28	−$200	$764 - 200 = 564$	8	$564 \cdot 8 = \$4512$
June 29-July 9	$120	$564 + 120 = 684$	11	$684 \cdot 11 = \$7524$
Total				$19,336

The sum of the total amounts owed each day of the month is $19,336. There are 30 days from June 10 to July 10.

Find the average daily balance.

$$\text{Average daily balance} = \frac{\text{sum of the total amounts owed each day of the month}}{\text{number of days in the billing period}}$$

$$= \frac{19,336}{30}$$

$$\approx \$644.53$$

Calculate the finance charge by using the simple interest formula.

$$I = Prt$$
$$I = 644.53(0.025)(1)$$
$$I \approx 16.11$$

The finance charge on the June 10 bill is $16.11.

Try it

1. An unpaid bill for $640 had a due date of August 10. A purchase of $240 was made on August 14, and $85 was charged on August 27. A payment of $300 was made on August 20. The interest on the average daily balance is 1.5% per month. Find the finance charge on the September 10 bill.

Quiz Yourself 7.3A

1. Calculate the finance charge for a credit card with an average daily balance of $213.89 and a monthly interest rate of 1.5%.

2. A credit card account had a $567 balance on May 10. A purchase of $126 was made on May 18, and a payment of $200 was made on May 31. Find the average daily balance if the billing date is June 10.

3. A credit card account has a $745 on February 20, on a non-leap year. A purchase of $329 was made on February 25. A payment of $500 was made on March 10. The interest on the average daily balance is 1.5% per month. Find the finance charge on the March 20 bill.

Practice Sheet 7.3A

Solve.

1. Calculate the finance charge for a credit card with an average daily balance of $1675.24 and a monthly interest rate of 1.25%.

    1. _____

2. Calculate the finance charge for a credit card with an average daily balance of $2694.75 and a monthly interest rate of 1.4%.

    2. _____

3. A credit card account had a $932 balance on September 15. A purchase of $687 was made on September 23, and a payment of $250 was made on September 30. Find the average daily balance if the billing date is October 15.

    3. _____

4. A credit card account had a $932 balance on September 15. A purchase of $687 was made on September 23, and a payment of $250 was made on September 30. Find the average daily balance if the billing date is October 15. The interest on the average daily balance is 1.75%. Find the finance charge on the October 15 bill.

    4. _____

Answers

Try it 7.3A
1. $10.26

Quiz 7.3A
1. $3.21
2. $595.97
3. $12.55

Solutions to Practice Sheet 7.3A

1. $I = Prt$
$I = 1675.24(0.0125)(1)$
$I \approx 20.94$
The finance charge is $20.94.

2. $I = Prt$
$I = 2694.75(0.014)(1)$
$I \approx 37.73$
The finance charge is $37.73.

3. Make a chart.

Date	Payments or Purchases	Balance Each Day	Number of Days until Balance Changes	Unpaid Balance Times Number of Days
Sept 15-22		$932	8	$932 \cdot 8 = \$7456$
Sept 23-29	$687	$932 + 687 = 1619$	7	$1619 \cdot 7 = \$11,333$
Sept 30-Oct 14	−$250	$1619 − 250 = 1369$	15	$1369 \cdot 15 = \$20,535$
Total				$39,324

$$\text{Average daily balance} = \frac{\text{sum of the total amounts owed each day of the month}}{\text{number of days in the billing period}}$$

$$= \frac{39,324}{30}$$
$$\approx \$1310.80$$

The average daily balance is about $1310.80.

4. From Exercise 3, the average daily balance is about $1310.80.
$I = Prt$
$I = 1310.80(0.0175)(1)$
$I \approx 22.94$
The finance charge is $22.94.

Objective 7.3B – Calculate the monthly payment on an APR loan

Key Words and Concepts

What does APR represent? _____

The stated interest rate for most consumer loans is normally the APR, as required by the Truth in Lending Act. This rate is based on compound interest rather than simple interest.

Payment Formula for an APR Loan

The payment formula for an APR loan is given by

$$PMT = Ai\left(\frac{(1+i)^{nt}}{(1+1)^{nt}-1}\right)$$

where PMT is the payment
$\quad A$ is the loan amount
$\quad t$ is the number of years

$$i = \frac{r}{n}$$

$\quad r$ is the annual interest rate as a decimal, and
$\quad n$ is the number of payments per year.

Example

A home improvement store is offering a refrigerator for $1150, including taxes, and financing at an annual interest rate of 5.25% for 2 years. Find the monthly payment.

Solution

Use the payment formula for an APR loan.

$A = 1150$, $t = 2$, $r = 0.0525$, $n = 12$, and $i = \dfrac{0.0525}{12}$

If we substitute the information into the formula we have

$$PMT = Ai\left(\frac{(1+i)^{nt}}{(1+1)^{nt}-1}\right)$$

$$PMT = 1150\left(\frac{0.0525}{12}\right)\left(\frac{\left(\left(1+\frac{0.0525}{12}\right)\right)^{12\cdot2}}{\left(1+\frac{0.0525}{12}\right)^{12\cdot2}-1}\right)$$

This looks very complicated, so here are three steps to show the solution.

Step 1: Calculate i. Since payments are monthly, $n = 12$.

$$i = \frac{r}{n} = \frac{0.0525}{12} = 0.004375$$

Step 2: Calculate $(1+i)^{nt}$. Use i from Step 1; $n = 12$, $t = 2$.

$$(1+i)^{nt} = (1+0.004375)^{12\cdot2} = (1.004375)^{24}$$
$$\approx 1.110456265$$

Try it

1. A store is offering a sofa set for $1500, including taxes, and financing at an annual interest rate of 4.5% for 3 years. Find the monthly payment.

Step 3: Substitute the calculated values in the formula.

$$PMT = Ai\left(\frac{(1+i)^{nt}}{(1+1)^{nt}-1}\right)$$

$$PMT \approx 1150(0.004375)\left(\frac{1.110456265}{1.110456265-1}\right)$$

$$PMT \approx (5.03125)\left(\frac{1.110456265}{0.110456265}\right)$$

$$PMT \approx 50.58$$

The monthly payment is $50.58.

Reflect on it
- APR is the annual percentage rate.
- The payment formula is used to calculate monthly payments on most consumer loans, like car loans or home mortgage loans.

Quiz Yourself 7.3B

1. You make a purchase of $1200, including taxes, financed for one year at an annual percentage rate of 6.25%. Find the monthly payment.

2. Sherry purchases a condo and obtains a loan for $230,000 at an annual interest rate of 3.75% for 30 years. Find the monthly payment.

3. Sally purchases a used car and obtains a loan for $11,000 at an annual interest rate of 7.5% for 5 years. Find the monthly payment.

Practice Sheet 7.3B

Solve.

1. You make a purchase of $750 including taxes. You finance the purchase for one year at an annual percentage rate of 5.9%. Find the monthly payment.

2. You make a purchase of $825, including taxes, financed for one year at an annual percentage rate of 5.5%. Find the monthly payment.

1. _____

2. _____

3. Jason purchases a condo and obtains a loan for $200,000 at an annual interest rate of 3.5% for 20 years. Find the monthly payment.

4. Jay purchases a condo and obtains a loan for $150,000 at an annual interest rate of 4% for 15 years. Find the monthly payment.

3. _____

4. _____

5. Tim purchases a used car and obtains a loan for $12,000 at an annual interest rate of 6.5% for 3 years. Find the monthly payment.

6. Ernie purchases a used camper and obtains a loan of $30,000 at an annual interest rate of 5.3% for 6 years. Find the monthly payment.

5. _____

6. _____

7. Margo purchases kitchen appliances and obtains a loan of $4500 at an annual interest rate of 4.9% for 2 years. Find the monthly payment.

8. Mike purchases a used car and obtains a loan of $8500 at an annual interest rate of 4.5% for 4 years. Find the monthly payment.

7. _____

8. _____

Answers

Try it 7.3B
1. $44.62

Quiz 7.3B
1. $103.42
2. $1065.17
3. $220.42

Solutions to Practice Sheet 7.3B

1. $PMT = Ai\left(\dfrac{(1+i)^{nt}}{(1+1)^{nt}-1}\right)$

$PMT = 750\left(\dfrac{0.059}{12}\right)\left(\dfrac{\left(\left(1+\dfrac{0.059}{12}\right)\right)^{12\cdot1}}{\left(1+\dfrac{0.059}{12}\right)^{12\cdot1}-1}\right)$

$PMT \approx 64.52$
The monthly payment is $64.52.

2. $PMT = Ai\left(\dfrac{(1+i)^{nt}}{(1+1)^{nt}-1}\right)$

$PMT = 825\left(\dfrac{0.055}{12}\right)\left(\dfrac{\left(\left(1+\dfrac{0.055}{12}\right)\right)^{12\cdot1}}{\left(1+\dfrac{0.055}{12}\right)^{12\cdot1}-1}\right)$

$PMT \approx 70.82$
The monthly payment is $70.82.

3. $PMT = Ai\left(\dfrac{(1+i)^{nt}}{(1+1)^{nt}-1}\right)$

$PMT = 200,000\left(\dfrac{0.035}{12}\right)\left(\dfrac{\left(\left(1+\dfrac{0.035}{12}\right)\right)^{12\cdot20}}{\left(1+\dfrac{0.035}{12}\right)^{12\cdot20}-1}\right)$

$PMT \approx 1159.92$
The monthly payment is $1159.92.

4. $PMT = Ai\left(\dfrac{(1+i)^{nt}}{(1+1)^{nt}-1}\right)$

$PMT = 150,000\left(\dfrac{0.04}{12}\right)\left(\dfrac{\left(\left(1+\dfrac{0.04}{12}\right)\right)^{12\cdot15}}{\left(1+\dfrac{0.04}{12}\right)^{12\cdot15}-1}\right)$

$PMT \approx 1109.53$
The monthly payment is $1109.53.

5. $PMT = Ai\left(\dfrac{(1+i)^{nt}}{(1+1)^{nt}-1}\right)$

$PMT = 12,000\left(\dfrac{0.065}{12}\right)\left(\dfrac{\left(\left(1+\dfrac{0.065}{12}\right)\right)^{12\cdot3}}{\left(1+\dfrac{0.065}{12}\right)^{12\cdot3}-1}\right)$

$PMT \approx 367.79$
The monthly payment is $367.79.

6. $PMT = Ai\left(\dfrac{(1+i)^{nt}}{(1+1)^{nt}-1}\right)$

$PMT = 30,000\left(\dfrac{0.053}{12}\right)\left(\dfrac{\left(\left(1+\dfrac{0.053}{12}\right)\right)^{12\cdot6}}{\left(1+\dfrac{0.053}{12}\right)^{12\cdot6}-1}\right)$

$PMT \approx 487.33$
The monthly payment is $487.33.

7. $PMT = Ai\left(\dfrac{(1+i)^{nt}}{(1+1)^{nt}-1}\right)$

$PMT = 4500\left(\dfrac{0.049}{12}\right)\left(\dfrac{\left(\left(1+\dfrac{0.049}{12}\right)\right)^{12\cdot2}}{\left(1+\dfrac{0.049}{12}\right)^{12\cdot2}-1}\right)$

$PMT \approx 197.22$
The monthly payment is $197.22.

8. $PMT = Ai\left(\dfrac{(1+i)^{nt}}{(1+1)^{nt}-1}\right)$

$PMT = 8500\left(\dfrac{0.045}{12}\right)\left(\dfrac{\left(\left(1+\dfrac{0.045}{12}\right)\right)^{12\cdot4}}{\left(1+\dfrac{0.045}{12}\right)^{12\cdot4}-1}\right)$

$PMT \approx 193.83$
The monthly payment is $193.83.

Name: Class:
Module 7: Statistics and Probability

Application and Activities

Discussion and reflection questions:

1. Do you think it is a good idea to keep a balance on your credit card? Why or why not?

2. Do you think it is better to pay off loans, or put money into savings (if you had to choose one over the other)? Explain your reasoning.

Class Activity

Objectives for the lesson:

In this lesson you will use and enhance your understanding of the following:

- Calculating the future value of an investment using compound interest
- Calculating the monthly payment on an APR loan
- Calculating interest on a credit card bill

Jan recently found out she was left $20,000 inheritance by her grandfather. She needs to decide what her best option is for the money over the next 5 years.

1. Jan could put the money in a 5 – year CD (Certificate of Deposit) she found that pays 2.3% compounded quarterly.
   a. Find the amount of money in the account after 5 years.

   b. How much money would she earn in interest over the 5 years?

2. Jan found a car she wants to buy that costs $20,000. The car loan she can get is a 5-year loan with an annual interest rate of 6%.
   a. Find the amount of her monthly payments if she gets a loan for the entire amount.

   b. Multiply the payments by 60 (total number of payments she would make over 5 years) to find the amount she would end up paying for the car after 5 years.

    c.   Subtract the amount you found in part b from $20,000 to find the amount of interest she would pay over 5 years.

3.   Jan owes $4,800 in credit card debt.   She read the fine print and found out that if she does not pay the balance of her credit card off each billing cycle, then she pays 18% annual interest compounded daily on the average daily balance every billing cycle she does not pay it off.

    a.   If she does not borrow any more money on her credit card, or pay any of the balance off, what will be her new balance after 5 years?

    b.   How much would she owe in interest after 5 years?

4.   Discuss as a class, which would be best for Jan to do with the $20,000 and justify your decision.

Group Activity (2-3 people):

Objectives for the lesson:

In this lesson you will use and enhance your understanding of the following:

- Calculating the future value of an investment using compound interest
- Calculating the monthly payment on an APR loan

1.   Jan decided to pay off the credit card first, which leaves $15,200 left from her inheritance. Decide which of the three scenarios would be financially better for Jan, and justify your answer.

Scenario I: Jan puts all $15,200 towards the $20,000 car and takes out a car loan for the remaining balance on the car.

Scenario II: Jan puts all $15,200 in the 5-year CD and takes out a car loan for all $20,000.

Scenario III: Jan puts half of the $15,200, which is $7,600 in a 5-year CD and half, or $7,600 towards the car, getting a car loan for the balance of the car.